Biomechanics

Biomechanics

Editors
R.K. Saxena • P. Mishra

Anamaya Publishers
New Delhi

Editors
R.K. Saxena
Centre for Biomedical Engineering
Indian Institute of Technology Delhi
New Delhi - 110 016, India

P. Mishra
Department of Biochemical Engineering and Biotechnology
Indian Institute of Technology Delhi
New Delhi - 110 016, India

Copyright © 2005 Anamaya Publishers

ANAMAYA PUBLISHERS
F-230, Lado Sarai, New Delhi - 110 030, India
e-mail: anamayapub@vsnl.net
anamayapub@indiatimes.com

All rights reserved. No part of this publication may be reproduced, stored in a retrieval system, or transmitted in any form or by any means, electronic, mechanical, photocopying, recording or otherwise, in part or as a whole, without the prior written permission of the publishers.

All export rights of this book vest exclusively with Anamaya Publishers.
Unauthorised export is a violation of the Copyright Law and is subject to legal action.

ISBN 81-88342-49-1

Published by Manish Sejwal for Anamaya Publishers,
F-230, Lado Sarai, New Delhi - 110 030. Printed in India.

Foreword

It gives me immense pleasure to write the 'Foreword' for this volume. I am happy to learn that the 42 research papers contained in this volume provides a forum to exchange the current knowledge in the field of biomechanics.

Biomechanics is a multidisciplinary branch of biomedical engineering and has made significant contribution in the field of cardiovascular mechanics, sports sciences, rehabilitation engineering, and physiotherapy, etc. It has also contributed in understanding the various complex problems of human system.

I congratulate the editors for their efforts to bring out this volume in the present form.

R.S. SIROHI
Director
IIT Delhi, New Delhi

Preface

Biomechanics is one of the branches of engineering sciences in which mechanical principles are applied for understanding the physiology of various systems and ailments. During the last few years there has been a significant growth in the area of biomechanics through the development and incorporation of medical and engineering techniques. Intricacies involved in understanding biomechanics involve input from almost all branches of knowledge domains. This volume attempts to bring people of science, engineering and medicine together. Various areas of biomechanics such as cardiovascular mechanics, sports mechanics and ergonomics, modeling and simulation in biomechanics, rehabilitation and physiotherapy are discussed. The aim of this volume is to address the various issues of mass healthcare and suggest the management. Modern topics of healthcare have been presented with good combination of macro and micro level analysis of various problems related to developing countries.

We are thankful to the contributors and to various agencies, especially to ICMR, DST, DBT, AICTE, DRDO, CSIR and SBI for their generous support. We appreciate the help of M/s Anamaya Publishers for bringing out this volume in excellent form.

EDITORS

Contents

	Foreword	*v*
	Preface	*vii*
1.	A Two Dimensional Infinite Element Model of Temperature Distribution in Human Peripheral Regions due to Uniformly Perfused Tumors *K.R. Pardasani and Madhvi Shakya*	1
2.	Simulation of Accidental Fall on the Floor with the Help of a Three-dimensional Finite Element Model of Human Pelvis *Santanu Majumder, Amit Roychowdhury and Subrata Pal*	11
3.	Gripping Techniques for Tendons *Boon Ho Ng, Siaw Meng Chou and Krishna Vaibhav*	20
4.	Foot Sole Soft Tissue Characterization in Diabetic Neuropathy Using Texture Analysis *P. Manika, K.M. Patil, V. Balasubramanian, V.B. Narayanamurthy and R. Parivalavan*	26
5.	Comparative Study of Static Balance Related Variables among the Indian Adult Male and Female *Manoj Kumar Chaudhary and Dhananjoy Shaw*	34
6.	A Mathematical Model for Computation of Electrical and Mechanical Properties of a Human Wet Bone and Detection of Osteoporosis *N.S. Rao, G.R. Babu, L.N. Merugu, V.S. Mallela and P.K. Subramanya Kumar*	42
7.	Swimming Strengthens Brawn as Well as Brain *Milind Parle and M. Vasudevan*	49
8.	A Three-dimensional, Anatomically Detailed Foot Model: A Foundation for a Finite Element Simulation and Means of Quantifying Foot-Bone Position *Nimesh Prakash and Piyush Soni*	56
9.	MEMS Based Implantable Systems for Bio Medical Application *Mohan Kumaraswamy and S. Ravi Kumar*	66
10.	Biomechanical Analysis of Vascular Stenting: A Review *J. Raamachandran and K. Jayavenkateshwaran*	78
11.	Classification of Human Erythrocyte Aggregates Using Neural Networks *A. Kavitha and S. Ramakrishnan*	86

12. Stress Analysis of Human Elbow Joint Fitted with Linked and Unlinked Prosthesis 91
    R. Daripa, S. Majumder, A. Roy Chowdhury, S.R. Quadery and R. Kumar

13. A New Algorithm for Diagnosing Disease Through Microscopic Images of Blood Cells 100
    S. Soundarapandian, S. Mahesh, P. Subbaraj and R. Murugesan

14. Microbial Bio-fuel Cell: A Potential Application Tool in Biomedical Instrumentation 107
    T.G. Deepak Balaji and A. Siva Kumar

15. Biomechanical Study of Orthopaedic Implants 113
    R.K. Saxena

16. Two Layered Model for Experimental and Analytical Study of Drag Reduction in Glass Model of Stenotic Glass Tube 119
    V.K. Katiyar and Manoj Kumar

17. Thermodynamic Properties of Bone 125
    Rinku Singh, Reeva Gupta, L.M. Aggarwal, A. Koul and D.V. Rai

18. Stress Analysis of Human Hip Joint Using CT-Scan Data 131
    Dibyendu Chakraborty, Bidyut Pal and Subrata Pal

19. Visualization of Brain Deformation for Neurosurgery 140
    Akash Kumar Singh

20. Effect of Physical Exercise on the Relationship between Selected Kinanthropometric Variables and Percentage Height of Centre of Gravity of Male 144
    Dhananjoy Shaw and Seema Kaushik

21. Effect of Cyclic Impact Mechanical Stresses on Biochemical Properties of Weight Bearing Synovial Joint Articular Cartilage Chondrocytes in Alginate Matrix 155
    Garima Sharma, R.K. Saxena and P. Mishra

22. Neural Network Based Inverse Kinematics Analysis for Design and Control of Artificial Leg 164
    R.P. Tewari and Sachin Chaudhry

23. Biomechanics of Foot, Its Deformity and Conservative Treatment 170
    M.D. Burman and Ranjan Das

24. Finite Element Analysis of Balloon and Artery Interaction During Stent Deployment 176
    C.S. Ramesh and Prashant K. Marikatti

25. Evaluation of Various Biomaterials Used in Treating Maxillofacial Injuries 183
    Smriti Sharma, Sandeep Tiwari, Vivek P. Soni and Jayesh Bellare

26. Sports Biomechanics and Human Performance 186
    Dhananjoy Shaw

27. Role of Obesity in Genesis of Osteoarthritis 194
    Amrita Parle

28.	Simulation of the Coronary Blood Flow in the Domain of a Severe Stenosis Employing a Dispersed Two-phase Modeling Approach *Gladwin Philip and Nimesh Prakash*	200
29.	Beta-Alanine Protects Mice from Memory-Impairment *Milind Parle and Dinesh Dhingra*	208
30.	Finite Element Analysis of Composite Hip Prosthesis *M. Sivasankar, D. Chakraborty and S.K. Dwivedy*	214
31.	Some Biomechanical Aspects of Fitness Training for Olympic Weightlifters *Sudhir Kumar*	221
32.	Effect of Alcohol on Mechanical Properties of Bone *Reeva Gupta, A. Koul and D.V. Rai*	228
33.	Effect of Moderate Cyclic Load Stress on Biochemical Properties of Articular Cartilage of Bovine Synovial Joints *Ritu Rathi and R.K. Saxena*	234
34.	Effect of Physical Exercise on the Relationship between Selected Kinanthropometric Variables and Percentage Height of Centre of Gravity of Female *Seema Kaushik and Dhananjoy Shaw*	240
35.	Finite Element Modeling and Analysis of $L_{2/3}$ Functional Spine Unit *J. Manjusha, Venkatesh Balasubramanian and C. Sujatha*	251
36.	Effects of Noise on Blood Pressure and Heart Rate *V.K. Katiyar, A.K. Guptar and Jaipal*	258
37.	Review of EMG Controlled Prosthetic Hand Control *A.S. Arora, S.K. Soni, P.K. Pathak, Vinay Gupta and Santosh Kumar*	267
38.	Simulation of Normal Human ECG and Arrhythmias Using Advanced Electronics *Shahanaz Ayub*	274
39.	Effect of Physical Exercise on Three-Dimensional Centre of Gravity, Height, Weight and Ponderal Index of the Students of University of Delhi *Dhananjoy Shaw, Seema Kaushik and Indu Kaushik*	281
40.	Handmade Peristaltic Pump *Abhishek Bakshi and A. Bhanu Prakash*	289
41.	Effect of Translations Mobilization on the Rotational ROM of the Normal Shoulder Done at the End Range of Abduction *K. Ramanathan, Jince Thomas, Patitapaban Mohanty, Monalisa Pattnaik, Sharada Lakshmana Nayak and Raja Bhattacharya*	293
42.	Proprioceptive Neuromuscular Facilitation in Restoring Altered Pulmonary Mechanics Post Mid-sternotomy *Neha Gupta, V.P. Gupta, Sajad Ali, Jeyasundar, Sundar Kumar and Thiruvarangan*	298

Biomechanics

Biomechanics
R.K. Saxena and P. Mishra (Editors)
Copyright © 2005, Anamaya Publishers, New Delhi, India

1. A Two Dimensional Infinite Element Model of Temperature Distribution in Human Peripheral Regions due to Uniformly Perfused Tumors

K.R. Pardasani and Madhvi Shakya
Department of Mathematics and Computer Applications,
Maulana Azad National Institute of Technology, Bhopal - 462007, India

Abstract: This paper deals with mathematical modelling of two-dimensional problem investigating thermal influence of malignant tumors in peripheral regions of human body. A finite element model has been extended to infinite element domains to study temperature distribution. The peripheral region is divided into three natural layers, namely, epidermis, dermis and sub-dermal tissues. A uniformly perfused tumor is assumed to be present in the sub-dermal region of the human body. The domain is assumed to be finite along the depth and infinite along the breadth. The region is discretized using appropriate number of triangular finite elements to match with geometry and structure of the region. These elements are surrounded by infinite domain elements along the breadth. The mathematical model incorporates all the important parameters like thermal conductivity, blood mass flow rate and metabolic heat generation. The controlled and uncontrolled rate of metabolic heat generation has been taken into account for normal and malignant tumors respectively. A mirror symmetry has been assumed in the region. Appropriate boundary conditions has been incorporated. The boundary value problem has been transformed into discretized variational form. The Ritz finite element technique has been applied to obtain the solution. A computer program has been developed and simulated to obtain the numerical results. The graphs have been plotted and the thermal effect of tumors on temperature distribution in surrounding normal tissues has been studied.

Introduction
Perl gave a mathematical model of heat mass distribution in tissues. He combined the Fick's second law of diffusion and Fick's perfusion principle along with heat generation term to deduce the model. The partial differential equation derived by him is given by

$$\rho c \frac{\partial T}{\partial \tau} = \text{div}(K \text{ grad } T) + m_b c_b (T_b - T) + S \tag{1}$$

The effect of blood flow and effect of metabolic heat generation are given by the terms $m_b c_b (T_b - T)$ and S respectively, where c = specific heat of the tissue, T_b = blood temperature, m_b = blood mass flow rate, t = time, c_b = specific heat of blood, K = thermal conductivity, ρ = tissue density, T = unknown temperature, S = rate of metabolic heat generation.

If the body is exposed to the environment then the boundary condition due to heat loss at outer surface is given by:

$$-K\left(\frac{\partial T}{\partial n}\right) = h(T - T_a) + LE \qquad (2)$$

where h is heat transfer coefficient, T_a is the atmospheric temperature, L and E are respectively the latent heat and rate of sweat evaporation and $\partial T/\partial n$ is the partial derivative of T along the normal to the skin surface. Also the human body maintains its core temperature at a uniform temperature (37°C). Therefore, the boundary condition at the inner boundary is generally taken as:

$$T = T_b$$

where T_b is the body core temperature.

Equation (1) has been used by Perl [10] to solve a simple case of Eq. (1) for a spherical symmetric heat source embedded in an infinite, tissue medium. Cooper and Trezek [4] found an analytic solution of heat diffusion equation for brain tissue with negligible effect of blood flow and metabolic heat generation. Trezek and Cooper [20] obtained a solution for a cylindrical symmetry considering all the parameters as constant and computed the thermal conductivity of the tissue. Chao, Eisely and Yang [2] and Chao and Yang [3] applied steady state and unsteady state models with all the parameters as constant, to the problem of heat flow in the skin and subdermal tissues. Saxena [12] used similarity transformation to find exact solution to certain time dependent problems in the beginning. Saxena [13, 14] obtained an analytical solution to one-dimensional problem taking position dependent values of blood mass flow and metabolic heat generation. Later on Saxena and Arya [14] initiated the use of finite element method for solving the problem of temperature distribution in three layered and six layered skin and subcutaneous tissues. Arya and Saxena [1] also used finite element technique to determine the thermal conductivity of SST region. Saxena and Bindra [15, 16] used quadratic shape functions in variational finite element method to solve a one dimensional steady state problem.

Later Saxena and Pardasani extended the application of analytical and finite element approach to the problems involving abnormalities like malignant tumors. Saxena and Pardasani [17, 18] also studied the relationship between dermal blood flow and skin surface temperature. Further Pardasani [6, 7] investigated the heat and water distribution problem in skin and subcutaneous tissues. Pardasani and Adlakha [8, 9] investigated the temperature distribution in cylindrical and spherical organs of human body involving tumors. Jas [5] has also studied temperature distribution in cylindrical organs of human body involving tumors using finite element method.

Mathematical Model

The partial differential Eq. (1) in a two dimensional steady state case for heat flow in skin and subcutaneous tissues may be written as:

$$\frac{\partial}{\partial x}\left(K\frac{\partial T}{\partial x}\right) + \frac{\partial}{\partial y}\left(K\frac{\partial T}{\partial y}\right) + m_b c_b (T_b - T) + S = 0 \qquad (3)$$

The tumor is characterized by uncontrolled rates of metabolic heat generation. The normal tissues are characterized by self-controlled metabolic heat generation. In view of this the metabolic term S in Eq. (3) can be broken into two parts as given:

$$S = S_1 + W$$

where S_1 = self controlled rate of metabolic heat generation and W = uncontrolled rate of metabolic heat generation.

We consider a rectangular composite region with boundaries $x = p$, $x = q$ and $y = r$, $y = s$ with the following conditions.

$$\left.\frac{\partial T}{x}\right|_{x=p} = \varepsilon \tag{4}$$

$$-\left.\frac{\partial T}{x}\right|_{x=q} = \theta \tag{5}$$

$$T(x, y) = T_b \text{ at } y = r \tag{6}$$

$$-K\frac{\partial T}{\partial y} = h(T - T_a) + LE \text{ at } y = s \tag{7}$$

The outer surface of the skin is exposed to the environment and heat loss takes place from this surface. The last condition incorporates this heat loss by radiation-convection and evaporation at the outer skin surface. The values of ε and θ are taken depending on the outward normal flow from the boundaries at $x = p$ and $x = q$. The region is divided into sufficiently large number of two types of elements to match with geometry and to incorporate minute details of physiology. The triangular shaped finite elements and rectangular shaped infinite elements. Thus the elements size is quite small and therefore we take linear variation of temperature within each element.

The variational formulation of Eq. (3) for some general e^{th} element along with the boundary conditions Eqs. (4), (5), (6) and (7) is given by

$$I^{(e)} = \frac{1}{2}\iint_{\Delta^{(e)}}\left[K^{(e)}\left(\frac{\partial T^{(e)}}{\partial x}\right)^2 + K^{(e)}\left(\frac{\partial T^{(e)}}{\partial y}\right)^2 + (m_b c_b)^{(e)}(T_b - T^{(e)})^2 - 2(S_1 + W^{(e)})T^{(e)}\right]dxdy$$
$$+ \frac{1}{2}\int_{\tau^1}\left[(T^{(e)} - T_a)^2 + 2LET^{(e)}\right]dx + \int_{\tau^2}\varepsilon^{(e)}T^{(e)}dy + \int_{\tau^3}\theta^{(e)}T^{(e)}dy \tag{8}$$

where $\Delta^{(e)}$ is the region contained in the eth element.

Here the second integral of Eq. (8) is valid for elements adjoining the outer skin surface, $\tau_1^{(e)}$ being the boundary of the eth element exposed to the environment and it is zero for all other elements. In the same way third and fourth integrals are valid only for the elements adjoining the boundaries τ_1 and τ_2 respectively and taken equal to zero for the remaining elements. Following assumptions have been made for $K^{(e)}$ and $(m_b c_b)^{(e)}$:

$$K^{(e)} = K^{(e)}\left(\lambda_1^{(e)} - \lambda_2^{(e)} y\right), \quad (m_b c_b)^{(e)} = m^{(e)}\left(\phi_1^{(e)} - \phi_2^{(e)} y\right)$$

For triangular finite elements

$$T_i^{(e)} = c_1^{(e)} + c_2^{(e)} x + c_3^{(e)} y \tag{9}$$

where $T^{(e)}$ is equal to T_i, T_j and T_k at the nodes of the eth element. Thus we have

$$T_i^{(e)} = c_1^{(e)} + c_2^{(e)} x_i + c_3^{(e)} y_i \tag{10}$$

$$T_j^{(e)} = c_1^{(e)} + c_2^{(e)} x_j + c_3^{(e)} y_j \tag{11}$$

$$T_k^{(e)} = c_1^{(e)} + c_2^{(e)} x_k + c_3^{(e)} y_k \tag{12}$$

In matrix form this can be written as

$$T^{(e)} = P^{(e)} C^{(e)} \tag{13}$$

where

$$C^{(e)} = \begin{vmatrix} c_1^{(e)} \\ c_2^{(e)} \\ c_3^{(e)} \end{vmatrix} \tag{14}$$

$$P^{(e)} = \begin{vmatrix} 1 & x_i & y_i \\ 1 & x_j & y_j \\ 1 & x_k & y_k \end{vmatrix} \tag{15}$$

where $C^{(e)}$ is obtained from Eq. (13) and is given below

$$C^{(e)} = R^{(e)} T^{(e)}, \text{ where } R^{(e)} = P^{(e)-1} \tag{16}$$

The expression Eq. (9) may also be written as

$$T^{(e)} = P^T C^{(e)} \tag{17}$$

where $P^T = \begin{bmatrix} 1 & x & y \end{bmatrix}$.

On substituting the value of $C^{(e)}$ from Eq. (16) in Eq. (17) we have

$$T^{(e)} = P^T R^{(e)} T^{(e)} \tag{18}$$

The rate of metabolic heat generation is directly proportional to gradient of issue temperature. So when gradient of tissue temperature increases, rate of metabolic heat generation also increases and when gradient of tissue temperature decreases, rate of metabolic heat generation also decreases. So the rate of self controlled metabolic heat generation for different triangular elements is prescribed below:

(i) If $y_i = y_k$ and $y_j > y_k$, $S^{(e)} = s^{(e)} \left(\alpha^{(e)} - \beta^{(e)} y \right) \left[1 + q^{(e)} \left(\frac{T_i + T_k}{2} - T_j \right) \right]$

(ii) If $y_i = y_k$ and $y_k > y_j$, $S^{(e)} = s^{(e)} \left(\alpha^{(e)} - \beta^{(e)} y \right) \left[1 + q^{(e)} \left(T_j - \frac{T_i + T_k}{2} \right) \right]$

(iii) If $y_i = y_j$ and $y_k > y_j$, $S^{(e)} = s^{(e)} \left(\alpha^{(e)} - \beta^{(e)} y \right) \left[1 + q^{(e)} \left(\frac{T_i + T_j}{2} - T_k \right) \right]$

(iv) If $y_j = y_k$ and $y_k > y_i$, $S^{(e)} = s^{(e)} \left(\alpha^{(e)} - \beta^{(e)} y \right) \left[1 + q^{(e)} \left(T_i - \frac{T_i + T_k}{2} \right) \right]$

For rectangular infinite elements:

For $-1 \leq \xi \leq 1$, $-1 \leq \eta \leq 1$

$$x = M_1(\xi, \eta)x_1 + M_2(\xi, \eta)x_2 + M_3(\xi, \eta)x_3 + M_4(\xi, \eta)x_4$$

$$y = M_1(\xi, \eta)y_1 + M_2(\xi, \eta)y_2 + M_3(\xi, \eta)y_3 + M_4(\xi, \eta)y_4$$

where the mapping functions M_1, M_2, M_3 and M_4 are given by

$$M_1(\xi, \eta) = (1-\eta)(-\xi)/(1-\xi)$$

$$M_2(\xi, \eta) = (1+\eta)(-\xi)/(1-\xi)$$

$$M_3(\xi, \eta) = (1+\eta)(1+\xi)/[2(1-\xi)]$$

$$M_4(\xi, \eta) = (1-\eta)(1+\xi)/[2(1-\xi)]$$

$$T = N_1(\xi, \eta)T_i + N_2(\xi, \eta)T_j + N_3(\xi, \eta)T_k + N_4(\xi, \eta)T_l$$

where the shape functions N_1, N_2, N_3 and N_4 are given by

$$N_1(\xi, \eta) = (1-\eta)(\xi^2 - \xi)/4$$

$$N_2(\xi, \eta) = (1+\eta)(\xi^2 - \xi)/4$$

$$N_3(\xi, \eta) = (1+\eta)(1-\xi^2)/2$$

$$N_4(\xi, \eta) = (1-\eta)(1-\xi^2)/2$$

Let

$$I = \sum_{e=1}^{N} I^{(e)} \quad (19)$$

N = total number of elements.

Now I is minimized by differentiating it with respect to each of the nodal temperature and setting the derivatives equal to zero

$$dI/dT = 0 \quad (20)$$

Here dI/dT denotes the differentiation of I with respect to each nodal temperature as given below:

$$\frac{\partial I}{\partial T} = \left| \frac{\partial I}{\partial T_1} \quad \frac{\partial I}{\partial T_2} \cdots \frac{\partial I}{\partial T_i} \quad \frac{\partial I}{\partial T_j} \quad \frac{\partial I}{\partial T_k} \quad \frac{\partial I}{\partial T_l} \cdots \frac{\partial I}{\partial T_n} \right| \quad (21)$$

Also we define

$$T = \begin{bmatrix} T_1 & T_2 & \ldots\ldots T_i & T_j & T_k & T_l & \ldots\ldots T_n \end{bmatrix}' \qquad (22)$$

Here n = total number of nodal points.

Numerical Results and Discussion

Here a uniformly perfused tumor is assumed to be situated in the subdermal region of the body. The skin and subcutaneous tissue (SST) region of human body has been divided into 10 layers. The epidermis divided into two layers. The dermis and subdermal tissues are divided into four layers each. The innermost layer is the core consisting of bone, muscles, large blood vessels etc. The vertical cross section of skin and subdermal tissues with a solid tumor is shown in Fig. 1 which is discretized into 504 triangular finite elements and 10 rectangular infinite elements. The epidermal layer is discretized into 98 elements with 2 infinite elements. The dermal layer is discretized into 196 elements with 4 infinite elements. The subdermal tissues is discretized into 220 elements with 4 infinite elements and also include 24 elements of tumor region.

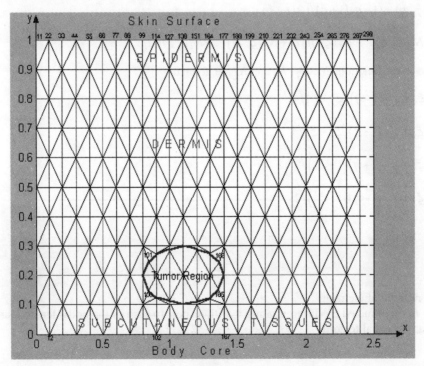

Fig. 1

Now using these assumptions and expressions, the integrals $I^{(e)}$ are evaluated and assembled a given:

$$I = \sum_{e=1}^{514} I^{(e)} \qquad (23)$$

The integral I is extremised with respect to each nodal temperature $T_i(i = 1(1)298)$ to obtain a following set of algebraic equations in terms of nodal temperature T_i ($i = 1(1)298$).

$$XT = Y \qquad (24)$$

Here X, Y and T are respectively the matrices of order 298×298, 298×1 and 298×1. A computer program has been developed for the entire problem in Microsoft Fortran Power station. Gaussian elimination method has been employed to solve the set of Eqn. (24) to obtain nodal temperatures, which give temperature profiles in each subregion.

The values of physical and physiological parameters have been taken from Cooper and Trezek [4], Saxena and Bindra [16] and Saxena and Pardasani [18] as given below:

$K_1 = 0.030$ Cal/cm-min °C $K_2 = 0.0845$ Cal/cm-min °C
$K_3 = 0.060$ Cal/cm-min °C $M = 0.003$ Cal/cm^3-min °C
$S = 0.0357$ Cal/cm^3-min $L = 579.0$
$h = 0.009$ Cal/cm^2-min °C

The following values have been assigned to physical and physiological parameters in each subregion.

Epidermis

$$\lambda_1^{(e)} = 1, \lambda_2^{(e)} = 0, K^{(e)} = 0.03, m^{(e)} = 0, W^{(e)} = 0, q^{(e)} = 2/(T_a + T_b), \alpha^{(e)} = 1, \beta^{(e)} = 0, s^{(e)} = S/16$$

Dermis

$$\lambda_1^{(e)} = 3, \lambda_2^{(e)} = 2.5, K^{(e)} = 0.03, \phi_1^{(e)} = 4, \phi_2^{(e)} = 5, m^{(e)} = m, W^{(e)} = 0, q^{(e)} = 2/(T_a + T_b),$$
$$\alpha^{(e)} = 3.8, \beta^{(e)} = 4.6, s^{(e)} = S$$

Tumor region

$$\lambda_1^{(e)} = 1, \lambda_2^{(e)} = 0, K^{(e)} = 0.036, \phi_1^{(e)} = 1, \phi_2^{(e)} = 0, m^{(e)} = m, \alpha^{(e)} = 3, \beta^{(e)} = 0, s^{(e)} = 0, W^{(e)} = S, q^{(e)} = 0$$

Figure 2 shows temperature profiles for $T_a = 15°C$ and $E = 0.0$ gm/cm^2 for a normal case (without tumor). Fig. 3 shows temperature profiles for $T_a = 23°C$, $E = 24 \times 10^{-3}$ gm/cm^2 for a normal case (without tumor) and an abnormal case (with tumor). The fall in temperature profiles is also more for higher rates of evaporation ($E = 0.24 \times 10^{-3}$ gm/cm^2-min) at same atmospheric temperature ($T_a = 15°C$). This may be due to increase in heat loss due evaporation at the skin surface. The temperature profiles fall down as we move away from body core to the skin surface. The numerical results for a normal case is comparable with those obtained by Saxena and Bindra (1984). An elevation in temperature profiles for skin and subdermal tissues with a tumor is observed while comparing the profiles for the normal and abnormal tissues. The maximum thermal disturbances are seen in the region between $y = 0.1$ cm and $y = 0.3$ cm. The points of change in the slopes of the curves are actually the junction of normal tissues and tumor region. This information is useful for justifying the boundaries of tumor periphery, tumor core and the regions affected by malignancy.

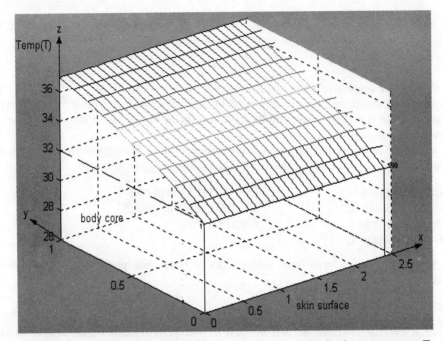

Fig. 2. Graph between temperature T(°C) and x, y (in cm) at atmospheric temperature $T_a = 15$°C and $E = 0.0$ gm/cm^2 for normal case.

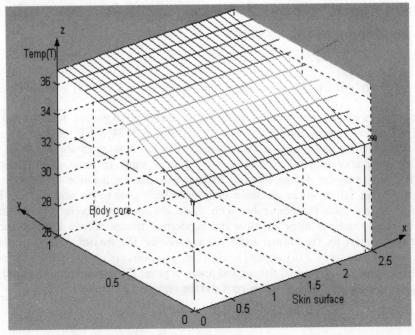

Fig. 3. Graph between temperature T (in °C) and x, y (in cm) at atmospheric temperature $T_a = 23$°C and $E = 0.0$ gm/cm^2 for normal case.

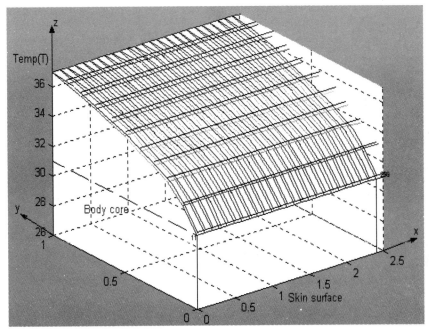

Fig. 4. Graph between temperature T (in °C) and x, y (in cm) at atmospheric temperature $T_a = 23$°C and $E = 0.24$ gm/cm^2 for normal and abnormal case.

References

1. Arya, D. and Saxena, V.P. 1981. Application of Variational Finite Element Method in the Measurement of Thermal Conductivity of Human Skin and Subdermal Tissues. *Proc. Nat. Acad. Sci., India 51(A), IV.* pp. 447-450.
2. Chao, K.N., Eisley, J.G. and Yang, W.J. 1973. Heat and Water Migration in Regional Skin and Subcutaneous Tissues. *Bio. Mech., ASME.* 69-72
3. Chao, K.N. and Yang, W.J. 1975. Response of Skin and Tissue Temperature in Sauna and Steam Baths. *Bio. Mech. Symp., ASME.* pp. 69-71.
4. Cooper, T.E. and Trezek, G.J. 1972. A Probe Technique for Determining the Thermal Conductivity of Tissue. *J. Heat Trans., ASME.* **94**: 133-140.
5. Jas. P. Finite Element Approach to the Thermal Study of Malignancies in Cylindrical Human Organs. *Ph.D. Thesis*.
6. Pardasani, K.R. 1988. Mathematical Investigations on Human Physiological Heat Flow Problems with Special Relevance to Cancerous Tumors. *Ph.D. Thesis*, Jiwaji University Gwalior. pp. 115-210.
7. Pardasani, K.R. 1988. Mathematical Investigations on Human Physiological Heat Flow Problems with Special Relevance to Cancerous Tumors. *Ph.D. Thesis*, Jiwaji University Gwalior. pp. 115-210.
8. Paradasani, K.R and Adlakha, N. 1993. Two-dimensional Steady State Temperature Distribution in Annular Tissue Layers of a Human or Animal Body. *Ind. J. Pure Appl. Math.* **24(8)**: 721-728.
9. Pardasani, K.R. and Adlakha, N. 1991. Exact Solution to a Heat Flow Problem in Peripheral Tissue Layers with a Solid Tumor in Dermis. *Ind. J. Pure Appl. Math.* **22(8)**: 679-682.
10. Perl, W. 1963. An Extension of the Diffusion Equation to include Clearance by Capillary Blood Flow. *Ann, NY. Acad. Sc.* **108**: 92-105.
11. Saxena, V.P. 1978. Application of Similarity Transformation to Unsteady State Heat Migration Problems in Human SST. *Proc. 6th Int. Heat. Trans. Conf.* **3**: 65-68.

12. Saxena, V.P. 1979. Effect of Blood Flow on Temperature Distribution in Human Skin and Subdermal Tissues. *Proc. 9th Nat. Conf. in Fluid Mechanics and Fluid Power.* pp. 156-161.
13. Saxena, V.P. 1983. Temperature Distribution in Human Skin and Subdermal Tissues. *J. Theo. Biol.* pp. 277-286.
14. Saxena, V. and Arya, D. 1981. Steady State Heat Distribution in Epidermis, Dermis and Subdermal Tissues. *J. Theor. Biol.* **89**: 423-432.
15. Saxena, V.P. and Bindra, J.S. 1984. Steady State Temperature Distribution in Dermal Blood Flow, Perspiration and Self Controlled Metabolic Heat Generation. *Indian J. Pure and Appl. Math.* **15(1)**: 31-42.
16. Saxena, V.P. and Bindra, J.S. 1989. Psuedo Analytic Finite Partition Approach to Temperature Distribution Problems in Human Limbs. *Int. J. Math and Mathematical Sciences.* **Vol. 12, No.2**: 403-408.
17. Saxena, V.P. and Pardasani, K.R. 1991. Effect of Dermal Tumor on Temperature Distribution in Skin with Variable Blood Flow. *Bull. Mathematical Biology.* **Vol. 53, No. 4**: 525-536.
18. Saxena, V.P. and Pardasani, K.R. 1989. Exact Solutions to Temperature Distribution Problem in Annular Skin Layers. *Bull. Calcutta Math. Soc.* **Vol. 81**: 1-8.
19. Trezek, G.J. and Cooper, T.E. 1968. Analytical Determination of Cylindrical Source Temperature Fields and their Relation to Thermal Diffusivity of Brain Tissues. *Thermal Problems in Biotechnology, ASME.* pp. 1-15.

2. Simulation of Accidental Fall on the Floor with the Help of a Three-dimensional Finite Element Model of Human Pelvis

Santanu Majumder[1], Amit Roychowdhury[1] and Subrata Pal[2]

[1]Department of Applied Mechanics, Bengal Engineering and Science University, Shibpur, Howrah - 711103, India

[2]School of Bioscience and Engineering, Jadavpur University, Kolkata - 700032, India

Abstract: Falling on the floor of residence or work place and on the playground is common phenomenon due to accidental slippage. In these situations, posterior part of the pelvis (both ichial tuberosities and coccyx of sacrum) is the most affected area. The objective of this study was to simulate this kind of accidental fall for a three-dimensional finite element model of human pelvis with the help of ANSYS LS-DYNA^{TM} (version 8.0) explicit dynamic analysis software. The complex geometry of the pelvis with part of the femurs was developed from the CT scan images of a male (58 yrs). CT scan images were processed with the help of MIMICS$^®$ (demo version 7.3), a medical image processing software. Cancellous bone was modeled with tetrahedral elements, having 9 dof (displacement, velocity, acceleration, three each). Cortical bone was represented by shell elements with 12 dof. Bones were assumed to be bilinear, inelastic. Also the floor, on which the pelvis would be impacting, was considered as rigid. A layer of soft tissue (skin-fat-flesh) was considered. To account for the full body weight, mass element (with same dof as tetrahedral element) was considered at 5th lumber (40 kg) and posterior femoral (25 kg) location. Full pelvis was assumed to hit the floor from a of height 88.1 cm with a velocity of 2.55 m/s and acceleration of 3.69 m/s^2 and analysis was carried out for 25 ms. Stress and strain (elastic and plastic) distribution was compared for cases with and without soft tissue layer. The simulation results were also compared with clinical findings for these falls.

Introduction

Accidental fall and slips are the cause of significant numbers of lower extremity injury cases, and are the major cause of disability in some work places. These account for 90% of pelvic and wrist fractures among the elderly populations (Grisso et al., 1991). Posterior part of the pelvis (both the ichial tuberosities and coccyx of the sacrum) is the most affected area in case of pelvic fracture, due to fall. Evidence suggests that the most important determinants of hip fracture risk during a fall are the body's impact velocity and configuration. Accordingly, protective responses for reducing impact velocity, and the likelihood for direct impact to the hip strongly influences fracture risk. The objective of this study was to simulate this kind of accidental fall for a three-dimensional finite element model of human pelvis with the help of ANSYS-LS DYNA^{TM} (version 8.0) explicit dynamic analysis software.

Method

The model of the pelvic-femur complex was generated from 98, 2.5 mm thick transverse CT scans (Fig. 1) of an old male (58 yrs). The CT images (in DICOM format) were reconstructed (Pal et al., 2004) after segmentation and thresholding, using medical image processing software, MIMICS® (demo version 7.3, Materialise Software, Malaysia). All these 98 CT images were stacked to produce the solid model (Fig. 2) and the output was taken in DXF format. Then with the help of Mechanical Desktop® (version 6.0, Auto Desk Inc.), this DXF formatted data was transferred to IGS format. Finally, with the help of ANSYS-LS DYNA™ software, this IGS data was translated and three-dimensional finite element model of human pelvis-femur complex were generated (Fig. 3) which consisted of sacrum; left and right ilium; left and right femur (proximal end).

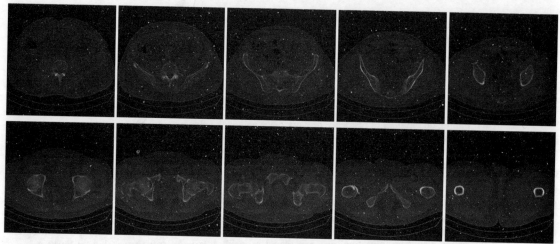

Fig. 1. 10 out of 98 CT scan images of pelvic-femur complex.

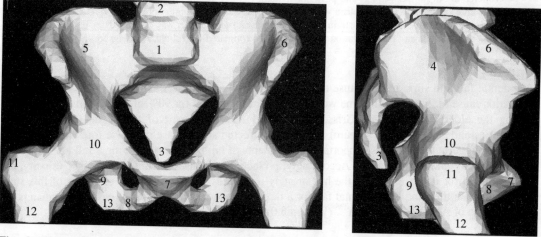

Fig. 2. Three-dimensional solid model of pelvic-femur complex, developed from 98 CT scan images 1: sacrum, 2: 5th lumber, 3: coccyx, 4: ilium (5) iliac fossa, 6: iliac spine, 7: pubic symphysis, 8: pubic ramus, 9: ichium, 10: acetabulum and femoral head, 11: greater trochanter, 12: proximal femur, 13: ischial tuberosity.

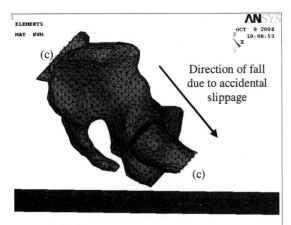 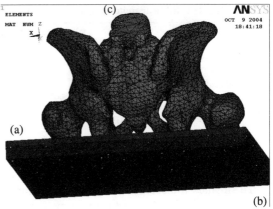

Fig. 3. Three-dimensional finite element model of human pelvic-femur complex with, (a) soft tissue layer, (b) rigid floor and (c) lumped mass, having 9,670 shell (cortical bone); 55,155 tetrahedral (cancellous bone); 2,851 tetrahedral (rigid floor); 6,677 tetrahedral (soft tissue); 57 mass elements, connected through 14,743 nodes.

During the accidental slippage on the floor, pelvis was assumed to hit the floor at an approximate angle of 45°. Hence a floor was modeled accordingly (Fig. 3). The pelvis is covered with soft tissue (fat, flesh and skin) and impact energy is absorbed to some extent by this soft tissue. To represent this effect, a soft tissue layer with 10 mm thickness was considered and it was placed just above the floor (Fig. 3), instead of covering the pelvis. It meant that the pelvis-femur complex will hit the soft tissue layer first and then it will come into indirect contact with the floor. This has reduced the huge CPU time of considering the bulk portion of the buttocks, abdomen and thigh. As we have considered only the pelvic-femur part, to account for the total inertia effect due to full body weight, we used uniformly distributed lumped masses (Fig. 3), attached to 5th lumber location (top surface of the sacrum, 25 kg) and femur locations (bottom surfaces of left and right femurs, 12.5 kg each).

Shell element and solid (tetrahedral) element were used to represent the cortical bone and cancellous bone of the pelvis-femur complex respectively. Soft tissue and rigid floor were modeled with solid (tetrahedral) elements. Mass elements were used to represent lumped mass, equivalent to body mass for upper and lower extremities. These finite elements are for explicit dynamic analysis.

The degrees of freedom (dof) for the ANSYS-LS DYNA^{TM} solid and mass elements are 9 (3 translational, 3 velocity, 3 acceleration) and for the shell elements, dof are 12 (3 translational, 3 rotational, 3 velocity, 3 acceleration) respectively. The pelvic FE model (Fig. 3) contained 9,670 shell elements, 55,155 tetrahedral elements and 57 mass elements. Hence the total 64,882 elements were connected through 11,711 nodes. The FE model of the floor and soft tissue contained 2,851 and 6,677 tetrahedral elements respectively, through 3,022 nodes. Hence in case of the pelvis-femur model with soft tissue and rigid floor (Fig. 3), the total 74,410 elements were connected through 14,743 nodes.

The material properties of pelvic bone were considered to be nonlinear, inelastic, isotropic and homogeneous throughout the pelvic model. We used the bilinear material features of ANSYS-LS DYNA^{TM} with yield stress and tangent modulus for the cortical and cancellous bone. The soft tissue was considered as nonlinear, elastic and homogeneous which was equivalent to the behavior of Mooney-Rivlin Rubber Elastic Model (Moes and Horvath, 2002) of ANSYS-LS DYNA^{TM}. All the properties required for this analysis are given in Table 1.

Table 1. Bilinear material properties, used for FE model of the pelvic bone and soft tissue

	Density ρ (g/cc)	Poisson's Ratio ν	Elastic Modulus E (MPa)	Yield Stress σ_y (MPa)	Tangent Modulus E_t (MPa)
Cortical bone	1.38423[*]	0.3[*]	17,000[**]	110[**]	900[**]
Trabecular bone	0.4491[*]	0.2[*]	380[**]	19[**]	0.01[**]
Soft tissue[***]	0.915[****]	0.4999[***]	-	-	-

[*] Keyak and Falkinstein (2003).
[**] Ozkaya and Nordin (1991).
[***] C_{10} and C_{01} (Mooney-Rivlin parameters) are 100 and 400 kPa respectively (Moes and Horvath, 2002).
[****] Nord and Payne (1995).

To simulate the accidental slippage situation on a finite element model, one needs to know the loading data during the impact between two objects (body and rigid floor). It was considered that the body was slipping on the floor with its CG at height of 88.1 cm (van den Kroonenberg et al., 1995) from the floor impacting the rigid floor with an acceleration of 3.69 m/s^2 (Hsiao and Robinovitch, 1998). It meant that the body would hit the floor with a velocity of 2.55 m/s. The same condition was applied to the pelvic model with the equivalent soft tissue layer and equivalent body mass (Fig. 3). To reduce the CPU time, the distance between the floor and pelvis was kept very low, so that the pelvis came in to contact with the rigid floor within a very short duration. Total dynamic analysis duration was taken as 25 ms.

Results and Discussion

We have used ANSYS® pre-processor (with LS DYNA™ options) to develop the finite element model of human pelvis-femur complex with soft tissue and floor. For getting the solution, the LS DYNA™ solver was used. Results were processed with ANSYS® post-processor and have considered various parameters like von-Mises stress, principal stresses, elastic and plastic strain and velocity.

We have compared the impact of pelvis without and with the equivalent layer of soft tissue (fat-flesh-skin) and considered the zone (right ischial tuberosity), which came into contact with the floor during impact. Von-Mises stress (SEQV) variations with time, for both the cases were considered and presented in the Fig. 4. It was clear that the due to the inclusion of the soft tissue layer, the peak value (207 MPa vs. 260 MPa) of the stress (SEQV) was lower. The peak for the soft tissue layer case occurred at 14.5 ms as compared to 8 ms for the case without the soft tissue. Also there was a delay in the rise of stress (3.5 ms vs 2.5 ms), as the pelvic bone came into contact with the floor at a later stage due to the compression of the soft tissue. Also the stress level came down to a level of 40 MPa at 21.3 ms (Fig. 4 (b)) as compared to 15 ms for the case without soft tissue layer (Fig. 4 (a)). Hence the total stressed duration was 17.8 and 12.5 ms for with and without soft tissue case respectively. There was a stress fluctuation (Fig. 4 (b)) in case of soft tissue in between 3.5 and 21.3 ms, as some amount of stress was absorbed. Also the principal stress variations (S1, S2, S3) with time depicted similar nature for both the cases (Fig. 4).

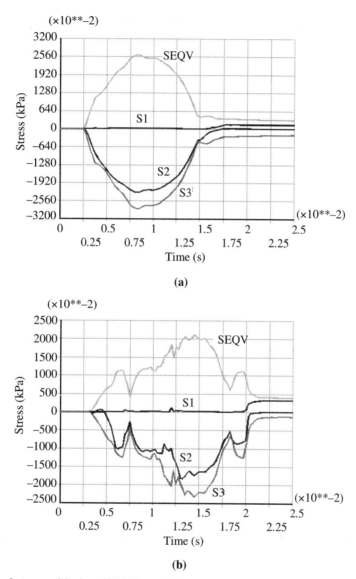

Fig. 4. Variations of stress with time (SEQV: von-Mises stress, S1, S2, S3: principal stresses) in the right ischial tuberosity zone for, (a) without (b) with soft tissue layer cases.

As we considered bone behavior as bilinear (elastic and plastic), the variations of von-Mises elastic (ELEQV) and plastic (PLEQV) strain with time are given in Fig. 5 for both the cases. The highest plastic strains (PLEQV) were 0.212 (at 17.5 ms) and 0.131 (at 22 ms) for without and with soft tissue cases respectively. Also the rise of plastic strain was less steep in case of the pelvis with soft tissue layer. It was also clear from the Fig. 5 that the highest elastic strain (ELEQV) was also lower for the soft tissue case (0.015 at 14.5 ms as compared to 0.021 at 8.4 ms).

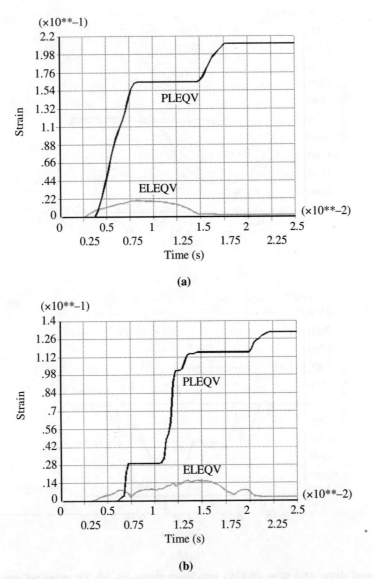

Fig. 5. Variations of strain with time (ELEQV: von-Mises elastic strain, PLEQV: von-Mises plastic strain) in the right ischial tuberosity zone for, (a) without (b) with soft tissue layer cases.

The variations of velocity components (VX, VY, VZ) with time (Fig. 6) for both the cases was also observed. Initial velocity (VZ) was assigned as −2.55 m/s. After impact, this component (VZ) started decreasing while the other components started picking up. For the soft tissue case, VZ component reduced to −0.6 m/s at 25 ms (Fig. 6 (b)) as compared to −1.25 m/s at 25 ms for the other case (Fig. 6 (a)).

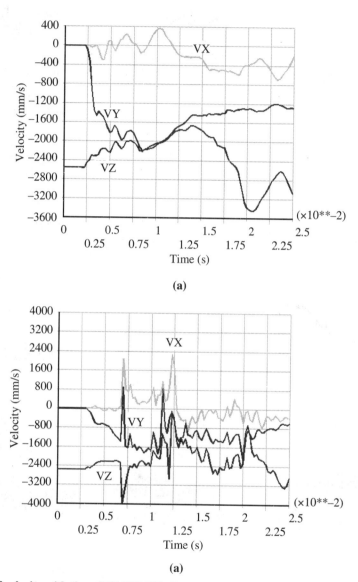

Fig. 6. Variations of velocity with time (VX, VY, VZ: Velocity along X, Y, Z axis direction respectively) in the right ischial tuberosity zone for, (a) without (b) with soft tissue layer cases.

The von-Mises stress contours (posterior, lateral and anterior views) at different instances of impact (5, 9, 16.5 and 20.5 ms) are given in Fig. 7. From these contours, it is clear that the ischial tuberosities, pubic region and sacro-iliac joint areas are heavily stressed. These are also in accordance with the radiographic observations from various clinical investigations (Spencer and Lalanadham, 1985; Tile, 1995; Young and Resnik, 1990).

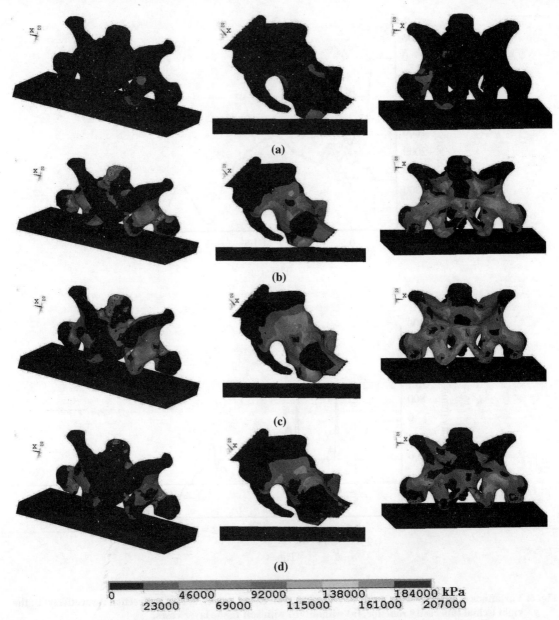

Fig. 7. Von-Mises stress contour (posterior, lateral and anterior views) for four different instances: (a) 5; (b) 9; (c) 16.5; (d) 20.5 ms.

Conclusion

In this investigation, we have developed a nonlinear, inelastic three-dimensional finite element (FE) model of human pelvis-femur complex, based on CT scan data of a male donor. The total FE model (with soft tissue and floor) contained 64,683 tetrahedral elements, 9,670 shell elements and 57 mass elements with 14,743 nodes. We simulated the condition of accidental slippage on the floor with the help of this FE model with and without considering an equivalent layer of soft tissue (skin, flesh and fat). Due to a fall from a height of 88.1 cm and hitting the floor with a velocity of 2.55 m/s, the stress distribution during 25 ms was found to be lower than the case without the soft tissue layer. This is due to the absorption of impulse by the soft tissue. Also the zones of higher stresses are similar to the radiographic observations of past clinical investigations. We considered the cortical shell thickness for the pelvic region to be as 1.5 mm. But it varied from 0.7 to 3.5 mm from location to location. Realistic consideration of cortical bone thickness and non-homogeneous distribution of cancellous bone properties throughout the pelvis will further reduce the peak stress. Hence it can be concluded that this FE model of pelvis-femur complex with soft tissue layer can simulate the fall condition properly and with the help of proper hip padding the severity of the injury can be brought down.

Acknowledgement

This study was supported by the grants (No. SR/S3/MECE/32/2002) from Department of Science and Technology, Government of India, New Delhi. The authors would like to thank Dr. Sujit Das and Mr. Subhash Chandra Jana, Department of CT Scan, The Calcutta Medical Research Institute (CMRI), Kolkata for their assistance to collect the CT scan data from CMRI. The authors also wish to acknowledge Dr. Jayanta Kumar Chakraborty, Assistant Professor, Department of Applied Mechanics, Bengal Engineering and Science University, Shibpur, for his assistance in selection of bone material properties.

References

1. Grisso, J.A., Kelsey, J.L., Strom, B.L., Chiu, G.Y., Maislin, G., O'Brien, L.A., Hoffman, S. and Kaplan, F. 1991. Risk Factors for Falls as a Cause of Hip Fracture in Women: The northeast hip fracture study group. *New England Journal of Medicine*. **324**: 1326-1331.
2. Hsiao, E.T. and Robinovitch, S.N. 1998. Common Protective Movements Govern Unexpected Falls from Standing Height. *Journal of Biomechanics*. **31**: 1-9.
3. Keyak, J.H. and Falkinstein, Y. 2003. Comparison of *in situ* and *in vitro* CT Scan-based Finite Element Model Predictions of Proximal Femoral Fracture Load. *Medical Engineering and Physics*. **25**: 781-787.
4. Van den Kroonenberg, A.J., Hayes, W.C. and McMahon, T.A. 1995. Dynamic Models for Sideways Falls from Standing Height. *Journal of Biomechanical Engineering*. **117**: 309-318.
5. Moes, C.C.M. and Horvath, I. 2002. Using Finite Elements Model of the Human Body for Shape Optimization of Seats: Optimization Material Properties. *International Design Conference - Design 2002*, Dubrovnik, May 14-17.
6. Nord, R.H. and Payne, R.K. 1995. A New Equation Set for Converting Body Density to Percent Body Fat. *Asia Pacific Journal of Clinical Nutrition*. **4**: 177-179.
7. Ozkaya, N. and Nordin, M. 1991. *Fundamentals of Biomechanics-equilibrium, Motion and Deformation, 1st Edn*. Van Nostrand Reinhold, New York. pp. 343-346.
8. Pal, S., Majumder, S., Sarkar, S., Roychowdhury, A. and Chatterjee, D. 2004. Reconstruction of 3D Images from CT Scans for Realistic Finite Element Analysis. *Proceedings of the 2nd International Conference on Medical Diagnostic Techniques and Procedures (ICMDTP)*, April 1-3, Indian Institute of Technology Madras, Chennai. p. 21.
9. Spencer, J.C. and Lalanadham, T. 1985. The Mortality of Patients with Minor Fractures of the Pelvis. *Injury*. **16**: 321-323.
10. Tile, M. 1995. *Fractures of the Pelvis and Acetabulum*. Williams and Wilkins, Baltimore.
11. Young, J.W. and Resnik, C.S. 1990. Fracture of the Pelvis: Current concepts of classification. *American Journal of Roentgenology*. **155**: 1169-1175.

Biomechanics
R.K. Saxena and P. Mishra (Editors)
Copyright © 2005, Anamaya Publishers, New Delhi, India

3. Gripping Techniques for Tendons

Boon Ho Ng[1], Siaw Meng Chou[1] and Krishna Vaibhav[2]

[1]School of Mechanical and Production Engineering,
Nanyang Technological University (NTU), Singapore

[2]Department of Mechanical and Industrial Engineering,
Indian Institute of Technology Roorkee, Roorkee - 247667, India

Abstract: Tensile properties of tendons were examined by many researchers in the past but no generally accepted properties have been adopted so far. Many inherent problems exist in the acquisition of tendon tensile properties. Due to the soft and acqueous nature of tendons, one of the main challenges in acquiring tendon properties lies in the gripping of tendon specimens. In this study, several gripping methods were attempted and tested including serrated jaw, sandpaper and cardboard lining, frozen and air-dried ends. These methods employed manual and pneumatic grips. It was found that 1 kN pneumatic grips (Shimadzu Co.) lined with cardboards provided an adequate grip without perceptible slip and damage to the tendons. Stress concentration at the grip specimen interface reduced substantially using this. Since this method employs no means, which may affect the properties of tendons, this method will possibly predict the true mechanical properties of tendons. An analysis of specimens failed at grip specimen interface versus those failed at mid-substance showed that tensile properties almost remain unaffected.

Introduction

Gripping is a major challenge faced in the testing of tendons. Besides the highly variable biological factors, which affect the tensile properties, many inherent problems exist in the acquisition of tendon tensile properties.

One of the main challenges in acquiring tendon properties lies in the gripping of tendon specimens. Slippage of specimens from the grips during tensile tests is common due to the soft and aqueous nature of tendons. Many past efforts were made to prevent the grip-specimen slip, namely using specially designed hydraulic or pneumatic clamps (Herzog and Gal, 1999; Matthews et al., 1996), serrated jaws (Herzog and Gal, 1999), waterproof abrasive paper, etc. All employed clamps with roughened grip surfaces. However, when slip is prevented, stress concentration may develop in the clamping region, resulting in premature failure of specimens.

Other holding methods have been developed to solve the problem of stress concentration by avoiding direct clamping of tendons. These include suturing, embedding, gluing and rolling methods. However, none of these are perfect for tendon gripping. Although the stress concentration at grip-specimen interface does not exist, problems such as specimen injuries, slippage and laxity emerged.

Special techniques such as cryo-jaws (Riemersma and Schamhardt, 1982) and air-dried ends (Haut, 1983) manage to solve the problems of slip as well as stress concentration. However, the results obtained by these techniques may not be able to represent the properties of living tissues because the properties of the tested specimens have been altered (Viidik, 1987).

An ideal grip should prevent slip without developing stress concentration in the grip-specimen interface. In this study, various gripping methods were evaluated for the best match of this criteria.

Materials and Methods

Thirty chicken flexor digitorum profundus (FDP) tendons were used in this investigation. Fresh chicken feet were obtained directly from an abattoir within three hours after sacrifice. All donors were about five weeks old weighing between 1.6 to 1.8 kg. Chicken feet were processed in non-sterilised environment and the FDP tendons of the second digit were retrieved and wrapped with saline-soaked gauze before they were sealed in airtight plastic bags. Tendons were stored at −40°C within six hours after death until required.

Prior to testing, the specimens were removed from the freezer and thawed in running water. Specimens were left in sealed plastic bag during the thawing process to avoid direct contact with water. The thawing process typically takes about 30 to 45 minutes.

Each specimen was dried thoroughly with tissue paper to remove the surface moisture and its weight was then measured by using Mettler digital weighing scale (model: AT400, D = 0.1 mg). The measured specimens were then kept in saline solution until time for testing. The volume of the specimen was calculated by assuming tendon density as 1120 kg/m^3 (Ker, 1981). The length of the specimen was measured and the average cross-sectional area was estimated by dividing the volume with length, assuming each tendon is cylindrical.

An Instron universal testing machine fitted with 1 kN load cell was employed for the tensile tests. The crosshead speed was set to 60 mm/min and the grip-to-grip length was set at 10 mm. Tendons were tested by using six gripping methods (namely, (1) plain-manual, (2) plain-pneumatic, (3) sandpaper, (4) cardboard, (5) frozen ends and (6) air-dried ends), and each gripping method was tested in quintuplicate. The mean values of ultimate tensile strength, ultimate crosshead strain and structural modulus were compared at 99% confidence level using the student's t method. Note that the structural modulus and ultimate crosshead strain differ from the actual elastic modulus and ultimate strain of the tendon as they are based on the crosshead motion of the grips. Plain manual makes use of standard manual grips provided by Instron for their machines. Specimens were clamped between two pieces of serrated metal. Plain pneumatic uses pneumatic grips (Shimadzu Co.), which were also serrated. However, these were powered by pneumatic force to provide a consistent 1 kN clamping force.

The rest of the methods make use of pneumatic grips. In the third method, sand paper was used in addition to the pneumatic grips while in the fourth method, the pneumatic grips were lined with cardboard. Sandpaper (grade: 400 Cw) and cardboard (thickness 0.4 mm) were cut into small pieces (40 × 50 mm), folded over the tendon ends and clamped for tensile test. The edges of the linings were carefully aligned to the edges of the clamp jaws to avoid uneven stress distribution as well as to ensure a constant length between the jaws.

The fifth method makes use of frozen ends. Both ends of a tendon were exposed to a stream of evaporating liquid nitrogen. Care was taken to avoid direct exposure of liquid nitrogen to the mid-substance. Both frozen ends were clamped as quickly as possible and tensile test was carried out immediately.

The last method makes use of air-dried ends. Specimens were prepared by wrapping up the mid-substance of specimens with saline soaked gauze while both ends were exposed to the air. The length of the wrapped portion was kept at 10 mm and the mid-substance was kept moist continuously. The drying process took about 1.5 h. Specimens were then tested by clamping the dried ends.

Most of the specimens failed at the grip-specimen interface. Thus, the results obtained may not be valid since rupture at the grip-specimen interface indicates that the specimens could have failed prematurely. To confirm this, an investigation was then conducted on 170 pieces of tendons. Specimens were tested by using pneumatic grips with cardboard lining and the results were grouped according to the failure location: failure at the grip-specimen interface (clamp failure) versus failure at the mid-substance (mid-failure). The preparation of these specimens were identical to those used for the evaluation of grips.

Results

The mean ultimate tensile strength of tendons for plain pneumatic grips was 21% lower than the plain manual grips (Table 1). However, the difference is not statistically significant. Compared to the plain manual grips, the utilization of lining, either sandpaper or cardboard, improved the ultimate tensile strength significantly by increment of 44 and 86% respectively. On the other hand, the frozen ends technique caused a 30% decrease compared to plain manual grips while the air-dried ends yielded similar ultimate tensile strength to plain manual grips with an insignificant increase of about 2%.

Table 1. Tensile properties of tendons tested with different grips

Grips	UTS (MPa)	Ultimate Cross-head Strain	Structural Modulus (MPa)
Plain-manual	36.57 ± 5.45	$0.26 \pm 0.04^+$	228.91 ± 24.57
Plain-pneumatic	$29.07 \pm 6.11^+$	$0.24 \pm 0.05^+$	190.13 ± 26.85
Sandpaper	$52.65 \pm 2.95^{*+}$	$0.30 \pm 0.02^+$	$287.56 \pm 9.40^*$
Cardboard	$67.94 \pm 9.55^{*+}$	$0.34 \pm 0.06^+$	$315.97 \pm 35.38^*$
Frozen ends	$25.44 \pm 6.11^+$	$0.33 \pm 0.04^+$	$143.33 \pm 20.87^*$
Air-dried ends	$37.45 \pm 2.22^+$	$0.28 \pm 0.04^+$	227.43 ± 32.60

Mean ± standard deviation
* $P < 0.01$ against plain manual grips
+ $P < 0.01$ against the results reported by Yamada (1970). (UTS = 43 ± 1.7 MPa; Ultimate Strain = 0.084 ± 0.0025)

All grips yielded similar results in ultimate crosshead strain, which ranges from 0.241 to 0.340 (Table 1). It should be noted here that it is the ultimate strain based on the crosshead motion of the jaws. The actual strain in the specimen should be less than this. For the relative comparison between different grips this value can be used. The results of structural modulus show similar trend to those of ultimate tensile strength (Fig. 1). Only the frozen ends and pneumatic grips with lining (both sandpaper and cardboard) were significantly different from the plain manual grips.

In the study on failure location, 72.35% (123 out of 170) of specimens failed at the grip-specimen interface while the rest (47 samples) failed at mid-substance (Fig. 2). No statistically significant difference was found in the ultimate tensile strength, ultimate crosshead strain and structural modulus between these two groups (Table 2).

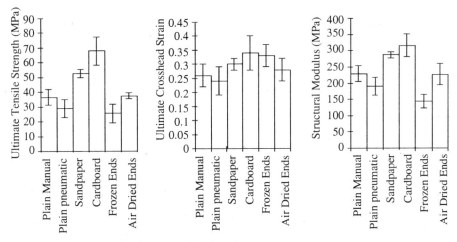

Fig. 1. Tensile properties of tendons versus gripping methods.

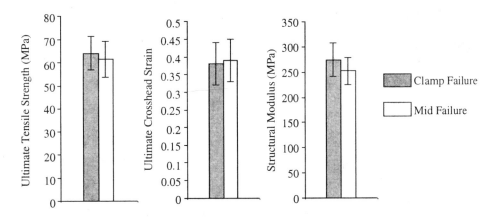

Fig. 2. Comparison of tensile properties for tendons failing at the clamp with those failing at the mid-substance (gripping with cardboard method).

Table 2. Comparison of tensile properties for tendons failing at the clamp with those failing at the mid-substance (gripping with cardboard method)

Tensile Parameter	Clamp Failure (1)	Mid-Failure (2)	Percentage Difference [(1) − (2)]/(1)
Occurrence	123	47	-
UTS (MPa)	63.89 ± 7.22	61.46 ± 7.70	3.80 %
Ultimate cross-head strain	0.38 ± 0.06	0.39 ± 0.06	2.63 %
Structural modulus (MPa)	274.59 ± 33.42	252.54 ± 26.19	8.03 %

Mean ± Standard Deviation.

Discussion

The ultimate tensile strength obtained by the plain manual grips (36.57 ± 5.45 MPa) is close to the value (43 ± 1.7MPa) reported by Yamada (1970) for the flexor tendon of domestic chicken (Table 1). There is no significant difference between these values. On the other hand, the ultimate tensile strength obtained by pneumatic grips with cardboard lining (67.94 ± 9.55 MPa) is significantly higher than the value reported by Yamada (1970) and also using the frozen ends and air dried ends in this study. The results indicate that the utilization of cardboard substantially reduced the stress concentration at the grip-specimen interface. It is predicted that cardboard, which is made up of coarse fibre, is compressible. It acts as a spongy material to dissipate stress at the grip-specimen interface.

Conclusions

It was found that pneumatic cardboard lining significantly improved the gripping of tendons. The utilization of cardboard lining significantly reduced, although could not refrain completely, the stress concentration at the grip-specimen interface. Pneumatic grips with cardboard lining are the best among the tested gripping methods with regard to the ease of use, slip prevention and minimum stress concentration at the clamping region. This method was found to be effective in gripping thin tendons. It was also found that the location of failure did not affect the tensile properties of tendons. The results of specimens failing at grip-specimen interface can also be treated as valid results.

Acknowledgement

The authors are grateful to Miss Goy Hsu Ann for her invaluable advice and SUPMAR Pvt. Ltd. for providing specimens.

References

1. Bader, D.L. and Bouten, C. 2000. Biomechanics of Soft Tissues. In: *Clinical Biomechanics*. Dvir, Z. (Ed.). Churchill Livingstone, New York. pp. 35-64.
2. Fung, Y.C. 1993. Bioviscoelastic Solids. In: *Biomechanics: Mechanical Properties of Living Tissues, 2nd Ed.* Fung, Y.C. (Ed.). Springer-Verlag, New York. pp. 242-320.
3. Haut, R.C. 1983. Age-dependent Influence of Strain Rate on the Tensile Failure of Rat-tail Tendon. *Journal of Tensile Engineering.* **105**: 296-299.
4. Herrick, W.C., Kingsbury, H.B. and Lou, D.Y.S. 1978. A Study of the Normal Range of Strain, Strain Rate and Stiffness of Tendon. *Journal of Biomedical Materials Research.* **12**: 877-894.
5. Herzog, W. and Gal, J. 1999. Tendon. In: *Biomechanics of The Musculo-skeletal System, 2nd Ed.* Nigg, B.M., Herzog, W. (Eds.). John Wiley & Sons Ltd, New York. pp. 127-147.
6. Ker, R.F., 1981. Dynamic Tensile Properties of the Plantaris Tendon of Sheep (*Ovis aries*). *Journal of Experimental Biology.* **93**: 283-302.
7. Matthews, G.L., Keegan, K.G. and Graham, H.L. 1996. Effects of Tendon Grip Technique (Frozen versus Unfrozen) on in vitro Surface Strain Measurements of the Equine Deep Digital Flexor Tendon. *American Journal of Veterinary Research.* **57(1)**: 111-115.
8. Mow, V.C., Hayes, W.C. (Eds.). *Basic Orthopaedic Biomechanics 2nd Ed.* Lippincott-Raven, Philadelphia. pp. 209-252.
9. Riemersma, D.J. and Schamhardt, H.C. 1982. The Cryo-jaw, A Clamp Designed for in vitro Rheology Studies of Horse Digital Flexor Tendons. *Journal of Biomechanics.* **15**: 619-620.
10. Rigby, B.J., Hirai, N., Spikes, J.D. and Eyring, H. 1959. The Mechanical Properties of Rat Tail Tendon. *Journal of General Physiology.* **43**: 265-283.

11. Smith, C.W., Young, L.S. and Kearney, J.N. 1996. Mechanical Properties of Tendons: Changes with sterilization and preservation. *Journal of Tensile Engineering.* **118**: 56-61.
12. Viidik, A. 1987. Properties of Tendons and Ligaments. In: *Handbook of Bioengineering.* Skalak, R., Chien, S. (Eds.). McGraw-Hill, New York. pp. 330-341.
13. Wang, X.T. and Ker, R.F. 1995. Creep Rupture of Wallaby Tail Tendons. *Journal of Experimental Biology.* **198**: 831-845.
14. Woo, S.L., Livesay, G.A., Runco, T.J. and Young, E.P. 1997. Structure and Function of Tendons and Ligaments.
15. Yamada, H. 1970. *Strength of Biological Materials.* The Williams & Wilkins Company, Maryland. pp. 99-105.

Biomechanics
R.K. Saxena and P. Mishra (Editors)
Copyright © 2005, Anamaya Publishers, New Delhi, India

4. Foot Sole Soft Tissue Characterization in Diabetic Neuropathy Using Texture Analysis

P. Manika[1], K.M. Patil[1], V. Balasubramanian[1], V.B. Narayanamurthy[2] and R. Parivalavan[2]

[1]Biomedical Engineering Division, Department of Applied Mechanics,
Indian Institute of Technology Madras, Chennai - 600036, India

[2]Diabetic Foot Clinic, Sundaram Medical Foundation, Anna Nagar,
Chennai - 600040, India

Abstract: Diabetes alters the properties of foot sole soft tissue resulting in increased tissue stresses and stress gradients at the skin-fat interface. This study attempts to analyze the textural characteristics of the foot sole soft tissue using ultrasound (US) images. Alteration in textural characteristics which possibly reflect the structural changes at the skin-fat interface with the progress of diabetes are noted. This is compared with normal feet to look for changes, which may signify as early predisposing factors in ulcer formation.

US images were acquired by placing a 7.5 MHz linear array scanner, over the dermal-epidermal junction of the soft tissue in parallel to skin surface. The US images and the foot sole hardness were measured using a Shore meter (in degree Shore) for both normal (40 feet areas) and diabetic subjects (20 feet areas). Textural features were derived from the US images using angular nearest neighbor gray level co-occurrence matrices.

Welch ANOVA test on the results showed that the mean values of the texture parameters namely, entropy, angular second moment (ASM) and inverse difference moment (IDM), were found to be significantly higher in normal as compared to diabetic feet in foot sole areas like the medial heel ($P < 0.001$; $P < 0.022$; $P < 0.003$), lateral heel ($P < 0.007$; $P < 0.033$; $P < 0.015$), first metatarsal ($P < 0.001$; $P < 0.015$; $P < 0.002$), lateral metatarsal ($P < 0.001$; $P < 0.003$; $P < 0.001$) and the big toe ($P < 0.007$; $P < 0.09$; $P < 0.049$). It is noted that there exists a high correlation between the foot sole hardness values and the textural features for both normal and diabetic subjects in all the above ulcer prone foot sole areas. Thus, different levels of the mean values of ASM, entropy and IDM texture features may signify different stages of progress of diabetes and possibly help the clinician in taking appropriate preventive measures.

Introduction

Diabetes continues to be one of the most common factors associated with lower-extremity amputation in post-industrialized and developing countries. Diabetic foot ulceration leads to life or limb threatening complications and is the single most common precursor to amputation. Work done by Birke et al. (1995) identifies diabetic foot ulcers as a component in 85% of lower-extremity amputations. It has also been found that the geometrical and material properties of the foot sole soft tissue change prior to plantar ulcers (Robertson et al., 2002). Such changes in the properties of the foot sole soft tissue result into increased tissue stresses and stress gradients (Thompson et al, 1999; Thomas et al., 2004). Earlier results reported (Charanya et al., 2004) establish that the foot sole hardness values (as characterized by shore levels) increase in certain foot sole areas prone to ulcer development and hence has been measured for each of such areas in our case. The changes in the thickness of the foot sole may alter the pattern of internal stresses in the foot. Study performed by Thompson et al. (1999) reports the skin-fat interface as a likely site of initial ulcer formation due to increased peak first principal stresses at this location with changes in its material properties. Therefore a reliable noninvasive method of early detection of such structural changes occurring at the skin-fat interface, for timely diagnosis about the existence of foot sole soft tissue ulcers, is clearly desirable. Diagnostic ultrasound has been a useful clinical tool for imaging organs and soft tissues in the human body, for more than two decades (Wells, 1977). This work attempts to perform differentiation between the normal and abnormal foot sole soft tissue based upon the examination of B-scan images. However, for diagnosing the progress of diabetes particularly in its early phase, the main obstacle is very subtle visual difference between sonograms of normal and abnormal foot sole soft tissue. One possible approach, as adopted in this study, can be found in using texture analysis to extract meaningful features or specific image properties and categorize or classify such sonograms. This is so because structural changes, as observed with the advancement of diabetes, leads to changes in acoustical properties of the foot sole soft tissue which can then be detected by ultrasound as a textural pattern different from the normal one.

The term *texture* is usually used to describe intrinsic properties of an object, such as granulation, roughness, coarseness etc. Formally, texture deals with a set of local neighbourhood properties. Hence, it can be stated that textures are like considering a group of pixels, sometimes called texture primitive or texture element and the texture described is highly dependent on the number of pixels considered (the texture scale); texture description is scale dependent. In the present study, the authors are concerned with the task of developing a set of texture measures which quantitatively indicate changes in ultrasound images inaccessible to unaided human appreciation.

Medical images e.g. ultrasound have been observed to convey useful diagnostic information through their visual texture. A survey of the usefulness of texture features to discriminate between normal-abnormal classes of radiographs is given in Sutton and Hall (1972); Chien and Fu (1974); Mir (1995). Work has also been carried out for texture analysis of ultrasound images as reported in Reath et al. (1985); Wagner et al. (1986); Garra et al. (1989).

In this study, spatial domain statistical techniques has been utilized because statistical methods have proven superior to frequency domain methods (Weszaka et al., 1976). Specifically, spatial grey level dependence method has been used in the present study as a foundational approach for establishing appropriate features that may be used to discriminate ultrasound image textures.

Methodology

Spatial Gray-Tone Spatial-Dependence Matrices (SGLDM)
The spatial gray level dependence method of texture description (Haralick et al., 1973) is concerned with the spatial distribution and dependence among the gray levels in a local area. It encodes repeated occurrences of some gray level configuration in the texture. Textural features are later derived from such angular nearest neighbour spatial dependence matrices.

A co-occurrence matrix is the joint probability of occurrence of gray levels 'a' and 'b' for two pixels with a defined spatial relationship in an image. This spatial relationship is defined in terms of inter-sample spacing d and orientation θ, where $\theta = 0, 45, 90$ and $135°$. Let I be the image to be analyzed of size $M \times N$. An occurrence of some gray level configuration can be described as a matrix of relative frequencies $P_{\alpha,d}(a,b)$, representing the number of times two pixels with gray-levels a and b occur in I separated by a distance d and orientation, from each other. Matrices are made symmetric, by using

$$P_{\alpha,d}(a,b) = P_{\alpha,d}(a,b) + P_{\alpha,-d}(a,b) \qquad (1)$$

The non-normalized definition of co-occurrence among pixels with their adjacency as function of angle and distance as shown in the diagram, can be represented formally as

$$P_{0,d}(a,b) = \left| \left\{ \begin{array}{l} \{(k,l),(m,n)\} \in D: \\ k-m=0, |l-n|=d, f(k,l)=a, f(m,n)=b \end{array} \right\} \right|$$

$$P_{45,d}(a,b) = \left| \left\{ \begin{array}{l} \{(k,l),(m,n)\} \in D: (k-m=d, l-n=-d) \\ \text{or } (k-m=-d, l-n=d), f(k,l)=a, f(m,n)=b \end{array} \right\} \right|$$

$$P_{90,d}(a,b) = \left| \left\{ \begin{array}{l} \{(k,l),(m,n)\} \in D: |k-m|=d, l-n=0, \\ f(k,l)=a, f(m,n)=b \end{array} \right\} \right|$$

$$P_{135,d}(a,b) = \left| \left\{ \begin{array}{l} \{(k,l),(m,n)\} \in D: (k-m=d, l-n=d) \\ \text{or } (k-m=-d, l-n=-d), f(k,l)=a, f(m,n)=b \end{array} \right\} \right|$$

where $|\{\ldots\}|$ refers to set cardinality. (2)

Textural Features Extracted from Gray-Tone Spatial-Dependence Matrices

Following the initial assumption that all texture related information is contained in gray tone spatial dependence matrices, we derived all the texture features from them. Using the SGLDM method, approximately two dozen co-occurrence features can be obtained (Haralick, 1978). In this study we restrict representation by six features for texture discrimination purpose. For each matrix, the following 6 normalized features were computed:

(1) $$\text{Angular Second Moment} = \sum_{i=1}^{N_g} \sum_{j=1}^{N_g} \left(p_{\alpha,d}(i,j) \right)^2$$

(2) $$\text{Entropy} = -\sum_{i=1}^{N_g} \sum_{j=1}^{N_g} p_{\alpha,d}(i,j) \log \left(p_{\alpha,d}(i,j) \right)$$

(3) $$\text{Contrast} = \sum_{n=0}^{N_g-1} n^2 \left\{ \sum_{\substack{i=1 \\ |i-j|=n}}^{N_g} \sum_{j=1}^{N_g} p_{\alpha,d}(i,j) \right\}$$

(4) $$\text{Inverse Difference Moment} = \sum_{i=1}^{N_g} \sum_{j=1}^{N_g} p_{\alpha,d}(i,j) \Big/ \left(1+(i-j)^2\right)$$

(5) $$\text{Correlation} = \sum_{i=1}^{N_g} \sum_{j=1}^{N_g} \left\{ \left((i-\mu_x)(j-\mu_y) p_{\alpha,d}(i,j)\right) \Big/ \sigma_x \sigma_y \right\}$$

(6) $$\text{Sum of squares} = \sum_{i=1}^{N_g}(i-\mu)^2 \sum_{j=1}^{N_g} p_{\alpha,d}(i,j)$$

where μ_x, μ_y, σ_x and σ_y are the means and standard deviations of the marginal distributions associated with $p_{\alpha,d}(i, j) = P_{\alpha,d}(a, b)/R$, R is a normalizing constant (here taken to be as the maximum value in the $P_{\alpha,d}$ co-occurrence matrix) and N_g is the number of gray levels in the image from which the gray level dependence matrices are extracted.

Data Acquisition

Foot Sole Hardness Measurement (Shore Level)

The instrument used for measuring the hardness of the foot sole is the shore meter, or durometer or hardness tester (as per ASTM-D 2240 standards). It is used to quantify the levels of hardness of soft materials such as rubber or soft tissue. The shore meter works on the principle of indentation using a truncated cone-tipped indenter. For each subject, normal and diabetic, measurements of hardness of the foot sole were carried out in all 10 standard foot sole areas as indicated in Fig. 1(a).

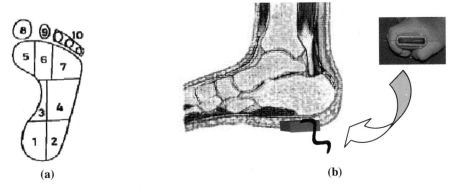

Fig. 1. (a) Foot sole areas, (b) The placement of an ultrasound probe over the foot sole soft tissue in parallel to the skin surface (held in medial to lateral direction).

Foot Sole Soft Tissue Image Acquisition

For the purpose of study, ultrasound images were recorded in B-mode by placing a 7.5 MHz linear array transducer over the dermal-epidermal junction of the foot sole soft tissue in parallel to the skin surface. As shown in Fig. 1(b), it was oriented either from lateral to medial or medial to lateral direction (depending upon the foot region scanned) for all foot sole areas prone to plantar ulcers namely, medial heel (area 1), lateral heel (area 2), first metatarsal (area 5), lateral metatarsals (area 7) and big toe (area 8). US images were recorded for normal as well as diabetic subjects.

Results

Data Set Description

Ultrasound images of the foot sole soft tissue used in this research were all captured from the same machine and then digitized with 480 × 576 pixels and 256 gray level resolution. Two sets of data have been taken: normal subjects (40 feet areas from 8 feet) and diabetic subjects (20 feet areas from 4 feet) since for each subject images in five-foot sole areas; area 1, area 2, area 5, area 7 and area 8 were recorded. As this study attempts to address the discrimination ability of image texture for detecting early changes occurring at the skin-fat interface, diabetic patients were considered in different stages of the disease. From each image, a region of interest of size 128 × 128 pixels was chosen. Figure 2 shows US images of foot sole soft tissue one each of normal and diabetic foot. Because the ultrasound images were taken by different clinicians, in order to eliminate effects of unequal gain settings, the mean of each image was removed before processing.

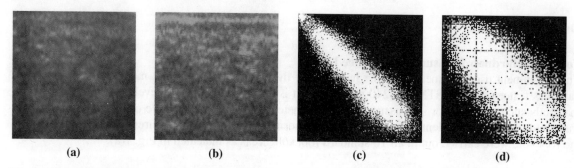

Fig. 2. (a) and (b) US images of foot sole soft tissue of a normal and a diabetic foot along with their respective gray-tone spatial dependence co-occurrence matrices in (c) and (d).

Comparison of Texture Features Between Normal and Diabetic Subjects at Early Phase of Neuropathy

Texture features on US images in each data set are computed using the SGLDM method and later examined for their sensitivity towards soft tissue changes that may occur with the progress of diabetes. As the texture changes in the US images are slow and visually non differentiable, different values of the inter-sampling distance d, varying from 2 to 10, have been used in the SGLDM method and an average of features for all d values are used for comparison.

Tables 1, 2 and 3 tabulate the values of texture measures namely, angular second moment (ASM), entropy and inverse difference moment (IDM), for different shore level values in all foot areas of both normal and diabetic subjects respectively.

Table 1. Comparison of normalized mean ASM feature values in the different ulcer prone foot areas for normal and diabetic subjects in standing position or loaded foot conditions

Foot Type	Foot Areas				
	1	2	5	7	8
Normal (Shore 20)	82.803	93.001	109.02	43.988	98.793
Diabetic (Shore 20)	-	-	1.1835	1.1488	-
Diabetic (Shore 25)	-	-	-	1.2646	1.1644
Diabetic (Shore 30)	-	-	-	1.3926	2.4149
Diabetic (Shore 35)	1.3141	2.2201	1.3737	-	-
Diabetic (Shore 40)	1.4868	3.3429	2.855	1.4189	-
Diabetic (Shore 45)	4.7145	-	-	-	-

Table 2. Comparison of normalized mean entropy feature values in the different ulcer prone foot areas for normal and diabetic subjects in standing position or loaded foot conditions

Foot Type	Foot Areas				
	1	2	5	7	8
Normal (Shore 20)	525.24	622.66	753.14	326.52	800.83
Diabetic (Shore 20)	-	-	63.742	32.241	-
Diabetic (Shore 25)	-	-	-	49.042	70.71
Diabetic (Shore 30)	-	-	-	67.878	181.56
Diabetic (Shore 35)	76.271	91.909	100.67	-	-
Diabetic (Shore 40)	121.7	122.78	196.35	96.549	-
Diabetic (Shore 45)	179.77	-	-	-	-

Table 3. Comparison of normalized mean IDM feature values in the different ulcer prone foot areas for normal and diabetic subjects in standing position or loaded foot conditions

Foot Type	Foot Areas				
	1	2	5	7	8
Normal (Shore 20)	41.755	31.121	44.605	44.002	46.564
Diabetic (Shore 20)	-	-	2.1475	1.6912	-
Diabetic (Shore 25)	-	-	-	2.3505	2.3883
Diabetic (Shore 30)	-	-	-	2.021	6.1446
Diabetic (Shore 35)	2.3996	3.3956	2.2485	-	-
Diabetic (Shore 40)	4.5952	4.1338	5.3029	2.9058	-
Diabetic (Shore 45)	5.8041	-	-	-	-

It is observed that there exists a significant texture difference between the normal and diabetic foot sole soft tissue in all foot areas. ASM is a measure of homogeneity of the image while entropy and IDM are highest for an image with equal spatial distribution of all its gray levels. Thus, higher values of ASM, entropy and IDM together imply uniformly distributed entries of large magnitude in the co-occurrence matrix computed from images of normal feet as compared to diabetic feet. It can also be seen for diabetic subjects that ASM, entropy and IDM values gradually increase with the increase in shore levels or the hardness of foot sole and causes a transition towards a slightly finer image texture. However even in cases of high shore levels in diabetic subjects, the texture parameters have an immense numerical difference from their values found in normal subjects.

Comparisons of mean ASM, entropy and IDM feature values are made between normal and diabetic subjects with different foot sole hardness levels by statistical tests to find the significant differences between them. Statistical studies, involving calculation of Welch ANOVA test followed by post hoc Dunnett 't' tests for parametric ASM, entropy and IDM data of all foot sole areas, are carried out separately for finding the 'P' values. The mean ASM, entropy and IDM values for diabetic subjects as compared to normal feet, are found to be significantly different in the foot sole areas, namely, area 1 ($P < 0.001$; $P < 0.022$; $P < 0.003$), area 2 ($P < 0.007$; $P < 0.033$; $P < 0.015$), area 5 ($P < 0.001$; $P < 0.015$; $P < 0.002$), area 7 ($P < 0.001$; $P < 0.003$; $P < 0.001$) and area 8 ($P < 0.007$; $P < 0.09$; $P < 0.049$).

Correlation between Textural Features and Foot Sole Soft Tissue Hardness

Figure 3 (a) to (c) represent the variations of ASM, entropy and IDM respectively, with different levels of hardness of foot sole in area 7 for 20 diabetic feet areas. The correlation coefficients (r) for all the above 3 are found to be high of the order of 0.91, 0.998 and 0.875. Table 4 represents coefficient of correlation between ASM, entropy and IDM, each individually with shore levels in all the specified areas of the foot.

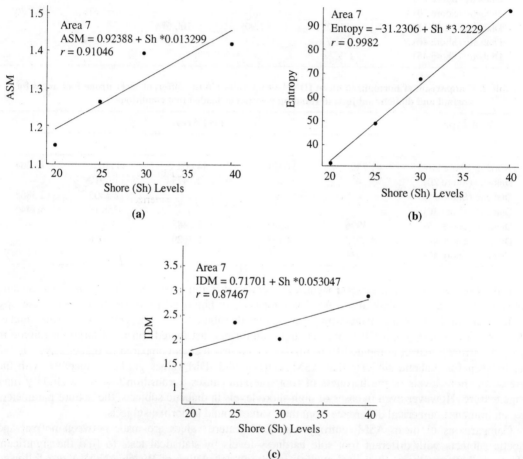

Fig. 3. Variations of ASM, entropy and IDM with levels of shore in area 7 for diabetic subjects from texture analysis of their foot sole soft tissue US images shown in (a), (b) and (c), respectively.

Conclusions

The mean values of ASM, entropy and IDM texture parameters are able to distinguish normal from diabetic feet in different foot sole hardness levels, in all specified areas of the foot and may serve to be good indicators of diabetes in its early phase. Also changes in the texture parameter values with varying shore levels suggest alterations in image properties possibly due to changes occurring in the foot sole soft tissue with the progress of diabetes. In this study, high correlation coefficients of the order of 0.8 or above, 0.9 or above, and 0.76 or above, were observed for shore values with each of ASM, entropy and IDM respectively.

References

1. Birke, J.A., Novick, A. Hawkins, E.S. and Charles, P.A. 1995. A Review of Causes of Foot Ulcerations in Patients with Diabetes Mellitus. *J. Prosthet. Orthot.* **14**: 13-22.
2. Charanya, G., Patil, K.M., Thoms, V.J., Narayanamurthy, V.B., Parivalavan, R. and Visvanath, K. 2004. Standing Foot Pressure Image Analysis for Variations in Foot Sole Soft Tissue Properties and Levels of Diabetic Neuropathy. *Innovation Technol. Biol. Med.* **25**: 23-33.
3. Chien, Y.P. and Fu, K.S. 1974. Recognition of X-ray Picture Patterns. *IEEE Trans. on Syst., Man. and Cyber.* **SMC-4**: 145-156.
4. Garra, B.S., Insana, M.F., Shawker, T.H., Wagner, R.F., Bradford, M. and Rusell, M. 1989. Quantitative Ultrasonic Detection and Classification of Diffuse Liver Disease Comparison with Human Observer Performance. *Invest. Radiol.* **24**: 196-203.
5. Haralick, R.M., Shanmugam, K. and Dinstein, I. 1973. Textural Features for Image Classification. *IEEE Trans. on Syst., Man. Cyber.* **SMC-3, 6**: 610-621.
6. Haralick, R.M. 1978. Statistical and Structural Approaches to Texture. *Proc 4th Int. Joint Conf. on Pattern Recognition.* pp. 45-69.
7. Mir, A.H., Hanmandlu, M. and Tandon, S.N. 1995. Texture Analysis of CT Images. *IEEE Engg. in Med. and Biol.* **14**: 781-786.
8. Reath, U., Schlaps, D. and Limberg, B. 1985. Diagnostic Accuracy of Computerized B-scan Texture Analysis and Conventional Ultrasonography in Diffuse Parenchymal and Malignant Liver Disease, *J. of Clin. Engng.* **13**: 87-99.
9. Robertson, D.D., Mueller, M.M., Smith, K.E., Commean, P.K., Pilgram, T. and Johnson, J.E. 2002. Structural Changes in the Forefoot of Individuals with Diabetes and Prior Plantar Ulcer. *J. Bone Joint Surg.* **84-A**: 1395-1404.
10. Sutton, R.N. and Hall, E.L. 1972. Texture Measures for Automatic Classification of Pulmonary Disease. *IEEE Trans. on Computers.* **C-21**: 667-676.
11. Thomas, V.J., Patil, K.M. and Radhakrishnan, S. 2004. Three-dimensional Stress Analysis for the Mechanics of Plantar Ulcers in Diabetic Neuropathy. *Med. Biol. Eng. Comput.* **42**: 230-235.
12. Thompson, D.L.L., Cao, D. and Davis, B.L. 1999. Effects of Diabetic Induced Soft Tissue Changes on Stress Distribution in the Calcaneal Soft Tissue. *Proc. XVII ISB Congress*, Calgury. pp. 8-13.
13. Wagner, R.F., Insana, M.F. and Brown, G. 1986. Unified Approach to the Detection and Classification of Speckle Texture in Diagnostic Ultrasound. *Opt. Engng*, **25**: 743-748.
14. Wells, P.N. 1977. *Biomedical Ultrasonics*. Academic Press, New York.
15. Weszaka, J.S., Dyer, C.R., Rosenfield, A. 1976. A Comparative Study of Texture Measures for Terrain Classification. *IEEE Trans. on Syst., Man. Cyber.* **SMC-6**: 269-285.

Biomechanics
R.K. Saxena and P. Mishra (Editors)
Copyright © 2005, Anamaya Publishers, New Delhi, India

5. Comparative Study of Static Balance Related Variables among the Indian Adult Male and Female

Manoj Kumar Chaudhary and Dhananjoy Shaw
Department of Natural/Medical Science, Biomechanics Laboratory, I.G.I.P.E.S.S.,
University of Delhi, New Delhi - 110018, India

Abstract: In order to attain perfection, excellence and better work-efficiency in the adult ages of human life, the researchers tempted and inspirited to undertake the study "A comparative study of static balance related variables among the Indian adult male and female". The adult people refers to the age ranging from 20 to 35 yrs, to accomplish the possibility of better utility in accordance with the literature already available. The study delimited to the objectives: (1) To study the trends of static balance related variables of adult male and female with the help of mean, standard deviation and bar diagram for each selected age categories. (2) To compare among the adult age group of male (i.e. 20-35 yrs) in regard to percentage height of center of gravity. (3) To compare among the adult age group of female (i.e. 20-35 yrs) in regard to percentage height of center of gravity. (4) To study the sex differences in each selected adult age categories i.e. 20, 21, 22, 23, 24, 25, 26, 27, 28, 29, 30, 31, 32, 33, 34 and 35 yrs in regard to percentage height of center of gravity. It was hypothesized that: (a) "Age is related to percentage height of center of gravity." (b) "There will not be any sex differences in regard to percentage height of center of gravity of adult in the age categories from 20 to 35 yrs." Keeping in the view the purpose of the study, a large number of male and female (10-15 in number) for each age category were randomly selected. The age categories in the percent study viz. 20, 21, 22, 23, 24, 25, 26, 27, 28, 29, 30, 31, 32, 33, 34 and 35 yrs refer to age ranging from 20 to 35 yrs. The following variables were selected to meet the purpose of study: (1) Height, (2) Weight, (3) Ponderal Index, (4) Height of center of gravity, and (5) Percentage height of center of gravity. The method followed in this study was the conventional "reaction board method" (Shaw 2000), which is the easiest way to determine the height of the center of gravity in a motionless condition for living body. The following statistical techniques were used to analyses the data: (1) Descriptive statistics (mean and standard deviation), (2) '1' ratio to compare between the sexes age wise, and (3) Analysis of variance to compare among the selected age groups of adulthood, independently for male and female.

Hypothesis was tested at 0.05 level of significance. On the basis of the finding, the following conclusions have been drawn: (1) By and large the height of male or female belonging to their selected age group from 20 to 35 yrs to be identical. (2) The weight of adult male or female belonging to selected age group exhibits a fluctuating trend. (3) The female having higher Ponderal Index was found in all the selected age group. (4) Male having higher height of center of gravity than female found in all the selected age group. (5) The percentage height of center of gravity of female was found to be higher than that of male in selected age group. (6) The comparison among the selected age group of male or female on the percentage height of center of gravity found to be statistically different. (7) The comparison between adult male and female of age group namely 30, 31, 32, 33, 34 and 35 yrs in regard to percentage height of center of gravity was found to be statistically different.

In order to attain perfection, excellence and better work-efficiency in the adult ages of human life, the researchers are tempted and inspirited to undertake the study "A Comparative Study on Height of Center of Gravity in Adulthood". The adult people refers to the age ranging from 20 to 35 yrs, to accomplish the possibility of better utility in accordance with the literature already available. The main intention of the researchers was to concentrate the study between 20 and 35 yrs of age to verify the sex differences in regard to balance related variables. It is well established that there are physical differences between the sexes specifically height and weight. Thus, while considering the nature of height and weight, changes during adulthood will definitely affect the height of the centre of gravity, so that a literature support will be possible to provide for the researchers of other fields, while dealing with the subjects age ranging from 20 to 35 yrs for their best professional practices as a basis of understanding of balance mechanism of such population. The reaction board method is well-established method for determining location of center of gravity. Shaw et al. (1998) in their study of "Validation of Indigenously Developed Reaction Board Method" concluded that the specimen as a whole ($N = 264$) have significant correlation ($r = 0.30$) between the weight and length at 0.05 level. The coefficient of correlation was found to be statistically significant ($r = 0.52$) in the medium length sample ($N = 34$). Whereas, the coefficient of correlation between length and weight in regard to the samples light weight ($N = 52$), middle weight ($N = 159$), heavy weight ($N = 53$), short length ($N = 35$) and tall length ($N = 195$) were found to be statistically insignificant at 0.05 level.

The relationship of location of center of gravity (CG) at the longitudinal axis determined by reaction board and as well as suspension method confirms the criterion validity. While specimen were grouped, in regard to weight and height were significantly correlated as CG was measured by both the systems. The analysis in regard to cross validity coefficient between reaction board method and suspension method for determining centre of gravity from bottom and top were found to be statistically significant at 0.05 level, with respect to all the classified group specimen or samples namely combined group ($r = 0.93$ and $r = 0.99$ for CG measured from bottom and top respectively), light weight group ($r = 0.88$ and $r = 0.93$ for CG measured from bottom and top, respectively), middle weight group ($r = 0.94$ and $r = 1.00$ for CG measured from bottom and top, respectively), heavy weight group ($r = 0.98$ and $r = 1.00$ for CG measured for bottom and top, respectively), short length group ($r = 0.87$ and $r = 0.89$ for CG measured from bottom and top, respectively) and medium length group($r = 1.00$ and $r = 0.99$ for CG measured from bottom and top, respectively) and tall length group ($r = 0.93$ and $r = 0.96$ for CG measured from bottom and top, respectively). Shaw et al. (1998) in their study of "Reliability of Reaction Board Method for Determination for Three-Dimensional Centre of Gravity" concluded that the reliability coefficient of centre of gravity in 'Z' dimension remain uninfluenced by the sex, conditioning, general physical education programmed and climatic conditions whereas 'Y' dimension observed little sex differences. It is concluded that the indigenously developed reaction board method is a reliable technique/protocol for determining three-dimensional centre of gravity irrespective of sex difference, growth, general physical education programmed and climatic conditions. Shaw et al. (1998) in their study of "Reliability of Reaction Board Method for Determination of Selected Segmental Mass" concluded that the conditioning, maturity and climatic changes having certain influences on the reliability limits of female subjects in regard to mass of arms but has no influence on the reliability limits on male subjects in regard to mass of arms. That the sex, conditioning, growth, general physical education programmed and climatic conditions do not influence the magnitude of coefficient of correlation of mass of trunk in all the considered groups. Ray and Sen (1983) studied the whole body CG and it was determined as a point of intersection of transverse (X-axis), frontal (Y-axis) and sagital (Z-axis) plane on 27 East Indian males. The CG in X-axis was 40.93% of total foot length from the posterior end of the feet, whereas, the CG in Y-axis was 48.43% of the foot breadth from the right side of the footmark. The respective values of CG in 'X' 'Y' and 'Z' axis in case of standing (feet held together) arm up condition were 40.96, 48.18 and 59.61%.

A significant change in the CG in Z-axis was observed with change in posture. The values were compared with western studies. It was observed that increased body height, body weight shifts the CG towards the head, whereas, the CG is shifted downward if height is unaltered but body weight is increased. Age has practically no influence over the change in the CG.

Objectives of the Study
The study was delimited to the following objectives:

- To study the trends of static balance related variables of adult males and females with the help of mean, standard deviation for each selected age categories.
- To compare among adult age group of male (i.e. 20-35 yrs) in regard to percentage height of CG.
- To compare among adult age group of female (i.e. 20-35 yrs) in regard to percentage height of CG.
- To study the sex differences in each selected adult age categories i.e. 20, 21, 22, 23, 24, 25, 26, 27, 28, 29, 30, 31, 32, 33, 34, 35 yrs in regard to percentage height of CG.

Hypothesis
It was hypothesized that:

1. Age is related to percentage height of CG.
2. There will be significant differences in regard to percentage height of CG of adult in the age categories from 20 to 35 yrs and will have sex differences in regard to percentage height of CG.

Selection of Variables
Keeping in mind the review of related literature, modern trends, scientific authenticity and administrative feasibility, the following variables were selected to meet the purpose of study:

1. Height (m)
2. Weight (kg)
3. Ponderal Index (m/kg)
4. Height of CG (m)
5. Percentage Height of CG (Percentage)

Collection of Data
Various scientific equipments were used to collect the data such as lever based weighing machine and anthropometric rod were used for weight and height respectively. For height of CG, measurement reaction board method was used. Adopting the standardized methods explained and recommended by Sodhi (1980), Das and Ganguly (1982), Clarke (1987), Mathews and Fox (1981), Shaw, Kaushik and Kaushik (1998), Shaw et al. (1998) etc.

Determination of Centre of Gravity

Method
The method followed in this study was the conventional "reaction board method" in a motionless condition for living body.

Apparatus
Different instruments, including the fabricated one, used for the determination of the centre of gravity are described as follows:

Platform Balance: A lever based weighing machine within the range of 0-100 kg was used in this study for measuring the reaction weights and the whole body weights of the subjects. The balance was frequently checked and calibrated against the standard weights of 1-100 kg. The accuracy of the balance was 10 g.

Wooden Block: A wooden block of the same height as that of the platform of the balance was used for the support of the wooden board.

Wooden Board: A wooden board 182 cm long, 50 cm wide and 2 cm thick was used. One end of the board was placed squarely on lever based weighing scale. The other end of the board was supported on wooden block of the same thickness as the weighing scale to keep the board horizontal.

Reaction-Board Procedure: This board is supported horizontally on two "knife-edges," one of which rests on a block of wood while the other rests on the platform of a set of scales. After the initial reading on the scales R_1 has been noted, the subject lies in a supine position on top of the board with the sole of the feet pressed against a vertical fitting placed so that the surface in contact with the feet lies directly above one of the knife-edges. The new reading on the scale R_2 is then noted. Now, because the board was in equilibrium before the subject lay upon it, and because the board-plus-subject is in equilibrium when the subject does lie down, it must be concluded that whatever forces and moments are introduced by the presence of the subjects are themselves in equilibrium. The forces introduced are the reactions to the increases in weight borne by the platform of the scales and by the wooden block and the weight of the subject. Consider the moments of these newly introduced forces about the point A. The moment of the force $R_2 - R_1$ applied by the platform of the scales to the knife-edge resting upon it is 1.83 $(R_2 - R_1)$ Nm and acts in a counterclockwise direction. (Note: The horizontal distance between the "knife-edges" of the reaction board is 1.83 m, in this case). The moment of the weight force is Wx Nm, where x (m) is the unknown horizontal distance of the centre of gravity from the point A. This moment acts in a clockwise direction. Because the body is in equilibrium, these moments must be equal in magnitude, i.e.

$$x = 1.83(R_2 - R_1)$$

$$x = 1.83 \times \frac{(R_2 - R_1)}{W}$$

where W = body weight of the subject.

Percentage Height of Centre of Gravity: The percentage height of centre of gravity was computed using following formula

$$\text{Percentage Height of CG} = \{(\text{Height of CG})/\text{Height}\} \times 100$$

Body Weight: Weight as a composite measure of total body size was measured in kilograms to the nearest 100 grams.

Procedure: The subjects in light indoor clothing stood erect in the centre of the scaled platform, weight was recorded in kgs. After taking each measurement the scale again was set to zero.

Stature (Height): Stature as a major indicator of general body size and of bone length was measured in meters to the nearest centimeters.

Land Mark: Maximum distance from the point of vertex on the head to the ground.

Tool: A movable anthropometric rod.

Procedure: The measuring tool consisted of a vertical graduated rod and another movable device/rod, which is brought onto the head, while measuring stature against an up ring wall. The measurement was taken with individuals standing straight against the wall touching it with heel, buttocks, back and arms hanging naturally by the sides. An upward force under the mastoid level was exerted to raise the subject. The subject inhaled deeply and was instructed to "look straight ahead". The subject's head positioning in the Frankfurt horizontal plane, stature was recorded to nearest of 1/10th of a centimeter.

Ponderal Index (PI): The Ponderal Index (Shaw, 1998) was computed using

$$\text{Ponderal Index} = \frac{\sqrt[3]{\text{Weight}}}{\text{Height}} \times 100$$

Statistical Procedure: Following statistical techniques were used to analyze the data:

1. Descriptive statistics (mean and standard deviation).
2. 't' ratio to compare between the sexes.
3. Analysis of variance.

Hypothesis was tested at 0.05 level of significance.

Analysis of Data and Results of the Study

Table 1. Descriptive statistics of adulthood male and female on height (m)

Age	Mean ± S.D. (M)	Mean ± S.D. (F)	Age	Mean ± S.D. (M)	Mean ± S.D. (F)
20	1.71 ± .07	1.60 ± 0.08	28	1.62 ± .09	1.54 ± .06
21	1.69 ± .07	1.64 ± .07	29	1.68 ± .09	1.53 ± .23
22	1.71 ± .06	1.57 ± .06	30	1.67 ± .08	1.57 ± .07
23	1.69 ± .11	1.58 ± .06	31	1.67 ± .09	1.54 ± .10
24	1.70 ± .06	1.59 ± .07	32	1.66 ± .07	1.57 ± .08
25	1.68 ± .07	1.57 ± .07	33	1.69 ± .06	1.53 ± .03
26	1.70 ± .06	1.55 ± .04	34	1.68 ± .04	1.54 ± .08
27	1.65 ± .07	1.56 ± .08	35	1.67 ± .06	1.56 ± .09

In Table 1 to 5, $N = 50$ for each age category for male and female independently. M = Male; F = Female.

Table 2. Descriptive statistics of adulthood male and female on weight (kg)

Age	Mean ± S.D. (M)	Mean ± S.D. (F)	Age	Mean ± S.D. (M)	Mean ± S.D. (F)
20	62.98 ± 9.78	53.69 ± 10.15	28	57.33 ± 10.64	53.33 ± 12.50
21	62.23 ± 7.59	54.50 ± 10.59	29	60.96 ± 15.82	56.29 ± 8.41
22	65.91 ± 9.35	48.07 ± 3.95	30	60.55 ± 14.98	51.08 ± 10.17
23	50.83 ± 8.47	50.08 ± 6.23	31	61.67 ± 16.25	50.19 ± 11.22
24	60.83 ± 12.40	52.08 ± 8.50	32	60.61 ± 14.01	56.44 ± 11.06
25	57.93 ± 10.57	53.42 ± 8.93	33	61.500 ± 13.63	57.29 ± 15.04
26	64.86 ± 14.69	46.68 ± 7.22	34	65.500 ± 11.53	60.57 ± 8.92
27	57.20 ± 10.02	50.39 ± 11.39	35	58.87 ± 12.98	52.84 ± 17.19

Table 3. Descriptive statistics of adulthood male and female on PI

Age	Mean ± S.D. (M)	Mean ± S.D. (F)	Age	Mean ± S.D. (M)	Mean ± S.D. (F)
20	231.72 ± 9.04	233.89 ± 10.94	28	236.88 ± 13.57	242.76 ± 14.42
21	233.64 ± 8.64	232.13 ± 10.03	29	233.97 ± 13.90	236.17 ± 13.47
22	234.77 ± 8.01	232.03 ± 6.66	30	235.46 ± 15.09	238.03 ± 11.15
23	218.82 ± 3.04	233.29 ± 10.18	31	236.19 ± 13.79	239.36 ± 19.86
24	231.05 ± 13.53	234.55 ± 13.70	32	236.19 ± 13.79	239.36 ± 19.86
25	229.35 ± 10.27	239.45 ± 13.80	33	232.83 ± 14.42	250.25 ± 20.32
26	235.03 ± 15.48	230.98 ± 9.81	34	239.80 ± 18.12	254.77 ± 12.81
27	232.20 ± 10.44	236.07 ± 9.58	35	233.18 ± 16.31	243.87 ± 14.08

Table 4. Descriptive statistics of adulthood male and female on height of centre of gravity

Age	Mean ± S.D. (M)	Mean ± S.D. (F)	Age	Mean ± S.D. (M)	Mean ± S.D. (F)
20	1.00 ± .06	.92 ± .090	28	1.00 ± .17	.98 ± .07
21	.97 ± .05	.94 ± .04	29	1.04 ± .07	1.01 ± .06
22	.98 ± .06	.91 ± .09	30	1.04 ± .08	1.06 ± .23
23	1.03 ± .05	.97 ± .08	31	1.04 ± .07	1.00 ± .07
24	1.05 ± .09	1.02 ± .07	32	1.07 ± .19	1.00 ± .09
25	1.03 ± .07	1.57 ± 0.07	33	1.03 ± .07	.98 ± .05
26	1.06 ± .09	1.00 ± .06	34	1.02 ± .06	.96 ± .04
27	1.01 ± 0.48	.98 ± .08	35	1.05 ± .18	1.00 ± .06

Table 5. Descriptive statistics of adulthood male and female on percentage height of centre of gravity

Age	Mean ± S.D. (M)	Mean ± S.D. (F)	Age	Mean ± S.D. (M)	Mean ± S.D. (F)
20	58.28 ± 3.16	57.13 ± 4.51	28	60.91 ± 7.39	63.84 ± 4.49
21	56.92 ± 1.44	57.23 ± 1.28	29	62.41 ± 3.21	62.43 ± 4.20
22	57.14 ± 3.60	58.11 ± 5.41	30	61.80 ± 3.63	64.07 ± 3.05
23	61.26 ± 1.54	60.92 ± 4.37	31	62.22 ± 3.13	65.07 ± 5.25
24	62.03 ± 5.32	63.73 ± 3.33	32	61.75 ± 2.49	64.15 ± 4.52
25	61.61 ± 3.92	61.59 ± 4.33	33	61.27 ± 4.72	64.36 ± 3.76
26	62.48 ± 4.76	64.42 ± 4.34	34	60.07 ± 3.70	62.68 ± 2.20
27	61.51 ± 3.52	63.37 ± 5.05	35	61.09 ± 3.75	64.57 ± 4.02

Table 6. Analysis of variance of percentage height of centre of gravity of male adult

Source	Sum of Square	D.F.	Mean Square	F Ratio	Probability
Between	967.760	15	64.52	3.57*	9.437 e – 06
Within	6280.44	347	18.10		
Total	7248.20	362			

*Significant at .05 level.

Table 7. Analysis of variance of percentage height of centre of gravity of female adult

Source	Sum of Square	D.F.	Mean Square	F Ratio	Probability
Between	1972.80	14	140.91	8.115*	7.000 e – 14
Within	3802.85	219	17.37		
Total	5775.65	233			

*Significant at .05 level.

Table 8. Comparison Between the Adult Male and Female on Static Balance Related Variables (Percentage Height of Center of Gravity)

Age	Mean ± S.D. (M)	Mean ± S.D. (F)	Mean Diff.	Std. Error of Diff.	Cal. 't'	Probability
20	58.28 ± 3.16	57.13 ± 4.51	1.15	0.96	1.21	0.12
21	56.92 ± 1.44	57.23 ± 1.28	–0.31	0.54	–0.58	0.29
22	57.14 ± 3.60	58.11 ± 5.41	–0.97	1.62	–0.60	0.28
23	61.26 ± 1.54	58.11 ± 5.41	3.15	3.22	0.98	0.17
24	62.03 ± 5.32	73.63 ± 3.3	–1.70	1.19	–1.43	0.08
25	61.61 ± 3.92	61.60 ± 4.33	0.19	1.08	0.19	0.49
26	62.48 ± 4.76	64.42 ± 4.34	–1.94	1.30	–1.50	0.07
27	61.51 ± 3.52	63.37 ± 5.05	–1.86	1.68	–1.11	0.14
28	60.91 ± 7.39	63.84 ± 4.49	–2.93	2.63	–1.12	0.14
29	62.41 ± 3.21	62.43 ± 4.20	–0.02	1.64	–0.012*	0.50
30	61.80 ± 3.63	64.07 ± 3.05	–2.28	0.98	–2.32*	0.011
31	62.22 ± 3.13	65.07 ± 5.25	–2.85	1.74	–1.64*	0.057
32	61.75 ± 2.49	64.15 ± 4.52	–2.40	1.19	–2.02*	0.025
33	61.27 ± 4.72	64.36 ± 3.76	–3.09	2.23	–1.39*	0.94
34	60.06 ± 3.70	62.68 ± 2.20	–2.61	1.57	–1.66*	0.058
35	61.09 ± 3.75	64.57 ± 4.02	–3.48	1.19	–2.93*	0.002

* Significant at .05 level. M: Male; F: Female.

Conclusions

Within the limitations of the present findings, following conclusions have been drawn:

1. The height of male belonging to selected age groups, i.e. 20-35 yrs found to be identical.
2. The height of female belonging to selected age groups, i.e. 20-35 yrs found to be identical.
3. Male having higher height than that of female in all the selected age groups.
4. The weight of male belonging to selected age groups, i.e. 20-35 yrs found to be almost identical.
5. The weight of female belonging to selected age groups, i.e. 20-35 yrs found to be almost identical.

6. Male having heavier weight than that of female in all the selected age groups, i.e. 20-35 yrs.
7. The Ponderal Index of male belonging to selected age groups, 20-35 found to be almost identical.
8. Ponderal Index of female belonging to selected age groups, 20-35 found to be almost identical.
9. Female having higher Ponderal Index than that of male in selected age group, i.e. 20-35 yrs.
10. Height of CG of male belonging to selected age groups, i.e. 20-35 yrs found to be almost identical.
11. Height of CG of female belonging to selected age groups, i.e. 20-35 yrs found to be almost identical.
12. Male having higher height of centre of gravity than that of female in all the selected age groups.
13. Percentage height of CG of male is higher in 20, 23 and 25 yrs than that of their counterpart.
14. The percentage height of centre of gravity of female is higher in 21, 22, 24, 26, 27, 28, 29, 30, 31, 32, 33, 34 and 35 yrs than that of their counter part (male) in corresponding age group.
15. The percentage height of centre of gravity of female is by and large found to be higher than that of male in selected age groups.
16. The comparison among the selected age group of male adult (20-35 yrs) on percentage height of centre of gravity found to be statistically different (F = 3.57*) at 0.05 level of significance.
17. The comparison among selected age groups of female adults from (20-35 yrs) on percentage height of centre of gravity found to be statistically different (F = 8.1*) at 0.05 level of significance.
18. The comparison between the adult male and female in regard to percentage height of center of gravity of age groups 20, 21, 22, 23, 24, 25, 26, 27, 28 and 29 yrs were not found to be statistically different.
19. The comparison between the adult male and female in regard to percentage height of center of gravity of age groups 30, 31, 32, 33, 34 and 35 yrs were found to be statistically different at 0.05 level of significance.
20. The mean values of percentage height of centre of gravity were higher in female than that of male belonging to age group of 30, 31, 32, 33, 34 and 35 yrs.

References

1. Clarke, H.H. and Clarke, D.H. 1987. *Application of Measurement of Physical Education* 4th Edn. Prentice Hall, Englewood, New Jersey.
2. Das, R.N. and Ganguli, S. 1982. Mass and Center of Gravity of Human Body Segments. *The Journal of the Institution of Engineer (India).* **62: IDGE**: 67-72.
3. Fox, E.L. and Mathews, D.K. 1981. *The Physiological Basis of Physical Education and Athletics* 3rd Edn. Sunders College Publishing, Philadelphia.
4. Kaushik, S., Kaushik, R. and Shaw, D. 1995. Changes in the Location of Three Dimensional Center of Gravity as a Result of Six Weeks Conditioning Programme on Male Athletes. *Souvenir: 5th National Conference of NAPESS and GANSF*, India. p. 33.
5. Ray, C.G. and Sen, R.N. 1983. Determination of Whole Body Center of Gravity in Indians. *Journal of Human Ergol.* **12**: 3-4.
6. Shaw, D. 2000. Determination of Center of Gravity by Reaction Board Method. *Mechanical Basis of Biomechanics.* Sports Publication, Delhi. pp. 263-266.
7. Shaw, D. et al. 1998. Validation of Indigenously Development Reaction Board Method. In: *Advances in Biomechanics.* Kamal Raj Pardasani (Ed.). Proceedings of Sixth National Conference. Om Printers and Stationer. pp. 240-244.
8. Shaw, Dhananjoy et al. 1998. Reliability of Reaction Board Method for Determination of Selected Segmental Mass. In: *Advance in Biomechanics* Kamal Raj Pardasani (Ed.). Proceedings of Sixth National Conference Om Printers and Stationer. pp. 262-267.

Biomechanics
R.K. Saxena and P. Mishra (Editors)
Copyright © 2005, Anamaya Publishers, New Delhi, India

6. A Mathematical Model for Computation of Electrical and Mechanical Properties of a Human Wet Bone and Detection of Osteoporosis

N.S. Rao[1], G.R. Babu[2], L.N. Merugu[3], V.S. Mallela[1] and P.K. Subramanya Kumar[4]

[1]ISSCITEW, Hyderabad, India
[2]GNIST, Hyderabad, India
[3]DLRL, Hyderabad, India
[4]Department of Orthopaedics, Kamineni Hospitals Ltd., Hyderabad, India

Abstract: Mathematical modeling of human bone is a complex phenomenon since, the cybernetics of the bone needs to be studied and analyzed as an interdisciplinary area of specialization. In this article, model of a human bone is presented considering isotropic, orthotropic and anisotropic properties of bone composition. This model gives theoretical estimates on various electric, dielectric and elastic properties of a wet bone. The viscoelastic behaviour of the bone has been retained in the model. The model makes it interesting and unique in predicting the elastic, electric and dielectric properties of the wet bone.

The theoretical results on electric and dielectric properties obtained have been compared with available experimental results. With regard to elastic properties of the bone, the theoretical results are encouraging when compared with experimental results.

Osteoporosis is a common disease which results in loss of bone tissue due to increased porosity or decreased density. Osteopetrosis is a rare disease in which the bone forms as calcified cartilage so that Haversian systems do not fully form. The haversian canals become clogged with unorganized mineral and collagen so that the density is comparable to that of normal bone. Parameter sensitivity analysis has been carried out using this model predicting the above. The model could be extended to estimate other properties of the wet bone viz. chemical composition and related parameters.

Introduction

Mathematical modeling of human bone is a complex phenomenon since the cybernetics of the bone needs to be studied and analyzed as an interdisciplinary area of specialization. Hard tissue, mineralized tissue and calcified tissue are often used as synonyms for bones, when describing the structure of bone. Bone is an anisotropic inhomogeneous, nonlinear, thermoheologically complex viscoelastic material. It exhibits electromechanical effects, presumed to be due to streaming potentials both *in vivo* and *in vitro* when wet. In dry state, bone exhibits piezoelectric properties (Rao et al. 2002; CRC Press, 2001).

In this article, an attempt has been made to model complex human bone (wet/dry) to predict it's electrical, dielectric and mechanical properties. Bone density, Young's modulus, Shear modulus, Poisson's ratio are some of the important mechanical properties of the bone while conductivity, permittivity and impedance are some of the electrical properties. Dielectric constant as a function of frequency is also an important property of the bone.

The composition of the bone is shown in Fig. 1, where various constituents shown in percentage are by weight. The important ones among basic parts of the bone are:

1. The *Cortical* or *compact* outer portion.
2. The porous or *cancellous*/trabecular portion.
3. Marrow and other tissues.

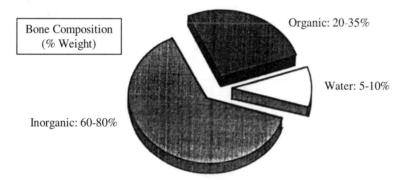

Fig. 1. Constituents of bone extracellular matrix.

Considering the nature and composition of bone material, anisotropy, orthotropy, transverse isotropy and isotropy relationships are taken into account for the model (CRC Press, 2001). The viscoelastic nature of the bone is also considered while modeling.

Development of Mathematical Model

Hook's law states that

$$\sigma_{ij} = (C_{ijkl})(\varepsilon_{kl}) \tag{1}$$

where σ_{ij} is the second rank stress tensor; ε_{kl} the second rank strain tensor; and C_{ijkl} is the fourth rank elasticity tensor.

$$\sigma_i = C_{ij}\varepsilon_j, \, i, j = 1 \text{ to } 6$$

where $[C_{ij}]$ are the stiffness coefficients. Inverse of C_{ij} (S_{ij}) are known as compliance coefficients.

The mechanical properties of the bone are classified under transverse isotropy, isotropy, orthotropy, anisotropy. The stiffness coefficients differ for each of these (Mehta, 1995). Thus, the computation of the $[C_{ij}]$ matrix becomes very important.

For transverse isotropic material the stiffness matrix $[C_{ij}]$ is given by (CRC Press, 2001)

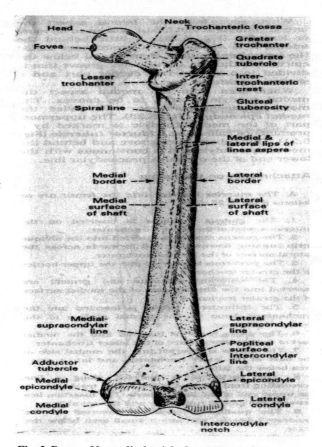

Fig. 2. Bones of lower limb: right femur, posterior aspect.

$$[C_{ij}] = \begin{bmatrix} C_{11} & C_{12} & C_{13} & 0 & 0 & 0 \\ C_{12} & C_{11} & C_{13} & 0 & 0 & 0 \\ C_{13} & C_{13} & C_{33} & 0 & 0 & 0 \\ 0 & 0 & 0 & C_{44} & 0 & 0 \\ 0 & 0 & 0 & 0 & C_{44} & 0 \\ 0 & 0 & 0 & 0 & 0 & C_{66} \end{bmatrix} \quad (2)$$

where $[C_{66}] = \frac{1}{2}[C_{11} - C_{12}]$. Of the 12 non-zero coefficients, only 5 are independent. Since bone is an orthotropic material, it requires that 9 of 12 non-zero elastic constants be independent.

$$[C_{ij}] = \begin{bmatrix} C_{11} & C_{12} & C_{13} & 0 & 0 & 0 \\ C_{12} & C_{22} & C_{23} & 0 & 0 & 0 \\ C_{13} & C_{23} & C_{33} & 0 & 0 & 0 \\ 0 & 0 & 0 & C_{44} & 0 & 0 \\ 0 & 0 & 0 & 0 & C_{55} & 0 \\ 0 & 0 & 0 & 0 & 0 & C_{66} \end{bmatrix} \quad (3)$$

$$\varepsilon_{ij} = S_{ij}\sigma_j, \quad ij = 1 \text{ to } 6.$$

where S_{ij}^{th} compliance is obtained by dividing the $[C_{ij}]$ stiffness matrix minus the ith row and j^{th} column by the full $[C_{ij}]$ matrix in terms of S_{ij}.

$S_{33} = 1/E_3$, where E_3 is the Young's modulus in bone longitudinal axis direction.

$E_3 \neq C_{33}$, since C_{33} and S_{33} are not reciprocals.

In order to solve this inequality, the problem may be approached from the measured values in mechanical tests such as uni-axial or pure shear. The resultant compliance coefficient may be expressed below. From the above, it is clear that stiffness matrix $[C_{ij}]$ is given by Eq. (2) for transverse isotropic material and by Eq. (3) for orthortropic material.

$$[S_{ij}] = \begin{bmatrix} 1/E_1 & -V_{21}/E_2 & -V_{31}/E_3 & 0 & 0 & 0 \\ -V_{12}/E_1 & 1/E_2 & -V_{32}/E_3 & 0 & 0 & 0 \\ -V_{13}/E_1 & -V_{23}/E_2 & 1/E_3 & 0 & 0 & 0 \\ 0 & 0 & 0 & 1/G_{31} & 0 & 0 \\ 0 & 0 & 0 & 0 & 1/G_{31} & 0 \\ 0 & 0 & 0 & 0 & 0 & 1/G_{12} \end{bmatrix} \quad (4)$$

For an orthotropic material, only 9 of the above 12 non-zero terms are independent due to symmetry of $[S_{ij}]$ tensor.

$$[V_{12}/E_1] = V_{21}/E_2; \quad V_{13}/E_3 = V_{31}/E_3; \quad [V_{23}/E_2] = V_{32}/E_3 \quad (5)$$

However, it is easily seen that for the transverse isotropic case

$$\begin{aligned} E_1 = E_2 &\Rightarrow V_{12} = V_{21} \text{ and } V_{31} = V_{32} = V_{13} = V_{23}; \\ G_{23} &= G_{31}; G_{12} = E_1/2*(1+V_{12}) \end{aligned} \quad (6)$$

Since the C_{ij} are simply related to the elastic moduli, such as Young's modulus E, Shear modulus G, Bulk modulus K and others (Saha and Williams, 1987), it is possible to describe the moduli along any given direction.

Bulk modulus (reciprocal of volume compressibility) is given by

$$1/K = S_{33} + 2(S_{11} + S_{12} + 2S_{13}) = [C_{11} + C_{12} + 2C_{33} - 4C_{13}]/\{C_{33}(C_{11}+C_{12}) - 2C_{13}^2\} \quad (7)$$

The other properties like speed of sound, conductivity and dielectric property are computed by using the standard equations

$$V^2 = E/\rho; \quad \sigma_i = C_{ij}\varepsilon_j; \quad \text{where } i, j = 1 \text{ to } 6 \quad (8)$$

Experimental Evaluation of Bone Impedance

The two methods which can be used for experimental evaluation of bone impedance are: Direct method and Differential method.

The drawback in the direct method is that at high frequencies, impedance measurement gets severely affected by the wiring and stray capacitances. Therefore, a differential technique was utilized to eliminate in this work.

Differential Method

The electrical equivalent circuit for the differential method is shown in Fig. 3.

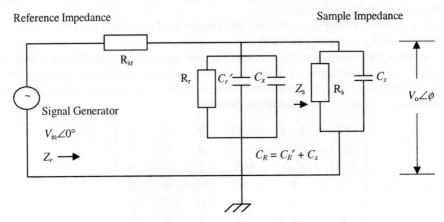

Fig. 3. Electrical equivalent circuit for differential method.

Here, a reference reading the amplitude ratio V_o/V_{in} and phase angle ϕ is noted with a gain-phase meter to estimate the stray capacitance. Such reference reading is necessary at each frequency of interest. From the reference reading (K, ϕ), the reference impedance $Z_r \angle \theta_1$ is calculated as

$$(Z_r \angle \theta_1) = R_{sr} K \angle \phi / (1 - K \angle \phi) \quad \text{where} \quad K = V_o/V_{in} \tag{9}$$

Once the reference impedance is calculated, the circuit's stray capacitance C_x is computed as

$$C_x = C_r' - C_r \tag{10}$$

$$C_x = (-\sin\theta_1/\omega Z_r) - C_r$$

where C_r' is the measured capacitance and C_r the capacitance of the reference circuit. On computing $Z_r \angle \theta_1$, a second reading (K', ϕ) is noted, with the bone sample across the reference circuit. The combined impedance of the parallel circuit is given by

$$Z' \angle \theta_2 = (R_{sr} K' \angle \phi)/(1 - K' \angle \phi) \tag{11}$$

$$Z_s \angle \theta = (Z_r \angle \theta_1 \, Z' \angle \theta_2)/(Z_r \angle \theta_1 - Z' \angle \theta_2) \tag{12}$$

From the known impedance, volume sample resistance R_s and the capacitance C_s are then computed as

$$R_s = |Z_s|/\cos\theta \quad \text{and} \quad C_s = -\sin\theta/\omega|Z_s|$$

Table 1 illustrates the computational procedure to measure the complex impedance $Z_s \angle \theta$, and how its resistive and capacitive components are obtained. The results of measurements made with a known impedance circuit, using direct method and the differential method indicated that the impedance measurements were significantly influenced by wiring and cable capacitance. Table 2 illustrates the computational procedure to measure the specific impedance using the differential technique from the known sample geometry and its resistance.

Table 1. The differential technique to measure the complex impedance

	Amp. Ratio (K)	Phase Angle (ϕ)	$Z_r \angle \theta_1 = \dfrac{R_{sr} K \angle \phi}{(1 - K \angle \phi)}$	$Z' \angle \theta_2 = \dfrac{R_{sr} K' \angle \phi'}{(1 - K' \angle \phi')}$	$Z_s \angle \theta = \dfrac{z_r \angle \theta_1 z' \angle \theta_2}{(z_r \angle \theta_1 - z' \angle \theta_2)}$	$R_s = \lvert z_s \rvert / \cos\theta$	$C_s = -\sin\theta / 2\Pi f \lvert z \rvert$
Without sample	0.6683	−16.8	15.99∠−45	-	-	-	-
With sample	0.6309	−16.3	-	14.27∠−41	-	-	-

Table 2. Computation of the specific impedance using the differential technique from known sample geometry and its resistance R_s and capacitance C_s

Specific Resistance	Specific Capacitance	Specific Reactance	Impedance Magnitude	Phase Angle	Specific Impedance
$R_{sp} = R_s A_x / L_x$	$C_{sp} = C_s (L_x / A_x)$	$X_{sp} = 1/\omega C_{sp}$	$Z_{SP} = \dfrac{R_{SP} \cdot X_{SP}}{(R_{SP}^2 + X_{SP}^2)^{0.5}}$	$\theta_{SP} = \dfrac{\tan^{-1}(R_{SP})}{X_{SP}}$	$Z_{SP} \angle \theta_{SP}$
KΩ cm	Pf-cm^{-1}	KΩ cm	KΩ cm	(°)	KΩ cm
15.84	13.97	113.8	15.69	−7.92	15.69/7.92

*Sample dimensions: Area = 1.1 cm × 0.59 cm = 0.649 cm^2; length = L_X = 4.51 cm; Measurement frequency = 100 kHz.

Sensitivity Analysis

The input parameters that correspond to the 6 diagonal elements of the stiffness matrix are varied by about ±3% to represent various patient's database. Based on these sensitivity analysis, shift in curves was observed. The results of this analysis were discussed with doctors in Kamineni Hospital. The results obtained in this analysis are encouraging.

Conclusion

- Generalized model developed to predict both electrical and mechanical properties of wet/dry bone.
- The results obtained with the above model have been validated with experimental data.

Acknowledgement

The authors acknowledge Sri Ganapathi Datta for his support in writing the code for the software development.

References

1. *Hand Book on Biomedical Engineering*, 2001. CRC Press, Boca Raton.
2. Mehta, S.S. 1995. Analysis of the Mechanical Properties of Bone Material Using Nondestructive Ultrasound Reflectrometry. *Dissertation Work*. University of Texas, Southwestern Medical Centre at Dallas.

3. Rao, N.S. et al. 2002. Overview Study of Dielectric Properties of Dry Bones. *Proceedings of National Conference on Sensors and Instrumentation.* Hyderabad. January 05-06, 2002.
4. Saha, S. and Williams, P.A. 1987. Electric and Dielectric Properties of Wet Human Cancellous Bone as a Function of Frequency. *13th Annual Meeting of the Society for Biomaterials.* New York. June 3-7, 1987.

Biomechanics
R.K. Saxena and P. Mishra (Editors)
Copyright © 2005, Anamaya Publishers, New Delhi, India

7. Swimming Strengthens Brawn as Well as Brain

Milind Parle and M. Vasudevan

Pharmacology Division, Department of Pharmaceutical Sciences, Guru Jambheshwar University, Hisar - 125001, India

Abstract: Alzheimer's disease affects physical as well as mental functions. The most common symptoms of Alzheimer's disease include loss of memory, impaired judgement, aphasia, and apraxia. The present study was undertaken to assess the benefits of swimming on the memory of rodents. The rats and mice of either sex were inserted in a specially designed swimming pool twice a day for 10 min for 30 consecutive days. Control group animals were not subjected to swimming during the above period. Swimming sessions were stopped in yet another group of animals after a period of 15 days. All these animals were tested for their memory scores on 1st, 16th and 31st day. Memory was tested employing elevated plus-maze and Williams Hebb maze. It was observed that both, rats as well as mice that underwent swimming regularly for 30 days showed sharp memories, when tested on three different exteroceptive models. The animals, which underwent swimming for 15 days only showed good memory on 16th day, which however, declined after 30 days. These findings indicate that regular swimming preserves brain function and probably prevents neurodegenerative processes.

Introduction

There has been a steady rise in the number of patients suffering from Alzheimer's disease (AD) all over the world. There are more than 25 million patients suffering from AD all over the world. Dementia is largely a hidden problem in India. Epidemiological studies of Indian population reveal that there are more than 2,500,000 patients suffering from AD (Dhingra et al., 2003). AD is a progressive, neurodegenerative, debilitating disorder manifested by loss of memory, impaired judgement, aphasia and apraxia. These personality disturbances interfere with one's profession, social activities and relationships. As the disease progresses, well learned memory-tasks get lost and the patients fail to recognize even close family members. The pathophysiology of AD features development of neuritic plaques and neurofibrillary tangles (Dhingra et al., 2004; Parle and Dhingra, 2004; Kozikowski and Tuckmantel, 1999). The slow death of brain cells particularly cholinergic neurons appears to be the main culprit for the development of AD (Dhingra et al., 2004). The treatment of AD is still a nightmare in the field of medicine. Therefore, neuroscientists all over the world are busy exploring the usefulness of alternative systems of medicine (e.g. nature cure, ayurveda, homeopathy etc). This prompted the authors to assess the effects of certain medicinal plants viz. *Glycyrrhiza glabra* and *Myristica fragrans* on memory of mice (Dhingra et al., 2004; Parle et al., 2004). Exercise probably regulates not only muscular activities but also brain functions. Lack of exercise constitutes one of the main risk factors for age-related diseases like hypertension, diabetes and Alzheimer's disease. Interestingly, dietary restriction (reduced calorie intake) has been shown to protect the brain from neuronal death and stroke (Mattson, 2000). Some studies have indicated that jogging is beneficial in preventing AD (Berchtold et al., 2001).

Objective

In the light of above, the present investigation was undertaken to delineate the effects of swimming on learning and memory of rodents employing various behavioral models.

Methods

Animals

Wistar rats (aged 16 months), weighing around 250 g and Swiss mice (aged 7 months), weighing around 25 g, (of either sex) were used in the present study. Animals were procured from the disease-free animal house of CCS Haryana Agriculture University, Hisar (Haryana, India). The animals had free access to food and water. Food given to the animals consisted of wheat flour kneaded with water and mixed with small amount of refined vegetable oil. The animals were acclimatized to the laboratory conditions for at least 5 days before behavioral experiments with alternating light and dark cycles of 12 h each. Experiments were carried out between 9.00 am and 4.00 pm on all the days. The experimental protocol was approved by the Institutional Animals Ethics Committee (IAEC) and care of laboratory animals was taken as per the guidelines of CPCSEA, Ministry of Forests and Environment, Government of India (Reg. No. 436).

Swimming Protocol

The rodents (rats and mice) were divided into 9 groups and each group comprised of a minimum of 6 animals. The swimming exposure to the rodents was for 10 min during each session and there were 2 swimming exposures on each day. The learning index and memory score (as determined by their TL/TRC) of all the animals was recorded on 1st, 2nd, 15th, 16th, 30th and 31st day.

Elevated Plus-maze

Elevated plus-maze served as the exteroceptive behavior model to evaluate learning and memory in rats and mice. The procedure, technique and end point for testing learning and memory was followed as per the parameters described by the investigators working in the area of psychopharmacology (Itoh et al., 1990; Reddy and Kulkarni, 1998; Dhingra et al., 2003; Parle and Dhingra, 2004).

Hebb-Williams Maze

Hebb-Williams maze is an incentive based exteroceptive behavioural model useful for measuring spatial working memory of rats (Parle and Singh, 2004). It consists of mainly three components. Animal chamber (or start box), which is attached to the middle chamber (or exploratory area) and a reward chamber at the other end of the maze in which the reward (food) is kept. All the three components are provided with guillotine removable doors. On the first day, the rat was placed in the animal chamber or start box and the door was opened to facilitate the entry of the animal into the next chamber. The door of start box was closed immediately after the animal moved into the next chamber so as to prevent back-entry. Time taken by the animal to reach reward chamber (TRC) from start box on 1st day reflected the learning index. The learning index was noted for each animal. Retention of this learning index (memory score) was examined 24 h after the first day trial. Each animal was allowed to explore the maze for 3 min with all the doors opened before returning to its home cage.

Statistical Analysis

All the results were expressed as mean ± standard error (SEM). Data was analyzed using one-way ANOVA followed by Dunnett's 't' test and student's paired 't' test. $P < 0.05$ was considered as statistically significant.

Results

Effect of Swimming Exposure on Time Taken to Reach Reward Chamber (TRC) of Rats Using Hebb Williams Maze

Time taken by the rat to reach reward chamber (TRC) from the start box on 1st day reflected the learning index, whereas, TRC of the next day (2nd day) indicated retention capacity (memory score) of animals. The rats, which were not exposed to swimming exercise (control group) showed TRC of 51.8 ± 3.0 (mean ± SEM) seconds when tested on 1st day and, whereas, TRC of 34.2 ± 2.7 seconds when tested on 2nd day. Interestingly, when the rats were exposed to swimming for 15 days (sub-chronic swimming exposure), there was a significant reduction ($P < 0.05$) in TRC of 15th day as compared to TRC of 1st day in the same group. Furthermore, the TRC of 16th day was also significantly reduced ($P < 0.05$) when compared to 2nd day TRC in the same group of rats (Table 1).

Table 1. Effect of swimming on TRC (Time taken to reach reward chamber) of rats using Hebb Williams Maze

Groups n = 6/group	Exposure to Swimming	TRC					
		Day 1	Day 2	Day 15	Day 16	Day 30	Day 31
Non swimming	Control 0 days	51.8 ± 3.0	34.2 ± 2.7	47.7 ± 3.4	33.3 ± 3.5	46.7 ± 5.2	32.2 ± 2.2
Sub-chronic swimming	15 days	53.7 ± 4.0	33.5 ± 2.4	38.8 ± 3.8[a]	25.2 ± 2.2[b]	40.8 ± 5.5[j]	28.3 ± 1.3[k]
Chronic swimming	30 days	50.8 ± 3.5	29.7 ± 1.2	36.7 ± 2.4[c,g]	20.8 ± .5[d,h]	32.5 ± 3.2[e,h,l]	16.2 ± 2.2[f,i,k]

Values represent Mean ± SEM, Data was analyzed using one-way ANOVA followed by Dunnett's 't' test and student's 't' test (paired), $P < 0.05$ was considered as statistically significant.

[a] $P < 0.05$ compared to TRC of 1st day in same group.
[b] $P < 0.05$ compared to TRC of 2nd day in same group.
[c] $P < 0.01$ compared to TRC of 1st day in same group.
[d] $P < 0.01$ compared to TRC of 2nd day in same group.
[e] $P < 0.001$ compared to TRC of 1st day in same group.
[f] $P < 0.001$ compared to TRC of 2nd day in same group.
[g] $P < 0.05$ compared to TRC of control group on same day.
[h] $P < 0.01$ compared to TRC of control group on same day.
[i] $P < 0.001$ compared to TRC of control group on same day.
[j] $P < 0.05$ compared to TRC of 15th day in same group.
[k] $P < 0.05$ compared to TRC on 16th day in same group.
[l] $P < 0.01$ compared to TRC on 15th day in same group.

This indicated that learning index and memory score of animals were remarkably improved after 15 days of swimming exercise. However, when the TRC was measured on the 30th day ($P < 0.05$) and 31st day ($P < 0.05$) of mice exposed subchronically to swimming (for 15 days), it was observed that there was significant increase in TRC as compared to corresponding TL of 15th day and 16th day. This indicated that when swimming exposure was halted suddenly after 15 days the process of neurodegeneration might have come into play and impaired the learning index and memory score of animals with the passage of time. Rats, which underwent chronic swimming exposure (for 30 days) showed excellent learning index and excellent memory as reflected by significant reduction of TRC value on 30th day (P < 0.001) and 31st day ((P < 0.001), when compared to the corresponding TRC of 1st day and 2nd day in the same group (Table 1). There was also a significant (P < 0.01) reduction in the 30th day TRC of rats exposed chronically to swimming (for 30 days) when compared to 30th day TRC of control group thereby indicating that chronic swimming exposure improved learning index. So also, there was a significant (P < 0.001) reduction in 31st day TRC (denoting memory) of rats exposed chronically to swimming (for 30 days), when compared to 31st day TRC of control group. This observation suggested that regular swimming schedule of 30 days prevented the brain damage possibly caused by neurodegenerative processes and improved learning index and retention capacity (memory) of animals. The effect of swimming on transfer-latency of mice using Elevated Plus Maze is shown in Table 2.

Table 2. Effect of Swimming on Transfer-Latency of mice using elevated plus maze

Groups n = 6/group	Exposure to Swimming	TRC Transfer Latency (TL)					
		Day 1	Day 2	Day 15	Day 16	Day 30	Day 31
Non swimming	Control 0 days	36.2 ± 6.4	26.3 ± 2.8	34.5 ± 2.7	23.2 ± 1.6	32.7 ± 3.6	22.3 ± 1.2
Sub-chronic swimming	15 days	37.3 ± 1.5	25.2 ± 0.5	29.3 ± 2.4^b	18.5 ± 2.2^a	31.5 ± 1.7^i	21.5 ± 1.4^h
Chronic swimming	30 days	35.7 ± 2.4	23.8 ± 1.3	$26.5 \pm 1.2^{b,e}$	17.8 ± 1.2^a	$21.7 \pm 2.0^{c,f,i}$	$11.2 \pm 1.5^{d,g,j}$

Values represent Mean ± SEM; Data was analyzed using one-way ANOVA followed by Dunnett's 't' test and student's 't' test (paired); P < 0.05 was considered as statistically significant.
[a] P < 0.05 compared to TL of 2nd day in same group.
[b] P < 0.01 compared to TL of 1st day in same group.
[c] P < 0.001 compared to TL of 1st day in same group.
[d] P < 0.001 compared to TL of 2nd day in same group.
[e] P < 0.05 compared to TL of control group on same day.
[f] P < 0.01 compared to TL of control group on same day.
[g] P < 0.001 compared to TL of control group on same day.
[h] P < 0.05 compared to TL of 16th day in same group.
[i] P < 0.01 compared to TL of 15th day in same group.
[j] P < 0.01 compared to TL of 16th day in same group.

Discussion

Alzheimer's disease (AD) affects physical as well as mental function (Teri et al., 1998; 2003), although, the relationship between physical health and AD has received little attention. Our results point out that integration of exercise schedule into the life style of AD patients is advantageous and worthwhile. This notion is supported by the studies of Teri et al. (1998), who has shown that gait, flexibility, body strength, and endurance was greatly improved in community-dwelling AD subjects and their care givers, when they participated in a structured exercise program. The most commonly prescribed medicines for the treatment of AD include either nootropic agents (such as piracetam) or anticholinesterases such as tacrine, rivastigmine or donepezil. However, all these drugs evoke several side effects limiting their use (Mashkovskii and Glushkov, 2001; Gauthier et al., 2003). AD is a progressive neurodegenerative disorder characterized by dementia, impaired judgement, aphasia, agnosia and apraxia. As this deadly disease progresses, the patient is unable to recall important events and sometimes even fails to remember the faces of close family members (Parle et al., 2004).

In the present study, the rodents were subjected to swimming (a pleasant exercise) for a period of 15 and 30 days. The authors observed that rats as well as mice that underwent swimming regularly for a period of 15 days (twice daily for 10 minutes) showed good learning index and memory score, when compared to control animals that did not swim. Interestingly, the learning index as well as memory score deteriorated significantly over the next 15 days, when the swimming exercise was halted suddenly after 15 days. This was perhaps due to the natural physiological process of forgetting, which might have come into play upon halting of swimming exposure to animals after 15 days. These findings suggested that an uninterrupted (regular) exercise schedule such as swimming probably helps in preventing the neurodegenerative damage of brain cells taking place gradually in aged rats and mice. Above findings underlined the importance of exercise in general and swimming in particular in preventing memory loss. Furthermore, the animals that continued to swim for a period of 30 days showed excellent learning index and memory score, when tested on the 30th and 31st day. There is a possibility that regular swimming for long periods not only arrested the neurodegenerative processes (responsible for dementia) but also stimulated the process of neurogenesis. This suggestion is in line with the reports of Van Praag et al. (1999), Neeper et al. (1995) and Kempermann et al. (1997). Voluntary physical activity on a running wheel apparatus doubled the number of surviving newborn hippocampal cells in adult mice (Van Praag et al., 1999). Although it is not known whether running has any influence on learning, it has been shown that physical activity facilitated recovery from injury (Johansson and Ohlsson, 1997) and improved cognitive functions (Fordyce and Wehner, 1993).

Exercise is known to induce numerous physiological alterations in vital organ systems of the body. The most important favorable changes include enhanced respiration, better utilization of oxygen by the muscles and increased blood flow to vital organs including brain. (Vijay Kumar and Naidu, 2002). Other benefits of exercise include higher energy levels, better stamina, retention of motor skills and improved sleep. Moreover, regular exercise results in a variety of adaptations that may be beneficial in attenuating the process of apoptosis (Leeuwenburgh et al., 1997; Powers et al., 1993; Sen et al., 1992). Gage et al. (1998) observed that new neurons continuously add to certain areas of the adult brain, such as the hippocampus and olfactory bulb. Kempermann et al., (1997) associated the improved learning ability of mice seen on exposure to an enriched environment to enhanced neurogenesis. Furthermore, release of trophic factors, responsible for progenitor cell survival (Ray et al., 1997), synaptic strength (Schuman, 1999), long-term potentiation (Patterson et al., 1992), and memory (Fischer et al., 1987), were all improved after exercise (Neeper et al., 1995). At the cellular level, wheel running enhanced the firing rate of hippocampal cells in a manner that correlated with the running velocity (Czurko et al., 1999). Ahmadiasl et al. (2003) showed that increased physical activity in adult rats significantly enhanced spatial learning performance in Morris water maze. Both running and living in an enriched environment doubled the number of surviving newborn cells and improved water maze performance (Russo-Neustadt et al., 1999). Over the past decade, a number of studies on humans have shown the benefits of exercise on brain health and function, particularly in aging population (Ivy et al., 2001). Maintaining brain health and plasticity throughout life is an important public health goal. The present research findings recommend that a well planned exercise schedule would certainly help aged people with or without AD for improving their cognitive functions. Daily swimming may do much more than keep you just physically fit. It might also prevent the deterioration of brain cells that can lead to AD.

Conclusion

In conclusion, our results emphasize the role of regular physical exercise, particularly swimming in the maintenance and promotion of brain functions. The underlying physiological mechanism for improvement of memory appears to be the result of enhanced neurogenesis.

References

1. Ahmadiasl, N., Alaei, H. and Hanninen, O. 2003. Effect of Exercise on Learning, Memory and Levels of Epinephrine in Rats Hippocampus. *Journal of Sports Science and Medicine.* **2**: 106-109.
2. Berchtold, N.C., Kesslak, J.P., Adlard, P.A. and Cotman, C.W. 2001. Estrogen and Exercise Interact to Regulate Brain-derived Neurotrophic Factor mRNA and Protein Expression in the Hippocampus. *The European Journal of Neuroscience.* **14**: 1992-2002.
3. Czurko, A., Hirase, H., Csicsvari, J. and Buzsaki, G. 1999. Sustained Activation of Hippocampal Pyramidal Cells by 'Space Clamping' in a Running Wheel. *The European Journal of Neuroscience.* **11**: 344-352.
4. Dhingra, D., Parle, M. and Kulkarni, S.K. 2003. Effect of Combination of Insulin with Dextrose, D (-) Fructose and Diet on Learning and Memory in Mice. *Indian Journal of Pharmacology.* **35**: 151-156.
5. Dhingra, D., Parle, M. and Kulkarni, S.K. 2003. Medicinal Plants and Memory. *Indian Drugs.* **40**: 313-319.
6. Dhingra, D., Parle, M. and Kulkarni, S.K. 2004. Memory Enhancing Activity of *Glycyrrhiza glabra* in Mice. *Journal of Ethnopharmacology.* **91**: 361-365.
7. Fischer, W., Wictorin, K., Bjorklund, A., Williams, L.R., Varon, S. and Cage, F.H. 1987. Amelioration of Cholinergic Neuron Atrophy and Spatial Memory Impairment in Aged Rats by Nerve Growth Factor. *Nature (London).* **329**: 65-68.
8. Fordyce, D.E. and Wehner, J.M. 1993. Physical Activity Enhances Spatial Learning Performance with Associated Alternation in Hippocampal Protein Kinase C Activity in C57BL/6 and DBA/2 Mice. *Brain Research.* **619**: 111-119.
9. Gage, F.H., Kempermann, G., Palmer, T.D., Peterson, D.A. and Ray, J. 1998. Multipotent Progenitor Cells in the Adult Dentate Gyrus. *Journal of Neurobiology.* **36**: 249-266.
10. Gauthier, S., Emre, M., Farlow, M.R., Bullock, R., Grossberg, G.T. and Potkin, S.G. 2003. Strategies for Continued Successful Treatment of Alzheimer's Disease: Switching cholinesterase inhibitors. *Current Medical Research and Opinion.* **19**: 707-714.
11. Itoh, J., Nabeshima, T. and Kameyama, T. 1990. Utility of an Elevated Plus Maze for Evaluation of Nootropics, Scopolamine and Electro Convulsive Shock. *Psychopharmacology.* **101**: 27-33.
12. Ivy, A.S., Rodriguez, F.G. and Russo-Neustadt, A. 2001. The Effects of NE and 5-HT Receptor Antagonists on the Regulation of BDNF Expression during Physical Activity. *Society for Neuroscience Abstracts.* **253**: 213.
13. Johansson, B.B. and Ohlsson, A. 1997. Environment, Social Interaction, and Physical Activity as Determinants of Functional Outcome after Cerebral Infarction in the Rat. *Experimental Neurology.* **139**: 322-327.
14. Kempermann, G., Kuhn, H.G. and Gage, F.H. 1997. More Hippocampal Neurons in Adult Mice Living in an Enriched Environment. *Nature (London).* **386**: 493-495.
15. Kozikowski, A.P. and Tuckmantel, W. 1999. Chemistry, Pharmacology and Clinical Efficacy of the Chinese Nootropic Agent Huperzine A. *Accounts of Chemical Research.* **32**: 641-650.
16. Leeuwenburgh, C., Hollander, J., Fiebig, R., Leichtweis, S., Griffith, M. and Ji, L.L. 1997. Adaptations of Glutathione Antioxidant System to Endurance Training are Tissue and Muscle Fiber Specific. *American Journal of Physiology.* **272**: 363-369.
17. Mashkovskii, M.D. and Glushkov, R.G. 2001. Drugs for the Treatment of Alzheimer's Disease. *Pharmaceutical Chemistry Journal.* **35**: 179-182.
18. Mattson, M.P. 2000. Neuroprotective Signaling and the Aging Brain: Take away my food and let me run. *Brain Research.* **886**: 47-53.
19. Neeper, S.A., Gomez-Pinilla, F., Choi, J. and Cotman, C. 1995. Exercise and Brain Neurotrophins. *Nature (London).* **373**: 109.
20. Parle, M. and Dhingra, D. 2004. Ascorbic Acid: A promising memory-enhancer in mice. *Journal of Pharmacological Sciences.* **93**: 129-135.

21. Parle, M. and Singh, N. 2004. Animal Models for Testing Memory. *Asia Pacific Journal of Pharmacology.* **16** (In press)
22. Parle, M., Dhingra, D. and Kulkarni, S.K. 2004. Neurochemical Basis of Learning and Memory. *Indian Journal of Pharmaceutical Sciences.* **66**: 371-376.
23. Parle, M., Dhingra, D. and Kulkarni, S.K. 2004. Neuromodulators of Learning and Memory. *Asia Pacific Journal of Pharmacology.* **16** (In Press).
24. Parle, M., Dhingra, D., and Kulkarni, S.K. 2004. Improvement of Mouse Memory by *Myristica fragrans* Seeds. *Journal of Medicinal Food.* **7**: 157-161.
25. Patterson, S.L., Grover, L.M., Schwartzkroin, P.A. and Bothwell, M. 1992. Neurotrophin Expression in Rat Hippocampal Slices: A stimulus paradigm inducing LTP in CAI evokes increases in BDNF and NT-3 mRNAs. *Neuron.* **9**: 1081-1088.
26. Powers, S.K., Criswell, D., Lawler, J., Martin, D., Lieu, F.K., Ji, L.L. and Herb, R.A. 1993. Rigorous Exercise Training Increases Superoxide Dismutage Activity in Ventricular Myocardium. *American Journal of Physiology.* **265**: 2094-2098.
27. Ray, J., Palmer, T.D., Suhonen, J., Takahashi, J. and Gage, F.H. 1997. In: *Isolation, Characterization, and Utilization of CNS Stem Cells.* Christen, Y. and Gage, F.H. (Ed.). Springer, Berlin. pp. 129-149.
28. Reddy, D.S. and Kulkarni, S.K. 1998. Possible Role of Nitric Oxide in the Nootropic and Antiamnesic Effect of Neurosteroids on Aging and Dizocilpine induced Learning Impairment. *Brain Research.* **799**. 215-229.
29. Russo-Neustadt, A., Beard, R.C. and Cotman, C.W. 1999. Exercise, Antidepressant Medications, and Enhanced Brain Derived Neurotrophic Factor Expression . *Neuropsychopharmacology.* **21**: 679-682.
30. Schuman, E.M. 1999. Neurotrophin Regulation of Synaptic Transmission. *Current Opinion in Neurobiology.* **9**: 105-109.
31. Sen, C.K., Marin, E., Kretzschmar, M. and Hanninen, O. 1992. Skeletal Muscle and Liver Glutathione Homeostasis in Response to Training, Exercise, and Immobilization. *Journal of Applied Physiology.* **73**: 1265-1272.
32. Teri, L., Gibbons, L.E., McCurry, S.M., Logsdon, R.G., Buchner, D.M., Barlow, W.E., Kukull, W.A., Lacroix, A.Z., McCormick, W. and Larson, E.B. 2003. Exercise Plus Behavioral Management in Patients with Alzheimer's Disease: A randomized controlled trial. *JAMA.* **290**: 2015-2022.
33. Teri, L., McCurry, S.M., Buchner, D.M., Logsdon, R.G., La Croix, A.Z., Kukull, W.A., Barlow, W.E. and Larson, E.B. 1998. Exercise and Activity Level in Alzheimer's Disease: A potential treatment focus. *Journal of Rehabilitation Research and Development.* **35**: 411-419.
34. Van Praag, H., Kempermann, G. and Gage, F.H. 1999. Running Increases Cell Proliferation and Neurogenesis in the Adult Mouse Dentate Gyrus. *Nature Neuroscience.* **2**: 266-270.
35. Vijay Kumar, K. and Naidu, M.U.R. 2002. Effect of Oral Melatonin on Exercise-induced Oxidant Stress in Healthy Subjects. *Indian Journal of Pharmacology.* **34**: 256-259.

Biomechanics
R.K. Saxena and P. Mishra (Editors)
Copyright © 2005, Anamaya Publishers, New Delhi, India

8. A Three-dimensional, Anatomically Detailed Foot Model: A Foundation for a Finite Element Simulation and Means of Quantifying Foot-Bone Position

Nimesh Prakash[1] and Piyush Soni[2]

[1]All Saints College of Technology, Bhopal - 462001, India

[2]Maharana Pratap College of Technology, Gwalior - 474006, India

Abstract: We generated an anatomically detailed, three-dimensional (3-D) reconstruction of a human foot from 286 computerized topographic (CT) images. For each bone, 2-D cross-sectional data were obtained and aligned to form a stacked image model. We calculated the inertial matrix of each bone from the stacked image model and used to determine the principal axes. Relative angles between the principal axes of the bones were employed to describe the shape of the foot, i.e., the relationships between the bones of the foot. A 3-D surface model was generated from the stacked image models and a detailed 3-D mesh for each bone was created. Additionally, the representative geometry of the plantar soft tissue was obtained from the CT scans, while the geometries of the cartilage between bones were obtained from the 3-D surface bone models. This model served dual purposes: it formed the anatomical foundation for a future finite element model of the human foot and was used to objectively quantify foot shape using the relationships between the principal axes of the foot bones.

This material is based on work done by the Department of Veterans Affairs, Veterans Health Administration, Rehabilitation Research and Development Service.

Introduction

Increasingly, musculoskeletal models of the human body are used as powerful tools to study biological structures, however, they frequently lack the geometric detail necessary to provide meaningful insights into biomechanical behavior. The lower limb, and in particular the foot, is of interest because it is the primary physical interaction between the body and the environment during locomotion. Just how variations in foot structure affect the interaction between the body and the environment is an ongoing research question. The work of Morag and Cavanagh (1999) has demonstrated that foot structure, as determined by two-dimensional (2-D) x-ray measurements, can affect plantar pressure. Objective measure of the 3-D relationships of the bones, rather than 2-D angular projections to cardinal planes, may provide further insight into foot-bone architecture.

Simplified biomechanical models of the foot have been generated but have lacked the necessary detail to accurately model biomechanical behavior. Phenomenological foot models have not provided anatomically detailed information about the structures within the foot (Meglan, 1991; Gilchrist, 1994). Others have developed models with better anatomical accuracy, including 2-D finite element models, but still lacked the detail required to study the individual motions of the bones of the foot (Nakamura et al., 1981; Patil et al., 1993, 1996, Lemmon et al., 1997, Ledoux and Hillstrom, 1999; Ledoux, 1999; Saucerman et al., 2000).

Several 3-D finite element models of the human foot have been developed, but most have been uniquely tailored to study the loading that a lower limb may undergo in automobile accidents (Beaugonin et al., 1997; Bedewi and Digges, 1999; Beillas et al., 1999; Dubbeldam et al., 1999; Tannous et al., 1996), while others have studied the effects of Hansen's disease (Jacob and Patil, 1999). These simulations range in fidelity and scope. The work of Beillas et al. (1999) is the most anatomically detailed model to date; it included osseous geometry obtained from computerized tomographic (CT) images, an approximation of cartilage geometry, foot and ankle ligaments, and plantar soft tissue properties. However, the model did not include anatomical toes, the plantar soft tissue was modeled coarsely, and 3-D cartilage models were not included.

Another means of studying foot biomechanics, namely quantitative measures of foot shape (neutrally aligned, pes planus (low arch), and pes cavus (high arch)), are limited by subjective error. Footprint indexes have been developed to describe the relationship between the footprint and the height of the arch (Staheli et al., 1987; Cavanagh and Rodgers, 1987; Hawes et al., 1992; Chu et al., 1995; Tareco et al.; 1999). However, these measurements are 2-D descriptions of 3-D phenomena, have not been correlated with foot type, and involve potential rater error. Various radiographic parameters have been employed to quantify foot morphology (Bordelon, 1980; Gould, 1982; Allard et al., 1982; Sangeorzan et al., 1993; Myerson et al., 1995; Saltzman et al., 1995; Dyal et al., 1997; Chen et al., 1998), but a comprehensive study correlating foot type with x-ray parameters has not been performed. X-ray measurements are also 2-D descriptions of 3-D phenomena and are similarly limited by rater error. Static alignment devices have also been used to quantify the amount of medial malleolar displacement during quiet stance (Rose, 1982; Welton, 1992; Thomson, 1994; Song et al., 1996). Although differences have been measured between foot types, the static alignment devices require an operator that places the device around a subject's limb and can only quantify rear foot position. Finally, CT images and the cardinal plane angular relationships between bones have been used to describe foot type (Grasso and Scarfi, 1993). However, the CT scans were not performed under weight-bearing conditions, measurements were only made in the two cardinal planes studied, and an observer was required to make the measurements, thus introducing further subjective error.

In summary, computational foot models have in general either lacked anatomical detail or been tailored for specific simulations. Further, all the aforementioned foot-type determination parameters require input from an observer, which introduces a subjective component. The purpose of this research is twofold. First, this paper will describe the development of a 3-D, anatomically detailed model of the human foot from CT scans. This model will serve as the foundation for a future finite element model of the human foot. Second, the geometrical data of the bones will be used to generate objective 3-D descriptions of bone position. The relationships between bones will be used in the future as a way of describing differences between feet of different architectures. Data will be presented from one foot to demonstrate the two purposes.

Methods

The CT images were acquired from the left cadaveric foot of a 67-year-old male donor. The specimen was obtained from the Department of Biological Structure, University of Washington. No gross deformities or significant degenerative changes were evident on anteroposterior (AP) or lateral radiographs. To prepare the foot for scanning, we thawed the specimen, dissected away the soft tissue around the tibia, reamed the tibial intramedullary canal, and threaded an acrylic rod into the tibia. The specimen was supported with an acrylic frame within a Hi-Speed Advantage CT scanner (General Electric Medical Systems, Milwaukee, WI) (Fig. 1). To hold the specimen in place, we loaded it with a nominal force within the frame. The threaded acrylic rod in the tibial shaft was cross-locked with an acrylic nut. The rod was held upright in the frame, and a second nut was used to apply minimal force to the foot and ankle with the tibia upright and the ankle at 90°, i.e., anatomical neutral position.

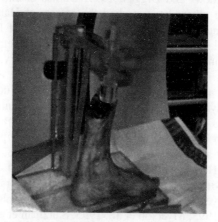

Fig. 1. Cadaveric foot supported in acrylic loading frame in CT scanner.

No attempt was made to load the foot to normal weight-bearing while it was being scanned. One future goal of this research was the development of a finite element foot model, which would require unloaded geometrical data to simulate initial conditions.

Another future goal was the quantification of differences in foot architectures in live subjects, which would require that the feet be loaded before scanning. For our purposes, i.e., obtaining the geometric data for a future finite element model and demonstrating the use of principal axes to describe foot-bone position, non-weight-bearing data were deemed adequate.

Frontal plane CT images of the specimen were acquired at 1-mm intervals, beginning posteriorly with the heel and proceeding anteriorly to the toes. The scans were taken with 512 × 512 pixels over a 206 mm × 206 mm area, for a dimensional accuracy of 0.4023 mm/pixel. For the entire foot to be scanned, 286 slices were required. The frontal plane was chosen as the optimal scanning orientation because it avoided potential difficulties in data processing that can occur when a bone appears as multiple discontinuous regions in the same CT image.

Each bone was represented as a series of 2-D outlines from the CT images. The CT data was downloaded to a Macintosh G3 PowerPC workstation (Apple Computer, Inc., Cupertino, CA) and visualized with the National Institutes of Health (NIH) Image 1.61 software (National Institutes of Health, Bethesda, MD). The 286 images were combined into a stack. For each slice, contours outlining the cortical shell of a particular bone were derived with NIH image's density slice option and automatic outlining tool (Fig. 2). A threshold of 137 (on an 8-bit scale) was optimal for contrasting bone. In rare instances that the border between two bones was sub-optimally delineated with the automatic outlining tool, manual pixel-level user input was required to define the borders. Once the edge of the bone of interest was determined, the stack was advanced to the next slice and the process repeated until every slice containing the particular bone had been examined.

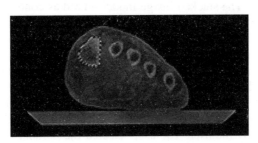

Fig. 2. A single CT slice with contour of first metatarsal cortical shell outlined.

A custom segmentation macro for NIH image (developed by Randal P. Ching) was created to obtain a description of the boundary of the object. The data, consisting of the x-y coordinates of each point that described the shape of the boundary of each slice, was saved to a separate file. This process was repeated for all of the bones of the foot; the sesamoids were included with the first metatarsal.

A 3-D stacked image model for each bone was created from the 2-D data. Each bone was represented by a set of files describing each slice of the particular bone of interest. To combine the slices into one file as a stacked image model, a custom software program (PolyLines 1.9, developed by Randal P. Ching) was used. PolyLines sequentially read each file of 2-D data (i.e., the x-y coordinates of the boundary of each slice) and combined the slices into a stacked image model by incorporating the space between slices as the global z-axis distance. To visualize the stacked image model, in the drawing exchange format (DXF), we converted the file to a Rotator file with a DXF-to-Rotator converter (http://raru.adelaide.edu.au/rotater/, developed by Craig Kloeden). The Rotator file was viewed with the Rotater 3.5 software package (http://raru.adelaide.edu.au/rotater/; developed by Craig Kloeden) (Fig. 3). The stacked image model was examined from all sides to look for irregularities in the surface, with corrections (i.e., regenerating boundaries for a particular slice with NIH image) made as necessary.

Fig. 3. A stacked image model of first metatarsal, demonstrating 2-D slices stacked together to form a 3-D object.

Fig. 4. 3-D surface model of first metatarsal.

A 3-D surface model of each bone was generated from the stacked image model. Each bone's DXF file was imported into form•Z 3.5 (auto•des•sys, Inc., Columbus, OH), a 3-D form synthesizer, to create a 3-D surface model. The stacked image model served as control lines for the generation of a controlled mesh 3-D surface (Fig. 4). A broken Bézier-controlled mesh-smoothing algorithm was used to generate the surface. All faces on the surface were then triangulated. After the triangulated mesh was created, it was exported in stereolithography (STL) format.

A 3-D surface mesh of each bone, which would be a foundation for a finite element model, was created from the surface models. The final software package used in the model development process was TrueGrid 1.4 (XYZ Scientific Applications, Inc., Livermore, CA), a finite element preprocessor and mesh generator. This software imported the 3-D STL file and enabled the mesh shape and density to be interactively generated by the user. The user selected the number of nodes as well as the initial position of the control nodes on the existing surface. The mesh was projected to the 3-D surface (Fig. 5). The number of nodes was selected to produce a mesh that was sufficiently detailed to accurately model biomechanical behavior without being too detailed such that future finite element simulation times would be rendered intractable. As before, the bony image was rotated to identify defects in the generated mesh; if defects were found, the control nodes and the mesh density could be adjusted. Furthermore, several diagnostic measurements could be conducted on the mesh (e.g., the orthogonality and the aspect ratio of each element) to quantify the quality of the mesh. A total of 7,022 four-noded shell elements were used to generate the surface mesh, ranging from 902 for the talus and 720 for the calcaneus to 32 for the fourth distal phalange (Fig. 6). We repeated this protocol for each of the bones of the foot. A mesh representative of the plantar soft tissue was also generated in this manner (Fig. 7). The borders of these tissues included the dorsal aspect of the foot bones, as well as the medial, lateral, and dorsal aspects of the foot. The same procedures were employed, except for the threshold level in the NIH image, which was adjusted to range from 133 to 185. The plantar soft tissue was represented by 2,112 eight-noded hexahedral elements.

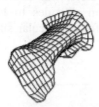
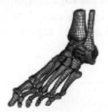

Fig. 5. Mesh representing surface geometry of first metatarsal.

Fig. 6. Mesh representing surface geometry of all of bones of foot.

Fig. 7. 3-D surface model representing plantar soft tissue.

The 3-D cartilage bodies were generated from the 3-D surface models of the bone, since the cartilage borders were not readily viewable in the CT scans. Once the bone models had been exported into form•Z, representative cartilage bodies were generated by creating a solid volume around the joint of interest. The bones were subtracted from the solid volume, and the remaining shape was trimmed to match the contours of the joint of interest (Fig. 8). Note that separate layers of cartilage were not created for each bone, rather one 3-D object represented all of the cartilages between the two bones. As with the bony objects, the cartilage objects were saved in STL format and exported to TrueGrid for mesh generation. The meshes ranged from 47 eight-noded hexahedral elements for the ankle-joint cartilage to 9 elements for the calcaneocuboid cartilage. A model of the entire foot, containing bones, plantar soft tissue, and cartilage was generated (Fig. 9).

Fig. 8. Cartilage between first metatarsal and medial cuneiform.

Fig. 9. Model of foot, containing bones, plantar soft tissue, and cartilage.

In addition to converting the 2-D slices into 3-D stacked image models, PolyLines were used to determine the center of volume and the inertial matrix of each bone via the parallel axis theorem. The bones were assumed to be of homogeneous density. This resulted in an inertial matrix that was based solely on the geometry of the bones and not on the mass density. While this "inertial" matrix would be inadequate for kinetic analysis, it was sufficient for describing the shape of the foot based on the volume and geometry of the bones.

The eigenvectors of the inertial matrix established the principal axes of each bone (Fig. 10). To remove the subjectivity associated with manual measurements on x-rays as well as the limits of 2-D projections of 3-D osseous geometry, relative angles were determined between bones from the principal axes. The talus and its associated joints are involved with frontal, sagittal, and transverse plane rotations within the foot, and the remainder of the foot often moves relative to the talus. Therefore, the positions of all bones of the foot were determined relative to the talus. For example, the direction cosine matrix between the first metatarsal principal axes (a 3 × 3 matrix M in the global coordinate system) and the talar principal axes (a 3 × 3 matrix T in the global coordinate system) was determined by multiplying T by M. Using established trigonometric relationships, an Euler angle description (z-y-x) from the direction cosine matrix was calculated. These three angles described the rotation of the first metatarsal relative to the talus; the rotations were made about the moving reference frame of the first metatarsal. The first rotation was about the z-axis, the second rotation was about the y-axis, and the third rotation was about the x-axis. (Note that the principal axes were determined before the TrueGrid surface mesh generation and thus were not subject to the potential additional levels of error introduced in those steps.) This procedure was done for the following bones: first metatarsal, second metatarsal, calcaneus, navicular, and cuboid.

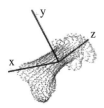

Fig. 10. Principal axes of first metatarsal.

Results

Anatomically accurate representations of the osseous and soft tissues of the foot using the reconstruction algorithms were generated. Data from CT images were processed for the generation of 3-D surface models of the bones, plantar soft tissue, and cartilage (Figs. 4, 7 and 8). The individual bones of the foot that were developed separately were combined with the soft tissues into a model of the entire foot (Fig. 9). Diagnostic analysis confirmed that the mesh quality was suitable for future finite element analysis.

The model provided objective, quantitative measures of the relative positions of the foot bones. The relative angles between bones from the principal axes for each bone (Fig. 10) were calculated. As an example, the three Euler angles (z-y-x) that described the transformation from the first metatarsal to the talus were −89.6°, 8.4°, 15.6° (Figs. 11 and 12). For each bone, the z-axis represented the axis about which the moment of inertia was smallest, the x-axis was the axis about which the moment of inertia was largest, and the y-axis was the cross product of the first two. The z-axis of the talus was the "long" axis, progressing from the center of mass anteriorly through the approximate center of the talar head. The y-axis was directed medially and the x-axis was directed superiorly. For the metatarsal, the z-axis was also the long axis, while the y-axis was directed dorsally and the x-axis directed laterally. Relative angles between the talus and other foot bones were also calculated (Table 1).

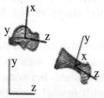

Fig. 11. Sagittal plane relationship of principal axes of talus and first metatarsal.

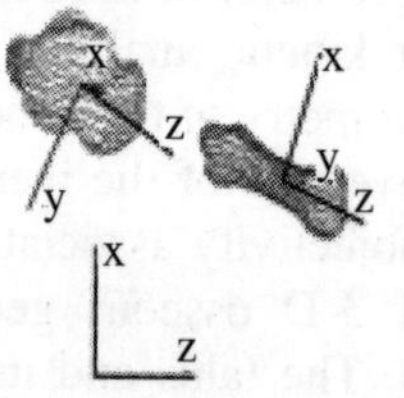

Fig. 12. Transverse plane relationship of principal axes of talus and first metatarsal.

Table 1. Euler angle rotations describing relationship between several bones of foot and talus

Bony Relationship	Alpha (z axis)	Beta (y axis)	Gamma (x axis)
First metatarsal to talus	−89.6	8.4	15.6
Second metatarsal to talus	−117.4	23.4	10.0
Calcaneus to talus	−71.6	34.3	−17.3
Navicular to talus	−134.4	−62.7	56.8
Cuboid to talus	−158.2	59.3	−21.2

Each relative angle describes a rotation of a bone about the talus, this relationship was determined for five bones, but only the first metatarsal was discussed. For clarification, it may help to think of these two bones as they sit in the cardinal planes. The long or z-axes of both bones (about which there is less inertia) were directed approximately anteriorly. The x-axes (about which there is the greatest amount of inertia) were not similarly directed; the talar x-axis was directed superiorly, while the first metatarsal x-axis was directed laterally.

One can clarify the Euler angle description by picturing the two bones rotated so that their principal axes are coincident; from this position, the Euler angles described the angular rotations that the first metatarsal must take relative to the talus to achieve its position in the cadaveric specimen. A rotation of −89.6° about the z-axis describes a rotation of the first metatarsal about the long axis of the first metatarsal (which was coincident with the long axis of the talus at this point). Thus, the axes with the most inertia differed between the bones by almost 90°. The 8.55° rotation about the y-axis indicated that the first metatarsal was plantar flexed relative to the talus. The 15.50° rotation about the x-axis demonstrated an external rotation of the first metatarsal relative to the talus. These two rotations, describing roughly the sagittal and transverse plane relationships, were clinically relevant for the two bones in question.

Discussion

Procedures for generating an anatomically detailed computer model from CT scans and for quantifying the relationships between the bones of the foot have been developed. Using custom software together with several commercial packages, CT images were processed so that surface models for individual bones, plantar soft tissue, and cartilage were generated. These geometries were used to create a 3-D anatomically detailed foot model that would serve as the basis for a finite element foot model. For each individual bone, the principal axes were calculated from the inertial matrix. These axes were used, which were determined objectively, to determine relative angles between bones. The angles describing the first metatarsal relative to the talus were presented as an example.

The study had several limitations. One important consideration was the amount of load applied to the foot while it was scanned. Because computer simulations typically start with unloaded initial conditions, scanning a minimally loaded specimen might be ideal for generating anatomical data for finite element modeling. However, when comparing the principal axes between different foot types, one should load specimens to physiological levels, because relative angles of foot bones would change when the foot is loaded. In future, when additional feet from live subjects are studied and contrasted for differences, they would be loaded with an acrylic frame during data collection. However, for this paper, the unloaded protocol was sufficient for demonstrating the utility of the method.

Additionally, the geometry of the bodies representing the cartilage was not obtained from actual cartilage. Instead, the cartilage models were created such that they filled the space between the bones. However, since the cartilage was not easy to distinguish in the CT images, our methodology provided the best geometrical representation possible within the limits of our system.

Certain steps in the reconstruction process did introduce limited amounts of subjective error. Although an attempt was made to quantify foot-bone position objectively, small amounts of subjectivity, e.g., the setting of the threshold level and the occasional need to perform manual pixel-level corrections with NIH Image, were impossible to avoid. However, the threshold level was constant for most CT images and the number of manual pixel-level corrections was minimal. Finally, while the single cadaveric foot specimen used in this study was thought to represent a typical adult foot and was deemed free of deformity or disease, the results discussed here do not represent a cross section of the population or the average results from a particular foot type.

The calculated relative angles between the talus and the first metatarsal provide information that is similar to the data obtained currently from planar radiographs. Two of the CT measurements, the plantar flexion and external rotation of the first metatarsal, describe angles that provide information similar to clinical x-ray measurements, i.e., the lateral talometatarsal angle and the transverse talometatarsal angle, respectively. However, the measurements were constructed in a different manner, i.e., the x-ray parameters were 2-D measurements made by one subjectively drawing lines connecting certain points on x-rays, while the CT parameters are 3-D measurements generated objectively from the osseous geometry. Thus, while a one-to-one correspondence between the parameters may not exist, both methods may describe similar trends for a particular foot type e.g., flatfeet will have characteristic x-ray measurements as well as CT parameters between the first metatarsal and the talus. However, unlike the radiographic data, the CT data will provide an objective 3-D description of the relationship between bones.

The soft tissue and osseous mesh created in the current study provides a foundation for future finite element analysis of foot biomechanics. Generated with largely automated reconstruction algorithms, the mesh possesses anatomical details, which are not described in previously mentioned existing foot models. These include more refined bones and plantar soft tissues, anatomically accurate toes, and cartilage, all of which may be critical in accurately simulating foot behavior.

Conclusion

The methodology discussed in this paper lays the foundation for the development of a finite element model of the foot as well as for future work on quantifying differences in foot shape between different foot types. The 3-D shapes of the bones, cartilage, and plantar soft tissue obtained from the CT scans will provide the necessary anatomical detail to begin finite element foot modeling. The relative angles between bones, as calculated from the principal axes, allow for objective determination of the relationships between the bones of the foot. This will be a new way to quantify differences between foot types.

Acknowledgment

This research was supported by the Department of Veterans Affairs, Veterans Health Administration, Rehabilitation Research and Development Service and examined thoroughly by Dr. K.C. Singhai and Dr. A.R. Sidiqui, Director.

References

1. Allard, P., Sirois, J.P., Thiry, P.S., Geoffroy, G. and Duhaime, M. 1982. Roentgenographic Study of Cavus Foot Deformity in Friedreich Ataxia Patients: Preliminary report. *Can. J. Neurol. Sci.* **9(2)**: 113-17.
2. Beaugonin, M., Haug, E. and Cesari, D. 1997. Improvement of Numerical Ankle/foot Model: Modeling of deformable bone. *41st Stapp Car Crash Conference*, Orlando (FL). pp. 225-37.
3. Bedewi, P.G. and Digges, K.H. 1999. Investigating Ankle Injury Mechanisms in Offset Frontal Collisions Utilizing Computer Modeling and Case-study Data. *43rd Stapp, Car Crash Conference*, San Diego (CA). pp. 185-202.
4. Beillas, P., Lavaste, F., Nicolopoulos, D., Kayventash, K., Yang, K. and Robin, S. 1999. Foot and Ankle Finite Element Modeling using CT-scan Data. *43rd Stapp Car Crash Conference*, San Diego (CA). p. 217-42.
5. Bordelon, R.L. 1980. Correction of Hypermobile Flatfoot in Children by Molded Insert. *Foot Ankle Int.* **1(3)**: 143-50.
6. Cavanagh, P.R. and Rodgers, M.M. 1987. The Arch Index: A useful measure from footprints. *J. Biomech.* **20(5)**: 547-51.
7. Chen, Y.J., Huang, T.J., Hsu, K.Y., Hsu, R.W. and Chen, C.W. 1998. Subtalar Distractional Realignment Arthrodesis with Wedge Bone Grafting and Lateral Decompression for Calcaneal Malunion. *J. Trauma.* **45(4)**: 729-37.
8. Chu, W.C., Lee, S.H., Chu, W., Wang, T.J. and Lee, M.C. 1995. The Use of Arch Index to Characterize Arch Height: A digital image processing approach. *IEEE Trans. Biomed. Eng.* **42(11)**: 1088-93.
9. Dubbeldam, R., Nilson, G., Pal, B., Eriksson, N., Owen, C., Roberts, A., et al. 1999. A MADYMO Model of the Foot and Leg for Local Impacts. *43rd Stapp Car Crash Conference*, San Diego (CA). pp. 185-202.
10. Dyal, C.M., Feder, J., Deland, J.T. and Thompson FM. 1997. Pes Planus in Patients with Posterior Tibial Tendon Insufficiency: Asymptomatic versus symptomatic foot. *Foot Ankle Int.* **18(2)**: 85-88.
11. Gilchrist, L.A. 1994. A Computer Simulation of Human Gait. *Ph.D. thesis*, University of Waterloo, Waterloo.
12. Gould, N. 1982. Graphing the Adult Foot and Ankle. *Foot Ankle Int.* **2(4)**: 213-19.
13. Grasso, A. and Scarfi, G. 1993. Computed Tomography Demonstration of Calcaneal Morphotypes in Plantar Arch Abnormalities. *Radiol Med (Torino)*. **85(4)**: 384-88.
14. Hawes, M.R., Nachbauer, W., Sovak, D. and Nigg, B.M. 1992. Footprint Parameters as a Measure of Arch Height. *Foot Ankle Int.* **13(1)**: 22-6.
15. Jacob, S. and Patil, M.K. 1999. Three-dimensional Foot Modeling and Analysis of Stresses in Normal and Early Stage Hansen's Disease with Muscle Paralysis. *J. Rehabil. Res. Dev.* **36(3)**: 252-263.

16. Ledoux, W.R. and Hillstrom, H.J. 1999. A Graphics-based, Anatomically Detailed, Forward Dynamic Simulation of the Stance Phase of Gait, Emphasizing the Properties of the Plantar Soft Tissue. *17th Congress of the International Society of Biomechanics*, Calgary, Alberta. p. 110.
17. Ledoux, W.R. 1999. A Biomechanical Model of the Human Foot with Emphasis on the Plantar Soft Tissue. *PhD thesis*, University of Pennsylvania, Philadelphia.
18. Lemmon, D., Shiang, T.Y., Hashmi, A., Ulbrecht, J.S. and Cavanagh, P.R. 1997. The Effect of Insoles in Therapeutic Footwear: A finite element approach. *J. Biomech.* **30(6)**: 615-20.
19. Meglan, D. 1991. Enhanced Analysis of Human Locomotion. *Ph.D. thesis*, Ohio State University, Columbus.
20. Morag, E., Cavanagh, P.R. 1999. Structural and Functional Predictors of Regional Peak Pressures under the Foot during Walking. *J. Biomech.* **32(4)**: 359-70.
21. Myerson, M.S., Corrigan, J., Thompson, F. and Schon, L.C. 1995. Tendon Transfer Combined with Calcaneal Osteotomy for Treatment of Posterior Tibial Tendon Insufficiency: A radiological investigation. *Foot Ankle Int.* **16(11)**: 712-18.
22. Nakamura, S., Crowninshield, R.D. and Cooper, R.R. 1981. An Analysis of Soft Tissue Loading in the Foot-A preliminary report. *Bull Prosthet Res.* **18(BPR 10-35)(1)**: 27-34.
23. Patil, K.M., Braak, L.H. and Huson, A. 1993. Stresses in a Simplified Two-dimensional Model of the Human Foot-A preliminary analysis. Mech. Res. Commun. pp. 1-7.
24. Patil, K.M., Braak, L.H. and Huson, A. 1996. Analysis of Stresses in Two-dimensional Models of Normal and Neuropathic Feet. *Med. Biol. Eng. Comput.* **34(4)**: 280-84.
25. Rose, G.K. 1982. Pes Planus. In: Disorders of the Foot and Ankle: Medical and surgical management. Jahss, M.H. (Ed.). W.B. Saunders, Philadelphia (PA).
26. Saltzman, C.L., Nawoczenski, D.A. and Talbot, K.D. 1995. Measurement of the Medial Longitudinal Arch. *Arch. Phys. Med. Rehabil.* **76(1)**: 45-49.
27. Sangeorzan, B.J., Mosca, V. and Hansen, S.T. Jr. 1993. Effect of Calcaneal Lengthening on Relationships Among the Hindfoot, Midfoot, and Forefoot. *Foot Ankle Int.* **14(3)**: 136-41.
28. Saucerman, J.J., Loppnow, B.W., Lemmon, D.R., Smoluk, J.R. and Cavanagh, P.R. 2000. The Effect of Midsole Plugs Inserted into Therapeutic Footwear for Localized Plantar Pressure Relief. *24th Annual Meeting of the American Society of Biomechanics*, Chicago (IL). pp. 235-36.
29. Song, J., Hillstrom, H.J., Secord, D. and Levitt, J. 1996. Foot Type Biomechanics: A comparison of planus and rectus foot types. *J. Am. Podiatr. Med. Assoc.* **86(1)**: 16-23.
30. Staheli, L.T., Chew, D.E. and Corbett, M. 1987. The Longitudinal Arch: A survey of eight hundred and eighty-two feet in normal children and adults. *J. Bone Joint Surg. Am.* **69(3)**: 426-28.
31. Tannous, R.E., Bandak, F.A., Toridis, T.G. and Eppinger, R.H. 1996. A Three-dimensional Finite Element Model of the Human Ankle: Development and preliminary application to axial impulsive loading. *40th Stapp Car Crash Conference*, Albuquerque (NM). pp. 219-38.
32. Tareco, J.M., Miller, N.H., MacWilliams, B.A. and Michelson, J.D. 1999. Defining Flatfoot. *Foot Ankle Int.* **20(7)**: 456-60.
33. Thomson, C.E. 1994. An Investigation into the Reliability of the Valgus Index and its Validity as a Clinical Measurement. *Foot.* **4**: 191.
34. Welton, E.A. 1992. The Harris and Beath Footprint: Interpretation and clinical value. *Foot Ankle Int.* **13(8)**: 462-68.

Biomechanics
R.K. Saxena and P. Mishra (Editors)
Copyright © 2005, Anamaya Publishers, New Delhi, India

9. MEMS Based Implantable Systems for Bio Medical Application

Mohan Kumaraswamy and S. Ravi Kumar
Core Technologies Division, Embedded Group, HCL Technologies Ltd.,
Ambattur Industrial Estate, Ambattur, Chennai - 600026, India

Abstract: Today's Technology has advanced to a level where yesterday's dream is today's reality. One such dream of Dr. Richard Feynman was on miniaturized machines. Today Micro Electro Mechanical Systems (MEMS) is the realizable first step towards that dream. Silicon technology has advanced to cater to the requirements of Micro-Electro-Mechanical Systems. In this paper, we have targeted a MEMS device, which can work without a power source of its own. This device can be used to monitor the pressure inside the human body and the monitored pressure information will be used for further studies to learn about different behavioral patterns of human organs. The same MEMS device will be enhanced in future to support neuro-simulation enabling the tissues to be controlled based on the pressure information.

In this system, radio frequency waves are used to transfer power to the implantable device and to transfer data from the implantable device to the external device. In this paper, few best possible approaches for realizing this functionality and the pros and cons of these approaches are discussed. Finally, the chosen method is discussed in detail and the modes employed for eliminating the impact of misalignments are elaborated. Apart from discussing the monitoring portion, this paper also discusses about the possible future enhancements of this system, such as neuro-simulation.

Introduction

MEMS based implantable pressure sensor is a miniaturized pressure sensor, which can be used for measuring pressure anywhere inside a human being. This implantable device is a passive device, which operates by deriving power from radio frequency (RF) signals through inductive coupling. The change in transducer value due to pressure variations will be communicated to the external device placed near the body, through the same inductive coupling principle used for power transfer. This MEMS device could be fabricated using either double-sided lithographic technique or by wafer bonding two back-to-back placed MEMS devices. There will be an inbuilt analog core, which will convert the analog value of pressure to digital data and do a digital transfer. This will help us in eliminating the measurement errors due to misalignment between the implanted and external device. In addition to reception and storage of the monitored pressure data, the external device also acts as a power source for the internal device.

The rest of the paper discusses the following: (i) a brief overview of the system requirements and the specifications that this product needs to comply with, (ii) system architecture and system states with emphasis on various approaches considered for realizing this functionality, (iii) limitations addressed by this system, and (iv) challenges foreseen in this development. Finally, various enhancements and value additions possible in this system are listed followed by our acknowledgements to the valuable contributions of our colleagues and senior managers.

Product Requirements

Compatibility Considerations
The following are the major points to be considered during the design of implantable wireless passive devices:

1. Size depends entirely on the location where this device will be implanted.
2. Accuracy and reliability of the measurement.
3. Standards governing RF safety limits and user frequency bands.
4. Thermal and athermal effects that affect the tissue.
5. Loss of RF energy in tissues.
6. Power requirements of the implantable device.
7. Bio compatibility.
8. Storage of captured data either internally or externally.
9. Mobility of the patient

Specifications to the Device

RF Specification
RF devices primarily need a tansmitter/receiver and an antenna. Generally designing an antenna for transmission and reception involves an antenna size comparable to wavelength λ (where λ = velocity of light/frequency of RF communication, e.g. $\lambda/2$ for a dipole antenna). In order to minimize the antenna size we need to use GHz (gigahertz) range of RF waves. However, the losses in tissues increase as the frequency increases. Based on the different data available from the past study reports given in different standards, it is observed that a frequency of 20 MHz (megahertz) can have a skin depth penetration of 23.16 cm which meets the basic requirement of this system. Since RF transmission at this frequency is not feasible with small antennas and since there is a need to transfer power as well through wireless transfer, inductive coupling is considered. Since RFID (Radio Frequency Identification) frequency is a generally accepted frequency and its values of 13.56 MHz gives a better skin depth we can have this as one of the choices. However, there is a possibility of a clash with other RFID devices commonly used. Hence, the operating frequency is narrowed down to a lower frequency range of 10 MHz based on electromagnetic analysis results.

Standards
Federal Communications Commission (FCC) standards and Institute of Electrical and Electronics Engineers, Inc (IEEE) standards are the major standards, which govern the implantable RF device standards. Standards classify the RF exposure effects as thermal and athermal effects. Athermal effects are not currently included in any of the major available standards. Based on thermal effects, the maximum permitted exposure limits are specified as 0.4 W/kg for averaged power loss (whole body exposure) and 8 W/kg (localized exposure) for concentrated power loss in a small area.

Packaging

Silicon substrate is used for MEMS and microelectronics device fabrication. Medical grade silicon can be used as the package itself. The membrane separating the pressure sensor and the external environment would be made of the same medical grade silicon to ensure biocompatibility of the implanted device.

Other Parameters Considered for Specification

Power requirement is another major requirement considered in the development of this system. The pressure transducer and associated analog circuitry needs low power design. At the same time due to digital transfer of data there needs to be a power transistor switch, which can handle slightly higher power to avoid any damages due to other RF signals in the vicinity. Since the implanted device is a passive device, this entire power needs to be transferred from the external device. The design of power transfer circuit in the external device should be in such a way that the implantable device power requirement is met after losses in tissues. Since the monitored pressure needs to be stored for further analysis, suitable memory needs to be provided. Due to size constraints, this memory will be provided in the external device.

Patient mobility is another important factor that needs to be considered while deciding the battery requirements of external device. Based on the power requirements for the implanted/external device and the duration for which the data needs to be collected continuously, a suitable battery needs to be identified. The external device can be controlled by a microcontroller, carefully chosen in such a way that it can operate at different power levels or system states. The architecture is chosen in such a way that the power transfer and hence the monitoring of pressure would be done only when required. This can be realized by a time-triggered wakeup in the external device.

System Architecture

Overall System

Figure 1 provides an overall pictorial representation of the system. Here the implantable MEMS device (IMD) is the MEMS based pressure sensor, which can be implanted in any part of the body. The only constraint is that the medium in which the IMD is implanted should be a fluidic, non-acidic and non-corrosive medium. The applied pressure will be converted to an electrical signal by the transducer and will be communicated immediately to the external device through electromagnetic waves. Other modes like ultrasound were also considered for power and data transfer. However, since electromagnetic waves and their studies are long proven, this principle is adopted for our architecture.

Power required for the transducer to convert the pressure information into electrical signals and to transfer the collected data through RF is derived from the external device. These monitored pressure samples will be stored inside the external device memory for further usage such as data analysis by physicians for diagnosing the problems. The external device could also be used to give a neural simulation to release the sphincter muscles in certain cases in a controlled manner. This can be initiated from the external device automatically or with the help of the patient control switches. The stored information could be moved from external device to a personal computer, for further analysis and diagnostics, using reader as the interface unit. The necessity of having a reader depends on the size of the external device chosen. In case of larger external devices, a direct personal computer (PC) interface can be provided. Else, the same wireless interface employed between IMD and external monitoring device (EMD) will be used for communication between EMD and reader. The reader will subsequently transfer this collected data to the PC. In our paper, we have considered only the IMD implementation as the major design challenge. To make this paper complete, the external device implementation relative to the wireless power transfer and data communication are also dealt with. Rest of the details are left as trivial and can be implemented in any fashion the designer imagines. Only the underlying principle is discussed in detail in this paper. MEMS fabrication and process steps are not discussed in detail.

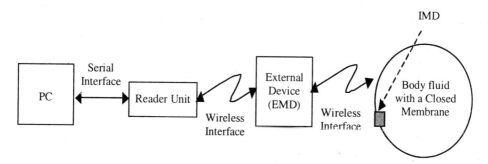

Fig. 1. Representative system.

System States

Since the system design needs to consider the entire life cycle of the product, there should be clearly defined system states and product lifetime for battery operated devices. Since the implantable device is a passive device, there are no lifetime limitations, except for the material (package and components) and transducer degradation time. The implanted device can be retained permanently inside the body if no harm is envisaged. The external device alone can be changed or recharged to get fresh set of data.

System States Considered

Based on the usage of the system, the following are identified as the system states:

Shelf State

This is the state in which the system will be maintained since its manufacture. If the device contains rechargeable or replaceable batteries, then the battery charge can be checked before it is deployed for monitoring the pressure. If it is a primary battery then it has a shelf life before which the EMD needs to be used. In case of primary batteries, the EMD can be put in permanent sleep state after manufacture to save power and a small circuitry can be introduced to trigger an interrupt. The microcontroller of EMD can wakeup based on this interrupt and look for valid communication or command from the user. In case of wrong trigger, the EMD will again go into sleep state. Since this is a trivial implementation, this is not discussed further in this document.

Activation State

The system switches to this state immediately after identifying a valid initiation from the user. This is a transition state in which it moves either to configuration state or data transfer state based on the command received from the user.

Configuration State
In this state, the configuration parameters and start monitor commands are issued to the EMD device. Once the start monitor command is issued to EMD after configuration, the EMD will be put into sleep mode for the set initial sleep time.

Implantation-Monitoring/Sleep
After the initial sleep time set during the configuration state, the EMD wakes up and enters this state. In this state, EMD powers the IMD device, reads back the monitored pressure data from IMD and stores the same. To conserve power, the EMD goes into sleep mode after receiving the measured data from IMD. EMD wakes up repeatedly after the set sample duration and collects the measured data, each time. This process is repeated until the required numbers of samples are collected. Once the desired numbers of samples are collected, EMD goes into shelf state, where it waits for the user to communicate and receive the stored pressure information.

Data Transfer State
In this state the data collected during implantation will be transferred to the PC through the reader unit. Once the data transfer is complete, EMD again goes back to shelf state.

One more state can be introduced based on the design studies done. If the design team is able to find a suitable rechargeable battery catering to the required power limits, then we can introduce a rechargeable solution. During charging, the system goes into battery charge state.

In addition, sufficient communication intelligence can be introduced during monitoring state to stop monitoring mode, if required, by placing the EMD in the reader unit. This could be used as an option to stop the monitoring mode and restart it later, if wrong initiations are done. This can also be used in test mode for measuring few samples before actual deployment of EMD, to check the proper functioning of the same. This verifies that there is no transportation damages.

Architectural Subsystem
Figure 2 shows the architectural subsystem block diagram.

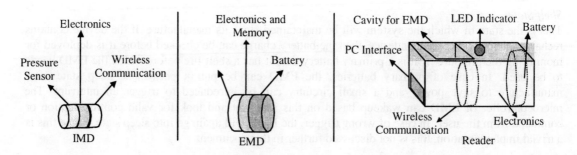

Fig. 2. System architectural block diagram.

Implantable Device (MEMS based)
Two options would be considered here. In the first approach, the implantable pressure sensor would be a capacitive pressure sensor, in which case the electronics would be completely eliminated. In the

second approach, the pressure sensor would use a piezo resistive transducer. The sensor output in this case will be digitized using a small analog circuitry before transmitting the same wirelessly.

Capacitive Pressure Sensor Approach
The resonant principle of the LC circuit (circuit containing an inductor and a capacitor) would be utilized to detect the capacitance value of the pressure sensor, which is proportional to the pressure variation. Variation in the measured pressure will vary the resonant frequency of the IMD, which in turn will alter the reflected impedance on the EMD. This impedance variation in the EMD will be directly detected and utilized for computing the pressure level. The IMD block would be implemented in MEMS. The rest of the circuit comprises the external device (Fig. 3).

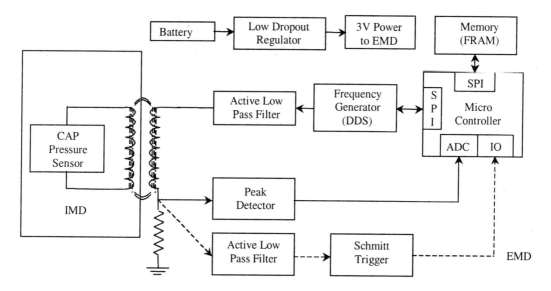

Fig. 3. Architectural block diagram using capacitive pressure sensor.

By the principle of electromagnetic induction, when the EMD coil generates an alternating electromagnetic field (caused due to the flow of alternating current) and if the IMD coil is placed in its vicinity, voltage gets induced in the IMD coil. The equivalent circuit of this is shown in Fig. 4.

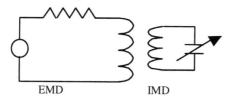

Fig. 4. Equivalent circuitry for single tuned reflected impedance.

When the impedance on the IMD coil (circuit) changes, it is "reflected" on the EMD side and manifests itself as a "voltage/current change". The change in the reactive impedance of the IMD circuit changes the natural resonant frequency of the IMD circuit. When the energizing frequency approaches the natural resonant frequency of the IMD, due to parallel resonance principle, the reflected impedance on the EMD side is maximum. Likewise, when the resonant frequency moves farther away from the energizing frequency, the reflected impedance on the EMD reduces. Hence, when the energizing frequency is equal to the resonant frequency of the IMD circuit, then the reflected impedance on the EMD increases to maximum and hence the current drawn in the EMD is minimal. This current drawn can be measured as a voltage drop across a resistance in series with the EMD coil and could be utilized to compute the pressure data.

Since circuit analysis shows that the amount of reflected impedance is also a function of the mutual inductance (which varies with misalignments between the two circuits), this needs to be eliminated for reliable pressure measurement. Various circuit choices have been considered to eliminate this fact. One approach, which was considered to address this effect, is to use two resonant IMD tank circuits. The frequency of these two circuits will be chosen in such a way that they are mutually exclusive and one tank circuit will not respond for the other tank circuits' frequency.

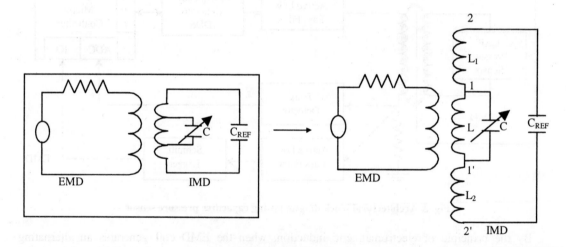

Fig. 5. IMD with two resonant circuits.

In this approach, when the tank circuit comprising of the capacitive sensor moves out of the resonance due to pressure variation, the other tank circuit cannot be totally eliminated in calculations and measurement. Hence, a small analog switching circuit is introduced that couples only one among the reference capacitor and the capacitive pressure transducer to the inductor at a time. This switching circuit switches from the reference capacitor to the pressure sensor capacitor, after a certain predetermined duration. This approach is given in Fig. 6.

Based on the permitted variation range of the capacitive transducer, a minimum and a maximum frequency of operation are fixed. The curves (Fig. 7) give an understanding of the curve movement with respect to the capacitance variation (pressure variation) and hence the resonant frequency variation.

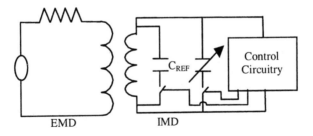

Fig. 6. IMD having control circuitry and a reference capacitor.

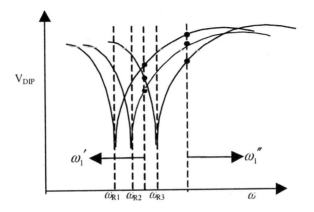

Fig. 7. V_{DIP} vs. frequency curve in a single-tuned circuit.

The graphs in Fig. 7 predict the values of the voltage drop with variation in frequency. The three curves correspond to three different capacitance values (min/typical/max). Using these curves, we can determine the energizing frequency range for the actual circuit. To determine the resonant frequency, the carrier frequency needs to be swept along the entire range. Since this needs more number of sampling frequencies and a huge power and measuring time, a different approach is adopted. In the new approach, a fixed energizing frequency (ω_1' or ω_1'') would be used. The voltage vs. capacitance curve (Fig. 8) derived from the V_{DIP} (voltage dip) vs. frequency curve (Fig. 7) can be used directly to measure the transducer capacitance. Here the reference capacitor value can be used to find the mutual inductance variation due to misalignment. This correction factor can be used to measure the actual capacitance of the pressure sensor. Since the misalignment parameter is measured using the reference capacitor, each time, just before the actual pressure sensor capacitance measurement, a small computation and interpolation technique can be used to eliminate the misalignment errors. Thus, two sampling per pressure measurement is sufficient.

In Fig. 8, the picture is read at the two different frequencies (ω_1' and ω_1''). For each of these frequencies the voltage vs. capacitance curves will be as follows, which gives a relation to decide on the design.

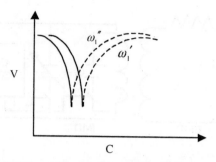

Fig. 8. Capacitance vs. V_{DIP} in a single-tuned circuit for various energizing frequencies.

Knowing the range of variations of the capacitance of the pressure sensor, the energizing frequency will be identified such that the voltage variations lie only along the dotted lines for all range of pressure sensor capacitance values. The reference capacitor value (C_{REF}) would be fixed within the range of variations of the capacitance of the pressure sensor.

Once the energizing frequency has been determined, this frequency is used to energize the EMD and hence the IMD circuit. The control circuitry includes suitable power extraction modules and control for switching between the two capacitors. After enough power has been transferred for the working of the control circuitry, it couples the reference capacitor to the inductor and the value of V_{DIP} is noted. This value is used for compensating for the variations in V_{DIP} due to mutual inductance variation attributed to misalignments. This information is stored for further computation. After this process is finished, the control circuitry switches from the reference capacitor to the variable capacitor (of the pressure sensor) and measures the value of V_{DIP}. Now, even though we do not eliminate the effects of variation in mutual inductance due to misalignments, they have certainly been compensated.

Piezo Resistive Pressure Sensor Approach

In the second approach, a piezo resistive pressure sensor would be used. The analog block will derive power from the carrier wave transmitted by EMD. This power will be utilized to energize the electronics and the pressure transducer. The analog block would be designed in such a way that the noise and temperature variation effects are fully eliminated. The analog block will also convert the measured analog pressure value into digital form and will transmit the same as a load modulated output employing a power transistor switch. The design is in such a way that the entire microelectronics and the pressure sensor will be on a single substrate, which would be manufactured using batch process. The equivalent circuit block diagram for this approach is given in Fig. 9.

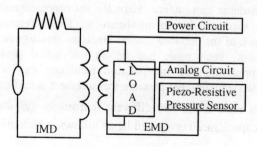

Fig. 9. Equivalent circuit for piezo resistive approach.

Figure 10 gives the load modulation waveforms for data transmission. When using the technique of load modulation, IMD load will be turned on/off (IMD coil load short/open circuited respectively) and hence the voltage read on the EMD circuit would be of two distinct levels. When the load is not turned on, the induced power will be available for the internal analog modules of IMD circuit. On the other hand, when the load is present, the IMD coil gets shorted and no power will be available for the analog block modules of IMD. This demands more power to be stored for energizing the analog circuitry during the load on period. To avoid these situations, load modulation is done in such a way that one level can be represented by a continuous no load condition and the other can be represented as a low frequency 0/1 pulsing (case b). This prevents higher power drain from the EMD during load on condition and also enables IMD to continuously derive power for energizing its analog circuitry.

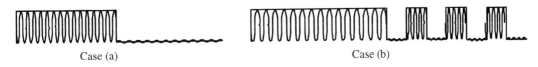

Case (a)   Case (b)

Fig. 10. Load modulation of digitized data.

Adopted Approach

In the first approach using capacitive pressure sensor, pressure data is transmitted in analog form. During the small gap of switching from reference capacitor to pressure sensor capacitor, there is a possibility of change in misalignments. Hence, this method might cause some difficulty in predicting the cause of measured pressure variations. Digital data transfer is preferable for RF data transfer for a reliable measurement in such cases. Based on this consideration, piezo-resistive approach is the most preferable method for our proto type. However, power requirements will be more in the piezo-resistive approach due to the involvement of analog circuitries whereas the power requirement for capacitive approach is minimal. The size required in the capacitive approach is smaller when compared with piezo-resistive approach. The circuit complexity is also less in capacitive approach.

Since each method has its own advantages, both methods are taken up for proto building and the final method will be chosen based on the test statistics.

In both the approaches, the implantable device is based on MEMS technology. Initial prototype development will be done using the available standard process steps. The test results of these prototypes would be utilized for finalizing one of the above-mentioned approaches. Further refinement would be done for batch process and biocompatibility as the next step.

External Device

The external device requirement for both the approaches of IMD is common. The block diagram of this is shown in Fig. 3.The external device will comprise a battery, which will act as the power source for both EMD as well as for the IMD through inductive coupling principle. The carrier wave generator and reflected impedance measurement circuits are part of the EMD. EMD will also contain sufficient memory to store the measured pressure data. The microcontroller chosen for EMD would be capable of operating at different low power modes and can be periodically woken up based on the sampling rate chosen for measuring pressure. EMD can be externally fixed closer to the IMD location and can be carried along with the patient. In case of measurements only in the clinic/hospital during patient checkup, it is not essential for the patient to carry the EMD. Moreover, in the clinic/hospital, during measurement, no extra effort is required other than just bringing the EMD closer to the patient near the location of implanted IMD. Only once, there is a need for an invasive operation for the patient to place the IMD, which then becomes part of the body and since it is a passive device, no product lifetime issues would be encountered.

Limitations Addressed

Electromagnetic Theory and Misalignments

There are many commercially available books and IEEE papers, which can be seen as reference for electromagnetic theory and misalignment principles. Since the misalignment is a major issue in RF communication, a good way to eliminate this is to fix the IMD and restrict its movement to have less predetermined misalignments. One way of doing this is by having an expandable cage to house the IMD. The predetermined misalignment requirements are considered well for the design of RF section. In case of EMD carried by patient, this restriction is mandatory. In case of monitoring in the clinic, there is no such constraint as the EMD can be reoriented based on the IMD location.

Challenges

The approach taken for fabrication of implantable device would be to use the readily available technology for MEMS based pressure sensor fabrication and CMOS (Complementary metal oxide semiconductor) compatible MEMS based design. As a first stage in our design, the inductive coil is external to the MEMS. Based on practical analysis, we would move towards a MEMS based coil as a second step. Biocompatible silicon packaging will be used where the coating of the MEMS final product itself will be a package. The outer layer on all sides will be processed to make it biocompatible. Fabricating this MEMS based coil and making this a lifetime biocompatible product would be a challenge, which needs exhaustive testing and analysis.

Road Map

The design for proto building is currently being carried out at HCL Technologies Ltd. The prototypes will be tested and a few process steps, if required, would be changed to make the process steps suitable for reliable batch fabrication of these devices. For implantable devices, the certification procedure needs to be followed before bringing the same into the market. Currently this device is targeted as a passive monitoring device through which accurate pressure data can be collected in real time and used for further analysis or diagnosis by the physicians.

Neuro simulation based on pressure information can be accomplished. There are studies on rats conducted in US, which shows that biodegradable micropatterned conduits pre-seeded with Schwann cells can be used for repairing sciatic nerve transactions. The study can be extended to grow nerve cells around the neuro simulation biocompatible electrodes through which controls can be simulated. An extensive study is needed with assistance of physicians, which is in its preliminary stage.

Acknowledgements

Both the authors wish to thank their General Manager, Mr. M. Venkatesan for the opportunity provided to present this paper and for giving his valuable comments during this study and this paper presentation. The authors also thank the reviewers and team members for their valuable contribution and for raising correct questions to improve the architecture, implement and refine the prototype.

References

1. ANSI/IEEE C95.1-1992. IEEE Standard for Safety Levels with respect to Human Exposure to Radio Frequency Electromagnetic Fields, 3 kHz to 300 GHz.
2. FCC Standards, Title 47 (Telecommunication).
3. Hayt, W.H. Jr., *Engineering Electromagnetics*, 5th Edition.
4. ICNIRP Guidelines for Limiting Exposure to Time-varying Electric, Magnetic, and Electromagnetic Fields.
5. Madou, M.J. *Fundamentals of Micro-fabrication: The science of miniaturization*, 2nd Edition.
6. Ramo, S., Whinnery, J.R. and Theodore. V.D. *Fields and Waves in Communication Electronics*.
7. Razavi, B. *Design of Analog CMOS Integrated Circuits*.
8. Stephen. D. Senturia, *Microsystem Design*.

10. Biomechanical Analysis of Vascular Stenting: A Review

J. Raamachandran and K. Jayavenkateshwaran
Department of Applied Mechanics, Indian Institute of Technology Madras, Chennai - 600036, India

Abstract: Vascular stents are scaffolding structures used to open up the clogged arteries or to prevent the recoiling of the atherosclerotic vessel wall after balloon angioplasty. The success rate of the stenting compared with the angioplasty increased the usage of the stents exponentially. The proliferation of the new vascular stent designs insists the need for proper analysis of the mechanical factors, which have a major impact on the efficacy of the procedure, to let the clinicians know the comparative merits and demerits of the different stent designs and also the regulatory authorities to prevent the substandard stents. Few studies explain that the mechanical interaction between the stent and the balloon with the artery have the potential to instigate stent related restenosis. Though a large number of trials study the long-term patency and the restenosis of different models, literature shows that the mechanical factors remain less explored. Moreover, the selection of stents for specific stenotic artery is done with minimal technical information provided by the manufacturer and the subjective information about the performance of particular stents, from the limited experience of individual clinicians. This study aims to present the state-of-the-art in biomechanical analysis of vascular stenting.

Introduction

Stents are flexible endovascular prostheses made from metal alloys designed as either metallic coils or slotted tubes. Most of the stents are balloon expandable and some are self-expandable. The stent is mounted on the catheter and is placed across the stenotic lesion and deployed with the help of fluoroscope and radio opaque markers. The deployment involves expansion of the stent circumferentially in apposition to the endothelial surface of the vessel.

The concept of scaffolding clogged or injured blood vessels is believed to be experimented first by Alexis Carrel, who placed glass tubes into the arteries of dogs, to treat traumatic injury of blood vessel. However, the stent era started with Charles T. Dotter in 1964, after he suggested the use of a pathway in a previously occluded blood vessel to maintain an adequate lumen. The word "stent" had been coined by him, in his report, in 1969 (Sigwart, 1997).

The vascular pathology may be just due to the degenerative changes, which results in loss of wall integrity, accumulation of material resulting in reduction of lumen diameter or external compression. The usage of stents triggered an incredible technological interest in areas like development of devices, the indications for the use of stents and the management after stenting.

The bothersome unresolved issue is the rate of recurrence of narrowing by intimal hyperplasia in stented patients. The restenosis has not yet been eliminated. A number of experiments and clinical data supports the concept that stent designs have a critical impact on thrombosis and hyperplasia. Variation in the design, without altering metal mass and material, may be associated with significant difference in the amount of thrombosis and restnosis in the rabbit model (Sigwart, 1997). It is possible to design "ideal" stents with highest long-term results, if the factors responsible for the platelet adhesion and smooth muscle cell proliferation can be determined.

The usage of stents has been increasing exponentially and the proliferation of new stent designs has also been started. Now it becomes a difficult task for the clinicians to select appropriate stent design to suit the occluded vessel and also for the regulatory authorities to prevent the substandard stent designs. Even though a large number of trials studied the long-term patency and the restenosis of different stent models, the mechanical factors, which influence the technical efficiency of the stenting technique, remain less explored. Moreover, the selection of the stents for specific stenotic artery is done, just with the minimal technical information provided by the manufacturer and the subjective information about the performance of the particular stents, from the limited experience of the individual clinicians (Tan et al., 2001). This study aims to review the biomechanical analysis of the stenting, available in the open literature.

The mechanical properties of different stent designs can be studied through two approaches:

1. Experimentation: Perform mechanical measurement directly on each stents in a simulated diseased artery, for various stent and the artery sizes (*ex vivo* study).
2. Numerical analysis: Standard numerical techniques, like finite element method (FEM), can be utilized to simulate the geometry, the material property, the loading and the boundary conditions, and the response of the different stents for the different conditions can be analyzed.

The experimentation needs to use the stent form different manufacturers and could be very expensive. Though the numerical simulation reduces the cost substantially, the reliability of the results totally depends on the material and geometry modeling and other assumptions in the simulation. Biomechanical modeling of stenting intervention is highly complex, because, it involves four components: the stent, the artery, the plaque and the balloon (for the balloon expandable stents). The overall system response is highly non-linear, due to the large deformation of every single component: geometrical non-linearity, and the material non-linearity (Auricchio et al., 2000).

Biomechanical Modeling of Stenting Using FEM

Free Expansion Modeling

It is possible to evaluate the behavior of the stent alone and to show possible unexpected response, by applying pressure on the internal surface of the stent and allowing it to expand freely. Free expansion of Palmaz stent shows, the distal strut touches the artery first, and then the central. It may lead to arterial lesions and may trigger the neointimal proliferation and also the chances of dissection. Auricchio et al. (2000) suggested a modified Palmaz stent design to avoid this effect with added distal struts. The expansion of the modified stent is more uniform thereby reducing chances of tissue damage.

Careful examination of *in vitro* experiments shows that the Palmaz stent expands almost uniformly except at the ends. So, it is reliable to consider an infinite prosthesis with uniform radial internal pressure, which allows working on a generic part of the structure. Based on the elastic-plastic model of stainless steel-3161, Dumoulin and Cochelin (2000) studied the degrees of radial and longitudinal recoil related to the elastic relaxation by removing the load on 2D and 3D models of the generic part of Palmaz stent. They also studied the stent resistance to crushing under external pressure.

Another study by Etave et al. (2001) studied two different stents: Palmaz-Schatz and Freedom stents, for the evaluation of, the pressure necessary for stent deployment, the intrinsic elastic recoil due to elastic deformation of the material used, the resistance of the stent to external compressive forces, stent foreshortening, stent coverage area, stent flexibility and stress and residual strain map. The analyses were performed for two different dimensions in each verity.

The comparison of the stents by free expansion modeling using finite element modeling was also done by Migliavacca et al. (2002). Here a 3D model of Palmaz-Schatz intravascular stent (Fig. 1) was developed to investigate the effects of the different geometrical features and the performance was also evaluated in terms of radial recoil, longitudinal recoil, foreshortening and dogboning. Parametric analyses were performed by varying the slot length l, the stent thickness s, and the metal to artery index (α_p/α_v) alternatively.

$$\text{Distal radial recoil} = \frac{R_{distal}^{load} - R_{distal}^{unload}}{R_{distal}^{load}}$$

$$\text{Central radial recoil} = \frac{R_{central}^{load} - R_{central}^{unload}}{R_{central}^{load}}$$

$$\text{Longitudinal recoil} = \frac{L^{load} - L^{unload}}{L^{load}}$$

$$\text{Foreshortening} = \frac{L - L^{load}}{L}$$

$$\text{Dogboning} = \frac{R_{distal}^{load} - R_{central}^{load}}{R_{distal}^{load}}$$

where L, R is length and radius, respectively.

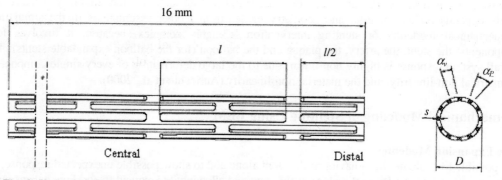

Fig. 1. Unexpanded geometry of Palmaz-Schatz, s thickness, α_p angle described by metallic surface, α_v angle described by slot, D outer dia.

The models resembling two other stents (Multi-Link Tetra and Carbostent) GEO1 and GEO2 were also considered (Fig. 2). The stronger expansion of distal zone with respect to the central zone during the stent implantation is called "dogboning", an undesirable effect.

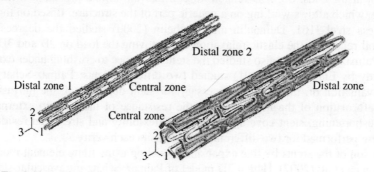

Fig. 2. Unexpanded geometry of GEO1 and GEO2 (Migliavacca et al., 2002).

GEO1: Corrugated ring patter (metal to artery ratio m/a = 0.341)
GEO2: Cellular geometry (metal to artery ratio m/a = 0.345)

The covered stents found its application in the proximal carotid and vertebral arteries to treat aneurysms. The covering material should be thin, flexible and should not migrate away from the stent. The study by Gu et al. (article in press) considers the covered microstents, with silicon covering of appropriate size. The material properties were adopted as same as that of Auricchio et al. (2000). The plastic behaviour of the microstent was assumed to be linear isotropic hardening between yield and ultimate stress. The cover had a linear elastic behaviour with Young's modulus of 2.47 MPa and Poisson's ratio of 0.3. The stent was modeled with 4 noded shell elements. To avoid migration of the covering, the stents were left uncovered for a small distance at both the ends, which would have trumpet like flares. Longitudinal shortening, elastic recoil, and the effect of the covering thickness were studied.

Modeling of Stent Insertion in Stenotic Artery

The 3D model of stent, plaque and the artery was developed by Auricchio et al. (2000), to analyze the revascularization of the stenotic arteries. A straight artery segment with parabolic plaque producing 53% stenosis and a tubular stent with rectangular slots which resembles Palmaz-Schatz stent were considered for the large deformation analysis using a commercial finite element code Abaqus (Fig. 3 (a)). The presence of symmetry allowed to consider only half of 30° of three dimensional segment of the whole model (Fig. 3 (b)). In this investigation the plaque and the artery were modeled as a homogeneous hyper-elastic isotropic incompressible material.

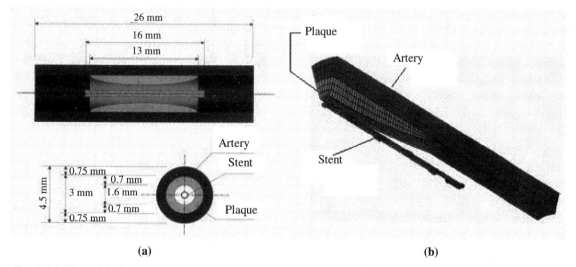

Fig. 3. (a) 3D model of the Artery, Plaque and Stent (b) 15° segment of the whole model (Auricchio et al., 2000).

On the other hand, Tan et al. (2001) considered two different models of stent, which were based on geometries, which could be constructed from the same basic pattern. He developed wire mesh models of Freedom stent (coil stent geometry) and Palmaz-Schatz stent (slotted tube geometry), and examined their mechanical characteristics in the loading and unloading regimes. The material properties for the stent were measured experimentally. The effect of the resistance force offered by the artery walls was included by incorporating an elastic foundation in the model. These stent models were analyzed for the different geometries by changing the number of parabolic convolutes. The hoop stress in the stent was calculated approximately by considering the stent mess as a cylinder of diameter D and thickness t, provided the amount of material in the cylinder of length h must be equal to the total amount of stent wire (length of wire = L, wire diameter = d)

$$\pi D h t = \pi \left(\frac{d^2}{4}\right) L$$

The circumferential stress in the wall of the equivalent cylinder is then given by

$$\sigma_\theta = \frac{PD}{2t} = 2P\left(\frac{D}{d}\right)^2 \left(\frac{h}{L}\right)$$

where P is the internal pressure.

So, for a particular material, the plastic pressure depends on the wire diameter, the yield strength and the stent diameter.

Hence, in the absence of the plaque geometry, the effect of the stent expansion is tested against an incompressible plaque. The plaque is represented as rigid body which is connected to a spring element through a gap element. The spring element was included just to incorporate the effect of irregular narrowing and to provide some circumferentially irregular arteries resisting elasticity.

A similar analysis in the wire mess representing Palmaz-Schatz stent shows that the deformation caused by the rigid arterial plaque in the mid, leads to the flaring of the edges of the stent, leads to the local vessel injury. Lower deployment pressure can cause an incomplete apposition of the struts against the vascular wall whereas the high deployment pressure can cause penetration of the stent struts into the adventitia.

The vascular trauma during the stent deployment causes more intimal hyperplasia than the balloon angioplasty. A thorough understanding of the factors involved in the vascular injury during stent deployment might allow to optimise the stent design. Stent deployment causes partial denudation of the endothelium in a pattern unique to each stent configuration. Rogers et al. (1999) addressed the hypothesis that the mechanism of endothelial cell denudation and therefore interstrut injury during stent deployment is balloon artery interaction. A 2D finite element model was developed to understand how the balloon artery interaction affects associated contact stress and thereby vascular injury. The model assumes a balloon membrane with no thickness, frictionless contact between balloon and artery, no slip between stent struts and luminal surface and no other substrates present between the balloon and the arterial wall. Nothing has been mentioned clearly about the material modeling of the balloon, artery and the stent materials in this study. Unlike the free expansion analysis of the stent, very few articles discussed about the role of arterial walls and the presence of plaque in the biomechanical modeling of stenting.

Stent Flexibility Analysis
Flexibility is an important mechanical property expected in vascular stents especially in coronary stents. It is the ability to turn or angle during delivery and to conform to the vessel wall after the deployment. Trackability, a property related to flexibility, is the ability of the system to advance distally around a guide-wire, while following the guide-wire tip along the path of the vessel. The flexibility of the tubular stent and the coil stent was studied by Etave et al. (2001) and it was explored

that the coil stents are more flexible than the tubular stents. A shell model of the tubular stent and wire mesh model of the coil stent were considered with a bending load at the center.

In the newer generation stents it is possible to recognize two different types of elements. Rings—sustain the vessel after the expansion and links—links the rings, and provides flexibility during delivery. Petrini et al. (2003) considered two 3D models of new generation stents (Cordis BX-Velocity and Sirius Carbostent) to measure the stent flexibility. Only a portion of the stent model was considered. The analyses were performed under displacement control by rotating the extremes at a fixed angle φ in both x-axis and y-axis in both unexpanded and expanded forms. Results were expressed in terms of bending moment (M) at the extremes as a function of the curvature index χ. The slop of the curve M-χ measures the stent stiffness, the reciprocal of flexibility.

$$\chi = \Delta\varphi / L$$

where L is length of the rings.

Experimental Analysis

Stents are expected to carry higher radial forces from the arterial walls and to exhibit minimum radial recoil. 17 coronary stents were compared experimentally for their radial forces by Regis et al. (1999) through *in vitro* experiments. In a different study by Regis et al. (2003), the trackability, flexibility, and conformability through the *in vitro* experiments where also assessed for the same set of stents. 23 coronary stents were compared experimentally for elastic recoil by Barragan et al. (2000).

7 new generation stents (3 balloon expandable and 4 self expandable) were compared in an experimental study by Dyet et al. (2000) for:

1. Radial strength: Force required for 50% reduction in stent diameter.
2. Flexibility: Force required per degree of flexion.
3. Radio-opacity: mm of aluminum required to render the stent invisible on the fluoroscopy.
4. Trackability: Ability of the stent on its delivery system to cross an iliac bifurcation.

Rogers et al. (1999) also tried to prove the endothelial denudation of arterial walls by the balloon wall interaction, as mentioned early, by *in vivo* experiments. Two distinct stent configurations, a slotted tube and a corrugated rings stent, with identical material were deployed in the iliac arteries of New Zealand white rabbits over 20 seconds at 8 atm. Endothelial staining was performed on the harvested arteries and the percentage area of endothelial denudation was calculated. Stent design in which the complex and closed struts (corrugated rings) existed, permitted 33% less injury in the space between each strut, than the stent design in which the inter-strut areas were simple and open.

Mathematical Modeling of Stent Geometry

The mathematical modeling of stent structure was first given by Tan et al. (2001) for two different stents (Freedom coil and Palmaz-Schatz tubular stent), which were used for their analysis as mentioned earlier. The basic structure of the wire frame that constitutes the stent was constructed on the surface of a cylinder of diameter D_i. The structure itself consists of paraboloidal loops arranged in a helical form winding along the surface of the cylinder with a helical climb angle α (Fig. 4).

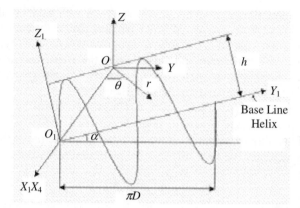

Fig. 4. Generation of stent geometry in (x, y, z) rectangular Cartesian coordinate system (Tan et al.).

If the stent wire diameter is d, then the factors influencing the design, other than the material properties are:

1. The diameter of the cylinder, D_i.
2. The angle of the loop subtended at the center of the stent cylinder, β.
3. The helical climb angle of the paraboloidal base helix, α.
4. The diameter of the stent in final expanded shape, D_0.
5. Stent wire diameter, d.
6. The initial distance between the peaks or trough from the paraboloidal baseline, h.
7. The initial or final elongated length of the stent is governed by other parameters because these parameters are interdependent.
8. The pressure required to facilitate expansion is also a function of parameters 1-4.

For a fixed value of d, the value of the parameters D_i, D_0 and β are determined by the initial and enlarged shape of the stent which will be affected by the initial and required conditions of the artery and its strength. Therefore, the only parameters which influence the design are the helical angle of climb, α, the distance between the peak and the trough in each loop of the paraboloidal helix from the helical base, h and the number of loops around the circumference of the cylinder n. The variables which need to be investigated in depth are h and α..

Conclusion

Various types of stents are currently available in the market. But most of them are balloon expandable stents and some are self-expandable stents. The stenting technique involves the artery, the plaque, the stent, the balloon and the interaction between these components. Even though it is believed that the interaction between the balloon and the artery has an important role on the efficacy of the stenting procedure, most of the analyses neglect the presence of the balloon in the simulation. To clearly understand the role of mechanical factors in in-stent restenosis, a simulation should consider all those four components and the interaction between them and also, well simulated experimental studies should be conducted. From the literature, it is understood that, the fatigue and radial strength of the stents should be explored to optimize the stent geometry.

References

1. Auricchio, F., Loreto, M.D. and Sacco, E. 2000. Finite-element Analysis of a Stenotic Artery Revascularization through a Stent Insertion. *Computer Methods in Biomechanics and Biomedical Engineering.* **00**: 1-15.
2. Barragan, P. et al. 2000. Elastic Recoil of Coronary Stents: A comparative analysis. *Catheterization and Cardiovascular Interventions.* **50**: 112-119.
3. Dotter, C.T. and Judkins, M.P. 1964. Transluminal Treatment of Arteriosclerotic Obstruction. **30**: 654-670.
4. Dotter, C.T., Buschmann, P.A.C., McKiney, M.K., Rosch, J. 1983. Transluminal Expandable Nitinol Coil Stent Grafting: Preliminary report. **147**: 259-260.
5. Duerig, T.W., Tolomeo, D.E. and Wholey, M. 2000. An Overview of Superelastic Stent Design. *Min. Invas. Ther. and Allied Technol.* **9(3/4)**: 235-246.
6. Dumoulin, C. and Cochelin, B. 2000. Mechanical Behaviour Modeling of Balloon-expandable Stents. *Journal of Biomechanics.* **33**: 1461-1470.
7. Dyet, J.F. and William, G. et al. 2000. Mechanical Properties of Metallic Stents: How do these properties influence the choice of the stents for specific lesions? *Cardiovascular and Interventional Radiology.* **23**: 47-54.
8. Etave, F. and Finet, G. et al. 2001. Mechanical Properties of Coronary Stents Determined by using Finite Element Analysis. *Journal of Biomechanics.* **34**: 1065-1075.
9. Gandhi, M.M. and Dawkins, K.D. 1999. Intracoronary Stents. *B.M.J.* **318**: 650-653.
10. Gu, L. and Santra, S. et al. Finite Element Analysis of Covered Microstents. Journal of Biomechanics. Article in press.
11. Han, R.O. and Schwartz, R.S. et al. 2001. Comparison of Self-expanding and Balloon-expandable Stents for the Reduction of Restenosis. *Am. J. Cardiol.* **88**: 253-259.
12. Holmes, D.R. et al. 1998. ACC Expert Consensus Document on Coronary Artery Stents. *JACC.* **32**: 1471-82.
13. Kastrati, A. and Julinda, M. et al. 2001. Restenosis after Coronary Placement of Various Stents Types. *Am. J. Cardiol.* **87**: 34-39.
14. Migliavacca, F. and Petrini, L. et al. 2002. Mechanical Behavior of Coronary Stents Investigated through the Finite Element Method. *Journal of Biomechanics.* **35**: 803-811.
15. Migliavcca, F. and Petrini, L. et al. 2003. Deployment of an Intravascular Stent in Coronary Stenotic Arteries: A computational study. *Summer Bioengineering Conference*, June 25-29, 2003.
16. Pache, J. and Kastrati, A. et al. 2003. Intracoronary Stenting and Angiographic Results: Strut thickness effect on restenosis outcome (ISAR-STEREO-2) trial. *J. Am. Coll. Cardiol.* **41**: 1283-8.
17. Petrini, L. and Migliavacca F. et al. 2003. Evaluation of Intravascular Stent Flexibility by Means of Numerical Analysis. *Summer Bioengineering Conference*, June 25-29, 2003.
18. Petrini, L. and Migliavacca, F. et al. 2003. Numerical Investigation of the Intravascular Coronary Stent Flexibility. Journal of Biomechanics. **37**: 495-501.
19. Regis, R. et al. 1999. Radial Forces of Coronary Stents: A comparative analysis. Catheterization and Cardiovascular Interventions. **46**: 380-391.
20. Regis, R. et al. 2003. Assessment of the Trackability, Flexibility and Conformability of Coronary Stents: A comparative analysis. *Catheterization and Cardiovascular Interventions.* p. 59.
21. Rogers, C. et al. 1999. Balloon-artery Interactions during Stent Placement: A finite element analysis approach to pressure, compliance, and stent design as contributors to vascular injury. *Circulation.* **84**: 378-383.
22. Ruygrok, P.N., Serruys, P.W. 1996. Intracoronaty Stenting: From concept to custom. *Circulation.* **94**: 882-890.
23. Sigwart, U. 1997. Stent: A mechanical solution for a biological problem? *European Heart Journal.* **18**: 1068-1072.
24. Stoeckel, D., Bonsignore, C. and Duda, S. 2000. A Survey of Stent Dsigns. *Min. Invas. Ther. and Allied Technol.* **11(4)**: 137-147.
25. Tan, L.B., Webb, D.C., Kormi, K., Al-Hassani, S.T.S. 2001. A Method for Investigating the Mechanical Properties of Intracoronary Stents Using Finite Element Numerical Simulation. *International Journal of Cardiology.* **78**: 51-67.
26. Tolomeo, D., Slater, T. and Wu, P. Predictive Modeling of Radial Strength for Superelastic Stents. *SMST-2000 Conference Proceedings.*
27. Trochu, F. and Terriault, P. 1997. Finite Element Stress Analysis of a Shape Memory Medical Stent. *SMST 97*, Asilomar, California, U.S.A.

Biomechanics
R.K. Saxena and P. Mishra (Editors)
Copyright © 2005, Anamaya Publishers, New Delhi, India

11. Classification of Human Erythrocyte Aggregates Using Neural Networks

A. Kavitha and S. Ramakrishnan

Department of Instrumentation Engineering, Madras Institute of Technology,
Anna University, Chromepet, Chennai - 600044, India

Abstract: Under physiological conditions the erythrocytes in static or slowly moving blood adhere face to face like piles of coins to form reversible cell-to-cell contact leading to formation of aggregates. Aggregation of erythrocytes which depends on their shape and deformability, is an important mechanism associated with cardiovascular blood flow. The essential factors influencing erythrocyte aggregation are low shear stress, erythrocyte membrane properties, macromolecular interaction between adjacent cells and physical and chemical properties of the suspending medium. Abnormal red cell aggregation has been found to be associated with several diseases, which include diabetes, malaria, cardiovascular malfunction, lacunar brain infarcts, essential hypertension, immunoglobulins and hematological disorders, local anesthesia and many others. The objective of this study is to develop an automated methodology to characterize the aggregation behavior of cells by analyzing its degree of aggregation using image processing, wavelet transforms and neural networks. In this three-step process, the images of various aggregate sizes are acquired and processed using various image-processing routines. An index is derived for the aggregates of different sizes using wavelet transforms and they are classified using an appropriate network chosen. The results demonstrate that precise classification of aggregates defining the intensity of aggregation could be made using this neural network based approach. This process of classifying aggregates automatically could be useful for assessing the erythrocyte mechanics in microvessels. The methodology and observations in identifying the appropriate network and classification of aggregates based on their sizes are discussed in detail.

Introduction

Erythrocyte aggregation is an important mechanism associated with cardiovascular blood flow. They adhere face-to-face in long columns under static or low flow conditions and are the prime determinants of blood flow in microcirculation (Rampling, 1987). It is well known that aggregation arises from bridging of RBC surfaces by macromolecules, particularly fibrinogen, which is present in plasma (Chien et al., 1970). Formation of aggregates represents the balance of energies at the cell surface. The essential factors influencing red blood cell aggregation are the low shear stress, erythrocyte membrane properties, macromolecular interaction between adjacent cells and physical and chemical properties of the suspending medium. They are observed in the axial regions of large vessels and under low flow conditions in smaller vessels (Ramakrishnan et al., 1999; 2001). Also, the forces holding the aggregates together are weak in normal blood and can be dispersed by low shear forces. But when flow forces are not strong enough to disperse, pathologically intensified aggregates occur. Erythrocyte aggregation increases blood viscosity and thus, affects the passage of the cells through micro vessels, especially in venules with low shear flow (Suzuki et al., 2001). Aggregates tend to form irreversible clumps in conditions arising due to pathological states and are capable of plugging arterioles and venules (Chien and Jan, 1973). Abnormal red cell aggregation has been found to be associated with several diseases and conditions which include diabetics, malaria, cardiovascular malfunction, lacunar brain infarcts, essential hypertension, and immunoglobulin and hematological disorders local anesthesia and many others (Ramakrishnan et al., 1999a; 1999b). Thus, aggregation of red cells is a major determinant of flow properties of blood in microcirculation (Schmid-Schönbein et al., 1990). Hence, studies on automated methods to assess size and index of aggregation are essential for clinical evaluation. In this study, we have adopted a three-step process using image processing, wavelet transforms and neural networks for characterizing the aggregation behavior of red blood cells.

Image processing is not new and techniques for the manipulation, correction and enhancement of digital images have been in practical use for many years. Two dimensional discrete wavelet transform is adopted (2-D DWT) in this work due to its fast computational algorithm for decomposition and the discrete nature of the image pixels. The wavelet transform has been attracting attention in diverse areas such as medical imaging, pattern recognition, data compression numerical analysis and image processing, especially for image analysis applications, since it provides information in both spatial and frequency domains. And, over the past few years, there have been numerous reports on the use of wavelets on medical imaging (Healy and Weaver, 1995). Daubechies wavelets are chosen for the analysis as they provide excellent results in the processing of images (Wang, 2001). A neural network is an information processing paradigm that is inspired by the way biological nervous systems, such as brain, process information. Although neural networks have been used extensively for engineering problems, their applications in biomedicine are of considerable value. They have been extensively used for recognition of a particular pathology in radiology, urology, laboratory medicine and cardiology (Dipti et al., 1996). Also, neural networks characterize local features, such as discontinuities in curvature, jumps in value or other edges (Rying et al., 2002), which make it suitable for defining the aggregation nature of erythrocytes. Thus, the objective of this work is to develop an automated methodology to characterize the aggregation behavior of cells by analyzing its degree of aggregation using image processing, wavelet transforms and neural networks.

Samples and Methods

The images for analysis were acquired from ten adult volunteers as described elsewhere (Ramakrishnan et al., 1999a; 2001). The images were then processed using certain image processing routines. They were resized, rescaled and segmented to detect transitions, which differed greatly in contrast from the background image. Changes in contrast were detected by calculating the gradient of the image using sobel operator. The binary gradient mask image obtained after edge detection was dilated and the borders cleared. The processed images were then subjected to wavelet transformations. The discrete wavelet transform provided multiresolution representations of an image and captured different details on the image as well as information on neighboring pixels. This information was contained in the wavelet coefficients which were in turn assigned as the components of the feature vector of a pixel (Yu and Erkströrm, 2003). Pixels in the wavelet transformed sub-images (i.e., wavelet coefficients) represented the characteristics of the pixels in the original image. The images were decomposed at level 5 using Daubechies-5 wavelets and the coefficients of approximation were estimated at level 1. An aggregation index was derived for aggregates of different sizes and bonding strengths. Difference between the standard deviation values of the histogram equivalents of the original image σ and the approximation coefficient σ' is defined as the aggregation index $-\Delta\sigma$ (AI).

$$\Delta\sigma = \sigma' - \sigma \tag{1}$$

Aggregates were classified into samples of lesser, medium and greater degree of aggregation based on their derived index values.

An index range of 0-10 was set with a target of −1, 10-20 with 0 and 20-40 with 1. Classification by perceptrons was adopted since they were specific in solving linearly separable classification problems. Perceptrons consisted of a single layer with the dotprod weight function, the netsum net input function, and the specified transfer function. Hardlim is the chosen transfer function for calculating the layer's output from its net input. Learnp is the perceptron weight/bias learning function adopted to calculate the weight change dW for a given neuron from the neuron's input P and error E according to the perceptron learning rule. The input neuron is chosen to be 1 with 3 output neurons. Fig. 1. shows the sequence of processes carried out in the study.

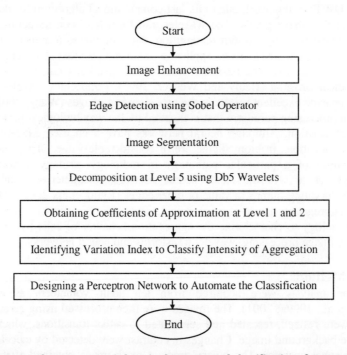

Fig. 1. Sequence of processes carried out in the automated classification of aggregates.

Results and Discussion

The index of aggregation $\Delta\sigma$(AI) is listed in Table 1. It was found that the index was less around 0-10 for samples with lesser degree of aggregation, moderate around 11-20 for samples with medium degree of aggregation and very high around 21-40 for cells with highest degree of aggregation. The target values for the network were set based on these index values.

Table 1 lists the standard deviation values of histogram equivalents for the original samples, approximation coefficient of the samples and difference in the deviations. In the neural networks based approach, results obtained using perceptron network were found to confirm the classification procedure effectively. Each category was distinctly identified with a slight variation in classification correctness with any test input value; the process could thus be automated. Due to closeness in the values of the training data of different classes the procedure had a slightly lesser percentage of correctness. To train the network adequately few hundreds of samples were required to generalize than to memorize, which could yield better performance in terms of correctness of classification.

Table 1. Standard deviation values of histogram equivalents for the original samples, the approximation coefficient of the samples and difference in the deviations.

Samples	σ	$\Delta\sigma'$	$\Delta\sigma$
1	52.94	71.42	18.48
2	54.16	60.94	6.78
3	62.41	78.01	15.6
4	55.35	71.92	16.57
5	53.71	85.57	31.86
6	54.77	57.7	2.93
7	59.78	72.8	13.02
8	57.93	74.2	16.09
9	60.38	72.44	12.06
10	64.06	81.54	17.48

Figure 2 shows the various values of the derived aggregation indices for samples of different degree of aggregation using Db 5 wavelets. It is very clearly seen that samples with less aggregation intensity have lesser AI values and a densely packed sample has a higher value of AI.

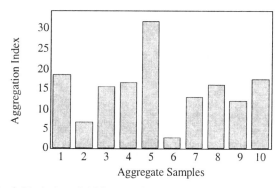

Fig. 2. Variation of AI for samples of various extents of aggregation.

Conclusions

Erythrocyte aggregation is the main hematological phenomenon in medicine having the greatest significance in contemporary daily healthcare (Schmid-Schönbein et al., 1990). It has been monitored by physicians as screening test for presence and/or severity of disease ever since the days of human pathology. Association and disassociation of red blood cells is a measure of red cell deformability. This deformability allows a reduction of bulk viscosity in large blood vessels and is the prime determinant of the very survival of red blood cells in microcirculation for 120 days (Weed, 1970). The rate and degree of erythrocyte aggregation depends on the suspending medium, physico-chemical properties and flow conditions and in the assessment of the mechanism of such dynamic transitions associated with aggregation behavior optical methods have been widely employed (Ramakrishana et al., 1990a). Different methods have been employed to assess the mechanics of formation and distribution of cellular networks. Under varied experimental and clinical conditions, there are only very few methods to evolve automated analysis. In the present work, an attempt has been made in this direction to automate the analysis using wavelets and neural networks. It is found that wavelets are useful in the classification as they are capable of reporting structural information at different locations within the image. Neural networks are most suitable as decision support systems in medicine and thus they are found to be well suited for assessing the extent of aggregation in samples. This wavelet and neural network based approach extends a possibility of automated classification of samples with various degrees of aggregation. Though the degree of aggregation is well classified, bonding strength and the intercellular spacing between adhered cells could not be evaluated using this method. Therefore, the work could be extended for analysis of population of cells in aggregate samples and the strength of bonding between cells, which are the major factors of significance.

Acknowledgement
One of the authors (SR) acknowledges DAAD, Bonn, Germany (DAAD Award 422/ind-4-tzn/95) and Dr. Sans Research Foundation in Biomedical Engineering, USA for their support.

References

1. Chien, S., Luse, S.A., Jan, K.M., Usami, S., Miller, L.H. and Fremount, H. 1970. Effects of Macromolecules on the Rheology and Ultrastructure of Red Cell Suspensions. *Proc. on Sixth Europ. Conf. on Microcirc.* Alborg, Ditzel, J. and Lewis, D.H., Kerger, B.S. (Eds.). pp. 29-34.
2. Chien, S. and Jan, K.M. 1973. Ultrastructural Basis of the Mechanism of Rouleaux Formation. *Microvasc. Res.* **5**: 155-166.
3. Dipti, I., Peter, B.S., Robert, J.A. and William, J.O. 1996. Artificial Neural Networks: Current status in cardiovascular medicine. *Journ. Am. Cardiol.* **28**: 515-521.
4. Healy, J.M. Jr. and Weaver, J.B. 1995. Two Applications of Wavelets and Related Techniques in Medical Imaging. *Ann. Biomed. Eng.* **23**: 637-665.
5. Ramakrishnan, S., Grebe, R., Singh, M. and Schmid-Schönbein, H. 1999a. Aggregation of Shape-altered Erythrocytes: An *in vitro* study. *Current Sci.* **77**: 805-808.
6. Ramakrishnan, S., Grebe, R., Singh, M. and Schmid-Schönbein, H. 1999b. Analysis of Risk Factor Profile in Plasmacytoma Patients. *Clin. Hemorheol. and Microcirc.* **20**: 20-25.
7. Ramakrishnan, S., Degenhardt, R., Vietzke, K., Grebe, R., Singh, M. and Schmid-Schönbein H. 2001. Influence of Immunoglobulin G and Immunoglobulin A on Erythrocyte Aggregation: A comparative study. *ITBM-RBM* **22**: 241-246.
8. Rampling, M.W. 1987. Quantization of Rouleaux Formation-A Useful Clinical Index. *Rev. Port. Hemorrheol.* **1**: 41-49.
9. Rying, E.A., Bilbro, G.L. and Lu, Jye-Chyi. 2002. Focused Local Learning With Wavelet Neural Networks. *IEEE Trans. on Neur. Net.* **13**: 304-319.
10. Schmid-Schönbein, H., Malotta, H. and Striesow, E. 1990. Erythrocyte Aggregation: Causes, consequences and methods of assessment. *Tijdchr NVKC.* **15**: 88-97.
11. Suzuki, Y., Tateishi, N., Cicha, I. and Maeda, N. 2001. Aggregation and Sedimentation of Mixtures of Erythrocytes with Different Properties. *Clin. Hemorheol. and Microcirc.* **25**: 105-117.
12. Wang, J.Z. 2001. Wavelets and Imaging Informatics: A review of the literature. *Journ. of Biomed. Informat.* **34**: 129-141.
13. Weed, R.I. 1970. Importance of Erythrocyte Deformability. *Ann. J. Med.* **49**: 147-151.
14. Yu, J. and Erkstörm, M. 2003. Multispectral Image Classification using Wavelets: A simulation study. *Patt. Recog.* **36**: 889-898.

Biomechanics
R.K. Saxena and P. Mishra (Editors)
Copyright © 2005, Anamaya Publishers, New Delhi, India

12. Stress Analysis of Human Elbow Joint Fitted with Linked and Unlinked Prosthesis

R. Daripa[1], S. Majumder[2], A. Roy Chowdhury[2], S.R. Quadery[3] and R. Kumar[1]

[1]N.I.O.H., Kolkata - 700090, India
[2]Department of Applied Mechanics, B.E. College (D.U.), Howrah - 711103, India
[3]B.R. Singh Hospital, E. Railway, Kolkata - 700014, India

Abstract: The human elbow is a complicated joint that allows two kinds of motion, i.e. Flexion and Extension of forearm as well as rotation of the forearm. The elbow bone, resembling a tubular structure, consists mainly of trabecular bone covered by high strength cortical bone of varying thickness in the form of a thin tube. This study was aimed at developing a realistic three-dimensional finite element model of the human elbow joint fitted with linked and unlinked implant, and finding out the behavior of the elbow under static load and analyzing the dynamic response. Static analysis was performed on the elbow model fitted with linked implant and dynamic analysis was performed on a separate model fitted with unlinked implant. All the modeling and analyses were carried out using ANSYS® software. The geometry of the three-dimensional finite element model was based on the dimensions of the original elbow joint, harvested from a 33 yrs old male, exhibiting the most frequent surface morphology of the trochlear notch [1]. The trabecular, cortical, bone cement and prosthesis were modeled with solid (10-noded tetrahedral elements) with 3 degree of freedom. After finite element mesh, the number of elements and nodes of the static model were 6187 and 10,567, respectively. The model for dynamic analysis consisted of 12,690 solid elements connected through 20,703 nodes. Impact load was applied simulating the sudden lifting of 10 kg load and impact duration of 10 ms to peak load. Static analysis was done considering straight elbow subjected to 20 kg load. Nodes of the distal humerus were kept fixed. In the dynamic analysis, displacements in the direction of impact and von-Misses stresses were analyzed with very small time steps. It was also observed that using plastic as prosthesis material, both the von-Misses stress in prosthesis and shear stress in bone cement decreased greatly. This result is significant as bone cement being brittle material is weak in shear and tension though strong in compression.

Introduction

The human elbow joint consists of three bones: radius, ulna and humerus, and three articulations: radio-humeral, radio-ulnar and humero-ulnar joints. It offers two kinds of motions (flexion-extension of forearm and rotation of forearm) that are unique within the human body. The elbow joint, which is used in many daily activities and gets damaged due to different diseases, becomes a potential area of biomechanical research. A total elbow replacement, or elbow arthroplasty, is used for those who have an elbow that has been damaged by trauma or arthritis.

The most common reason for an artificial elbow replacement is arthritis. The two main types of arthritis are degenerative and systemic. Degenerative arthritis is also called wear-and-tear arthritis, or osteoarthritis. Any injury to the elbow can damage the joint and lead to degenerative arthritis. Arthritis may not show up for many years after the injury. There are many types of systemic arthritis. The most common form is rheumatoid arthritis. All types of systemic arthritis are diseases that affect many, or even all, of the joints in the body. Systemic arthritis causes destruction of the joints' articular cartilage lining.

An elbow joint replacement may also be used immediately following certain types of elbow fractures, usually in elderly patients. Elbow fractures are difficult to repair surgically in the best of circumstances. In many elderly patients, the bone is also weak from osteoporosis. (People with osteoporosis have bones that are less dense than they should be.) The weakened bone makes it much harder for the surgeon to use metal plates and screws to hold the fractured pieces of bone in place long enough for them to heal together. In cases like this, it is sometimes better to remove the fractured pieces and replace the elbow with an artificial joint.

Patients with these conditions often suffer chronic pain and decreased joint function. The treatments that are generally adopted in such cases are largely unsatisfactory and unable to restore the joint mobility and stability. Total joint replacement (surgical intervention and reconstruction of the joint) with an artificial prosthesis is the only viable option for the patient. This study was aimed at designing total elbow joint fitted with linked and unlinked prosthesis and understanding the static and dynamic behavior of the artificial elbow joint, using a three-dimensional finite element approach.

Methodology

The major steps adopted were solid modeling, finite element mesh generation, selection of material properties and imposition of boundary conditions (loads and constraints). The detailed, geometrically accurate three-dimensional finite element model of the total elbow prosthesis (humero-ulnar) was developed (Fig. 1) using the dimensions of the original elbow joint, harvested from a 33 yr old male exhibiting the most frequent surface morphology of the trochlear notch [1] and no macroscopically visible damage. Solid modeling and mesh generation was done with the help of commercially available finite element modeling and analysis software, ANSYS® (ANSYS, Inc. Pennsylvania, USA).

Solid Modeling

For static analysis prosthesis was modeled (Fig. 1 (a)) as linked component, i.e. ulnar and humeral prosthesis was connected by external means. For dynamic analysis model (Fig. 1 (b)) was developed as unlinked implant, i.e. no built-in bearing mechanism between its components. In both the models, humeral and ulnar components have long, tapered stem designed to keep the implant stable as the elbow

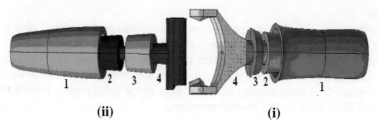

Fig. 1 (a). Model of (i) humeral and (ii) ulnar prosthesis for static analysis. 1: cortical bone, 2: bone cement, 3: trabecular bone and 4: prosthesis.

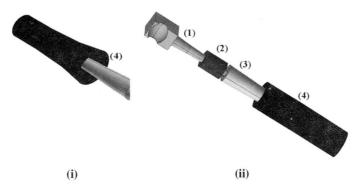

**Fig. 1 (b). Model of (i) humeral and (ii) ulnar prosthesis for dynamic analysis.
1: prosthesis, 2: trabecular bone, 3: bone cement and 4: cortical bone.**

flexes and extends. Linked implants are generally used in patients whose tissues are no longer strong enough to support the elbow joint. The model of this implant consisted of 182 key-points connected through 288 lines. The number of volumes and areas generated in the model was 8 and 160, respectively. The model of unlinked implant consisted of 217 key-points connected through 364 lines. The number of volumes and areas generated in the model was 9 and 186, respectively.

Mesh Generation

The finite element mesh was obtained by discretizing the solid model using a free mesh technique. At this stage, the number of elements, element types and element size should be chosen depending on the complexity of the geometry, desired problem size and analysis nature of the finite element model. In this study, solid element was chosen to mesh the complete system of bone-bone cement-prosthesis in both the models (static and dynamic case). From the anatomical point of view, the elbow bone contains mainly the tubular structure [2] of low-density trabecular bone, which is covered by the high strength cortical bone of varying thickness. Bone marrow is removed from the hollow cavities of humerus and ulna, and replaced by acrylic bone cement [3] for the purpose of fixation of the prosthesis. Among the various types of solid elements available in the ANSYS library, the 10-noded tetrahedron element was chosen to mesh the whole field of bone-bone cement-prosthesis (8 volumes in static model and 9 volumes in dynamic model). This particular type of element has mid-side nodes and three translational (UX, UY, UZ) degrees of freedom (dof). Element sizes were varied in different portions of the models and fine meshing was done in the critical areas where possibilities of stress concentration were high. The final three-dimensional model with linked prosthesis (Fig. 2) along with the associated bone cement, trabecular and cortical bone contained 6187 10-noded tetrahedral elements connected through the 10,567 nodes, whereas, unlinked prosthesis had 14,231 elements connected through 25,505 nodes (Fig. 3).

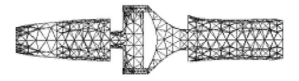

**Fig. 2. Front view of three-dimensional elbow model for static analysis with
6187 elements (10-noded tetrahedral) and 10,567 nodes.**

Fig. 3. Front view of three-dimensional elbow model for dynamic analysis with 14,231 elements (10-noded tetrahedral) and 25,505 nodes.

Material Properties

We used linear, elastic, isotropic material property throughout the elbow bone. Young's moduli of cortical bone, used in various finite element studies of elbow, varied from 8000 to 18,000 MPa. Felix et al. used Young's modulus of 8000 MPa in their finite element studies. Young's modulus of trabecular bone was also found to be varying between 300 and 500 MPa. Felix et al. used Young's modulus of 500 MPa for the trabecular bone in their study. In our study, we used Young's Modulus of cortical and trabecular bone as 18,300 and 300 MPa, respectively. Poisson's ratio was taken as 0.3. Young's Modulus and Poisson's ratio for prosthesis material (steel and plastic) was taken as 2,05,300, 1100 MPa and 0.3, 0.34, respectively. Investigators like Carter and Hayes [4] and Rice et al. found out the density of elbow bone based on CT scan. As their study was a standard one, we use the same values of material properties in our study. Hence, for our dynamic model, the density allocated for the elbow trabecular and cortical bone was 2 and 0.5 gm cm^{-3}, respectively. Density of bone cement was taken as 1.52 gm cm^{-3} [5]. Density of steel and plastic (UHMWPE) [6] was taken as 7.8 and 0.976 gm cm^{-3}, respectively. Bone cement was used to fix the prosthesis with the bone. The most widely used bone cement is based on PMMA, also called acrylic bone cement [3]. It is self-polymerizing and contains solid PMMA powder and liquid MMA monomer. Young's Modulus and Poisson's ratio of bone cement was taken as 2500 MPa and 0.34, respectively.

Boundary Conditions

Loading

We considered the case of elbow impact occurring due to sudden lifting of a load of falling on an outstretched hand. The magnitude of the impact force was considered as 10 kg, i.e.100 N. It was applied (Fig. 4) for impact duration (ramp time to peak load) of 10 ms, and again applied for two different materials of prosthesis viz. steel (316-L) and plastic (UHMWPE). For the linked prosthesis (Fig. 4 (a)) only static analysis was performed with magnitude of static load being 20 kg, i.e. 200 N. For the entire load cases (static and dynamic), loads were applied on few nodes to distal end of forearm and were acting outwards parallel to the axis of the ulna. In all the cases, load was acting as distributed load on nodes of distal forearm (Figs. 4 (a) and (b)). All the above dynamic analyses were performed for a fixed density of 2, 0.5, 1.52, 7.8 and 0.978 g/cc for the cortical bone, trabecular bone, bone cement, steel prosthesis and plastic (UHMWPE) prosthesis, respectively, i.e. for a fixed mass of elbow model.

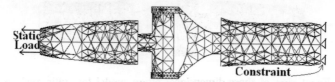

Fig. 4 (a). Location of nodes where load and constraints were applied on the static model.

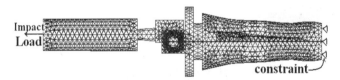

Fig. 4 (b). Location of nodes where load and constraints were applied on the dynamic model.

Constraints

As the impact load was applied to the distal end of the forearm, the constraints were to be applied on the distal end of the humerus. Few nodes at the cutting section of the distal humerus (Fig. 4), perpendicular to the long axis of the humerus were constrained in all the directions (UX, UY, UZ, ROTX, ROTY and ROTZ) for all the load cases as described earlier.

Results

Rigorous analyses were done for both the static and dynamic models, considering various load cases. Due consideration was given for both the prosthesis (humeral and ulnar) and bone cement, while analyzing the responses under static and dynamic loads. Static analysis of the elbow model (considering linked prosthesis) was done for the normal physiological loading. Analysis was done for 10 kg (i.e. 100 N) load applied statically on straight elbow. Again, analysis was done for two different materials of prosthesis viz. steel (316-L) and plastic (UHMWPE). Hence, the static model (Fig. 4(a)) was solved separately for the two different load cases. In this discussion, among all the stresses, importance was given to the von-Misses stress criteria for both the humeral and ulnar prosthesis and the shear stresses in bone cement [3]. Special observations were made for few zones of the artificial elbow model, like the articulating surface between humeral and ulnar prosthesis, and interfaces between prosthesis and bone cement, as well as bone cement and cortical bone.

At first, the von-Misses stress distribution pattern of the humeral and ulnar prosthesis for the two different cases of analysis is presented in Figs. 5 (a), (b) and 6 (a), (b). The stress contour map revealed that for the case of straight elbow lifting 10 kg load, the maximum von-Misses stress in ulnar prosthesis was 13.42 and 20.93 MPa for steel and plastic prosthesis, respectively. These values were 40.597 and 78.435 MPa, respectively, for humeral prosthesis. The zone of highest von-Misses stress was same for the two prosthesis materials.

(a)

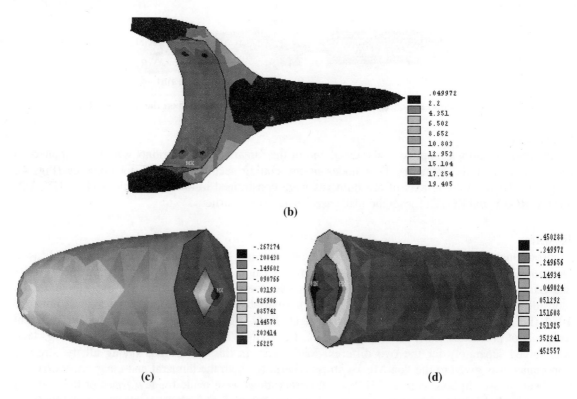

Fig. 5. von-Misses stress distribution in (a) ulnar prosthesis and (b) humeral prosthesis. Shear stress distribution in (c) bone cement of ulnar side and (d) bone cement of humeral side when prosthesis is made of steel.

The shear stress distribution in z-x plane for the bone cement of ulnar and humeral side for the two cases of study has been shown separately in Figs. 5 (c), (d) and 6 (c), (d), respectively. It has been observed that the maximum values varied between 0.3 and 0.5 MPa, and remain below 1 MPa. These maximum values were located mostly in the interfaces between prosthesis-bone cement and bone-bone cement region.

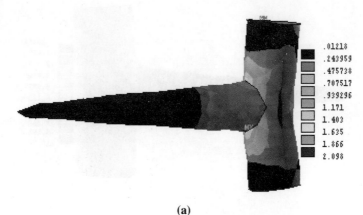

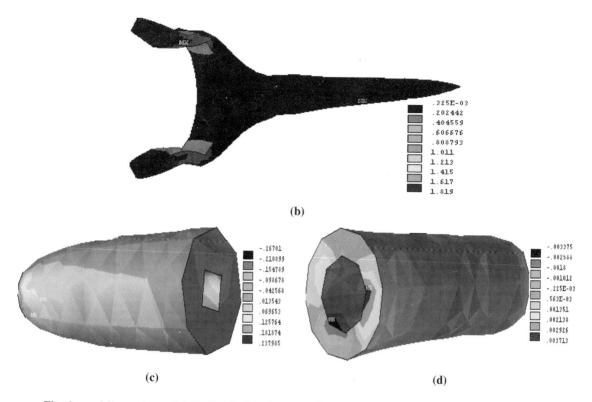

Fig. 6. von-Misses stress distribution in (a) ulnar prosthesis and (b) humeral prosthesis. Shear stress distribution in (c) bone cement of ulnar side and (d) bone cement of humeral side when prosthesis is made of plastic.

Again, two more static analyses were done changing the thickness of bone cement layer from 3 to 5 mm in order to study the effect of thickness of bone cement layer on stress value. It was observed that stress in both the humeral and ulnar prosthesis, as well as bone cement, reduced significantly.

Dynamic analysis was performed on a second model of the artificial elbow prosthesis that was developed as unlinked component, i.e. there was an external connection between the humeral and ulnar components of the prosthesis. This analysis was performed for a fixed impact load of 100 N and impact duration (ramp time to peak load) of 10 ms to peak load. The results are shown in Figs. 7 (a) to (f). Displacement-time history plots of the prosthesis model are presented for the two materials of prosthesis (Figs. 7 (e), (f)). Von-Misses stress-time history plots are presented in Fig. 7 (a) (humeral component) and Fig. 7 (b) (ulnar component) for plastic prosthesis. Von-Misses stress-time history plots for the steel implant are presented in Fig. 7 (c) (humeral component) and Fig. 7 (d) (ulnar component). Peak stress did not occur at the instant when impact load was maximum for all the cases. In general, von-Misses stress in prosthesis and shear stress in bone cement was lower for plastic prosthesis in dynamic analysis also as evident from plots shown in Figs. 7 (c) to (f).

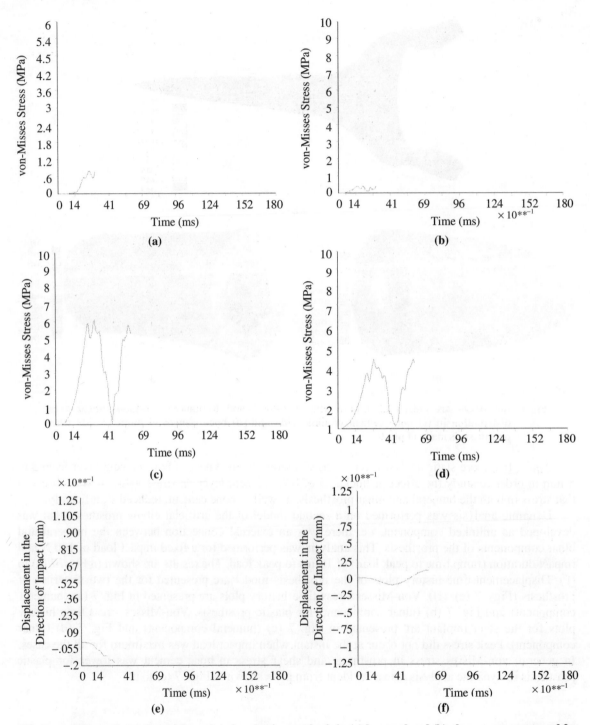

Fig. 7. von-Misses stress-time history plots for plastic prosthesis in (a) humeral and (b) ulnar component, and for steel prosthesis in (c) humeral and (d) ulnar component. Displacement-time history plots of entire model for (e) plastic and (f) steel prosthesis.

Conclusions

Regarding the stress distribution, as shown in Figs. 5 (a) to (d) and 6 (a) to (d), sufficiently high stress occurred at the articulating surface of humeral and ulnar prosthesis, during most of the analysis. Higher values of shear stress also occurred at the interface of cortical bone-bone cement and prosthesis-bone cement. But values of shear stresses were less than 1 MPa for most of the analysis, except a few. Values of von-Misses stress in most of the cases were found higher in humeral prosthesis as compared to ulnar prosthesis.

It is clear from these analyses that the interface of prosthesis and bone cement is a vulnerable zone in the artificial elbow system where very high stress was found for normal physiological loading condition.

It was also observed that the bone cement layer faces 1-2 MPa shear stress, which can be allowed. But it was found that if thickness of bone cement layer is decreased from 2.5 to 1.5 mm, then tensile and shear stress in bone cement increased dangerously. Bone cement being a brittle material is weak in tension and shear though strong in compression. Therefore, it is concluded that bone marrow part should be totally replaced by bone cement to keep the tensile stress in bone cement within allowable limit.

From the results obtained from the dynamic load cases, it can be concluded that if the impact force in daily activities or in accidents remains within 10-15 kg in 10 ms, then our designed prosthesis is not likely to fail or loose under impact load as peak stresses (< 10 Mpa in prosthesis and < 1 Mpa in bone cement) can be restricted below their respective compressive yield limit, i.e. 20 Mpa for bone cement, 23 Mpa for plastic and 700 Mpa for steel (316-L). Similar to the static case, it is also concluded here that using plastic as implant material for both the components of the artificial elbow, peak stresses can be further lowered in both the humeral and ulnar prosthesis, as well as shear stress in bone cement, in both the humeral and ulnar side. This is significant as bone cement being brittle material is weak in shear though strong enough in compression.

References

1. Tillmann, B. 1978. A Contribution to the Functional Morphology of Articular Surfaces Thieme, Stuttgart.
2. Verdonschot, N. and Huiskes, R. 1990. FEM Analysis of Hip Prostheses: Validity of the 2-D side-plate model and the effects of torsion. *Transaction of 7th Meeting of the European Society of Biomechanics.* p. 20A.
3. Saha, S. and Pal, S. 1984. Mechanical Properties of Bone Cement: A review. *J. Biomedical Materials Research.* **18**: 435-462.
4. Carter, D.R., Hayes, W.C. and Schurman, D.J. 1976. Fatigue Life of Compact Bone II: Effect of microstructure and density. *Journal of Biomechanics.* **9**: 211-218.
5. Indian Academy of Sciences. *Bull. Mater. Sci.* **Vol. 23, No. 2, April 2000**: pp. 135-140.
6. Pinchuk, L.S. 1998. Concept of Endoprostheses Approximation to Natural Joint. *Proc. of 1st Asia Int. on Tribology (Asiatrib. 98).* Beijing, China. pp. 834-837.

Biomechanics
R.K. Saxena and P. Mishra (Editors)
Copyright © 2005, Anamaya Publishers, New Delhi, India

13. A New Algorithm for Diagnosing Disease Through Microscopic Images of Blood Cells

S. Soundarapandian[1], S. Mahesh[2], P. Subbaraj[3] and R. Murugesan[4]

[1]Department of Mechanical Engineering, S.A.C.S., M.A.V.M.M. Engineering College, Madurai - 625301, India

[2]F.W. Technology Center, Bangalore, India

[3]Arulmigu Kalasalingam College of Engineering, Krishnankoil - 626190, India

[4]Madurai Kamaraj University, Madurai - 625021, India

Abstract: This paper outlines a simple, fast and reliable method for automatically diagnosing diseases through digitized images of blood cells that have been segmented to reveal the functions of the principal effectors. Organs and tissues that employ blood cells as principal effectors of their physiologic functions. Hematopathology is somewhat unique in its approach to the patient and the disease, in which many diseases are understood at molecular level and the function of the blood is relatively simple when compared to that of other organ systems. Erythrocyte, leukocyte and platelets of the blood cell plays a crucial role in resisting foreign particles such as bacteria and virus etc. In microscopic images, the diagnosis is based both on the evaluation of some general features of the blood cells such as color, shape, border, and the presence and aspect of characteristic structures. Perception of these structures depend both on magnification and image resolution. The introduction of vision system associated with image processing techniques enables the numerical and objective description of some pattern cells features. MS Visual C++ and Matrox Inspector image processing software are used for the experimental work. At preprocessing stage, which enhances the image of blood cells by the use of calibration, mapping and morphology is introduced. In the secondary stage, algorithm consists of blob analysis and second order edge detection. Experimental results using proposed technique gives way for new research area in biomechanics (modeling and simulation).

Introduction

Health is a positive state of physical and mental well being and not merely absence of disease. It has systems of checks and balances and mechanisms, which help preserve a state of normality in whatever environment the human body is placed, and whatever provocations it is subjected to. The terms 'health' and 'disease' often convey something different when used in different contexts. In complex organisms like man, the body is made up of millions of tiny units called 'cells'. Cells perform functions of varying degrees of specialization. Cluster of cells performing related functions are grouped together in the form of organs. The chief requirement of every cell is that its interior and immediate surroundings should be suitably warm, salty and acidic. The range of temperature, salt concentration, and acidity that is compatible with life is extremely narrow. The cells of the body are provided a fairly constant internal environment, or homeostatic condition. Derangement of homeostatic mechanisms is the fundamental basis of disease. Blood has been associated with life since time immemorial. We are alive only so long as our cells are alive. Our cells stay alive so long as they are supplied with nutrients and cleared of their waste products. Blood does this job for every cell of the body. But it is remarkable that it does so while moving in, and remaining essentially confined to an extensively branched tree of tubular structures called blood vessels. This is because there are portions of this tree, which come very close to every cell of the body. Although blood is only a part of the body fluids, the concentration in body fluids and concentration in blood can be used interchangeably. The reason is that blood and other fluid compartments of the body have constant exchange going on amongst them. Humoral immunity comes into play in response to invading germs, as well as some harmful chemical substances. The harmful chemical substances (toxins) may be those, which are produced by the invading germs. The main feature of the response is production of proteins (antibodies), which circulate in the blood stream and neutralize or help eliminate the invading agent.

Blood: The Vital Fluid

Blood is a liquid tissue. Human blood is the fluid circulated by the heart through the human vascular system. There are three cellular components of human blood: red blood cells, white blood cells, and platelets. Red blood cells transport oxygen to other cells of the body. Packed with hemoglobin (an iron-bearing protein) and shaped like plump disks with indented centers, red bloods cells are produced in bone marrow and have a life span of about 120 days. White blood cells (purple) protect the body from infection, attacking and destroying foreign particles like dust, pollen and viruses. White blood cells are far less numerous than the red cells, the ratio being one white cell to every 600 red cells. They are on the average, slightly larger than the red cells, and differ from them in three important respects. First they have nuclei. Second, they do not contain hemoglobin, and are therefore colorless, finally, some white cells can move and engulf particles or bacteria much like anemia. Platelets defend the body against excessive blood loss. Platelets (blue) flow freely in the blood in an inactive state but when an injury is sustained, platelets become sticky to plug the injured area. The blood also transports hormones from the different glands located throughout the body. These hormones act as messengers that depended on the blood for travel. Another function is to help us keep immune from disease that may enter the body. The clotting process not only stops you from bleeding to death, but also assists in stopping microbes from entering and causing infection. Years of research have gone into trying to avoid the problems of blood perishability and safety by developing blood substitutes. Most of these have focused on materials that will transport adequate amounts of oxygen to the tissues.

Experimental Setup

The technology used in solid state imaging sensors is based principally on charged coupled devices (CCD), which can convert the images from analog to digital format. Simple strategies can be adapted to grab the images from the electron microscope. The experiment apparatus consists of structured light source, electron microscope, CCD camera and vision system. The relative cells of the blood are first measured. Before the actual diagnosis is made, the real coordinates and world coordinates must be determined in the camera coordinate system by a calibration process.

The camera acquisition requires the knowledge of calibration analysis. The camera can be viewed simply as a device for mapping a 3D image into a 2D image. A detailed experimental study is carried out in diseased cell structure with the normal one. The real time experiment is to illustrate the actual implementation of the estimation method developed by spatial encoding technique. The innovative hardware and vision-processing algorithm with menu driven user interface was used to analyze the condition of the blood cells in the diseased cell structure. The experiment is planned taking into consideration the above requirements and detection of diseases through microscopic images of the diseased cell structure. The software facilitates capturing of high-resolution images in color or in black and white, comparison of built up model (template) with that of subjected model and the images can be sized and stored for future reference. The algorithm used is shown in Fig. 1.

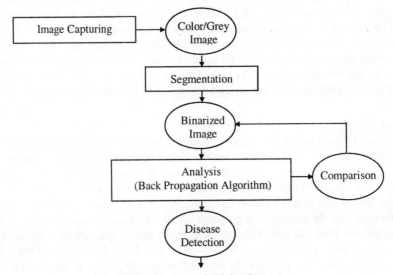

Fig. 1.

Results and Discussion

Blood cell recognition, disease diagnosis and some of the morphological methods are briefly reviewed in this paper. This includes experimental investigations and image processing algorithms that have been developed to diagnose diseases. To overcome the problems associated with blood cells, disease constituents and effectors during the diagnosis process, the image processing technique has been formulated with vision system to provide closed loop control. Several experiments were carried out to characterize the disease of the main modeling steps, i.e., establishing corresponding cells from a set of learning shapes and deriving a statistical model. We applied the modeling procedure to distinct the incoming image with the normal one. We tested our methods on 15 images of healthy and 30 images with disease. In our approach, we established the corresponding cells prior to pose and scaling estimation during statistical modeling. This assumes that the position of the cells does not change significantly with different pose and scaling, which is justified by the high accuracy of the model. To assess how well the statistical model generalizes to unseen objects, we measured how close a shape not part of the learning sample can be approximated with the statistical shape model. To compute the robustness of the method with respect to different parameter settings, the entire modeling was carried out for a range of parameter at a time while keeping the remaining parameters at their optimal values. The images used are presented here, which have exhibited a wide variety of lesions and confusing manifestations. On this difficult data set, our methods successfully detect the disease in most of the reliable cases.

Normal blood - dense area, segmented neutrophil, normal blood.

 Normal blood - thin area, normal blood.

 Normal blood - dense area, normal blood.

 Normal blood - side edge, large cells.

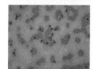 Blood - tail area, segmented neutrophils, tail area - mostly segmented neutrophils.

 Blood - thin area, lymphocytes, neutrophils, thin area - 2 lymphocytes, 1 band neutrophil, 2 disrupted neutrophils.

 Lymphocyte - normal blood.

 Lymphocyte - thin tail area blood.

 Red blood cells - normal. This photomicrograph shows normal red blood cells (RBCs) as seen in the microscope after staining.

 Red blood cells - tear-drop shape. This photomicrograph shows one of the abnormal shapes that red blood cells (RBCs) may assume, a tear-drop shape. Normally, RBCs are round.

Red blood cells - elliptocytosis. Elliptocytosis is a hereditary disorder of the red blood cells (RBCs). In this condition, the RBCs assume an elliptical shape, rather than the typical round shape.

Red blood cells - sickle cells. These crescent or sickle-shaped red blood cells (RBCs) are present with sickle cell anemia, and stand out clearly against the normal round RBCs. These abnormally shaped cells may become entangled and block blood flow in the small blood vessels (capillaries).

Megaloblastic anemia - view of red blood cells. This picture shows large, dense, oversized, red blood cells (RBCs) that are seen in megaloblastic anemia. Megaloblastic anemia can occur when there is a deficiency of vitamin B-12.

Malaria - photomicrograph of cellular parasites. Malaria is a disease caused by parasites. This picture shows dark stained malaria parasites inside red blood cells (a) and outside the cells (b). Note the large cells that look like targets. It is unknown how these target cells are related to this disease.

Red blood cells - target cells. These abnormal red blood cells (RBCs) resemble targets. These cells are seen in association with some forms of anemia, and following the removal of the spleen (splenectomy).

Segmented neutrophil, lymphocyte, thin area blood, one segmented neutrophil, 1 large lymphocyte. Very thin area.

Small platelets, megathrombocyte, band neutrophil, ten small platelets which are round and slightly swollen, 3 larger platelets and 1 megathrombocyte. A normal band neutrophil is in the field.

Plasma cell with four nuclei, acute myeloid leukaemia, one plasma cell with four nuclei. Its golgi area (light in color) is in its center. Four blasts surround it. They are medium in size, have a high nuclear/cytoplasmic ratio, blue cytoplasm and an evenly distributed nuclear chromatin with one or more prominent nucleoli.

Young megakaryocyte, AIDS patient marrow, one young megakaryocyte in process of nuclear division.

Plasmacytoid lymphocyte, severe arthritis with osteoporosis, one plasmacytoid lymphocyte. It is a classic example: large size, deep basophilic blue cytoplasm, nucleus centrally located with a dense chromatin with a few randomly distributed open or colourless areas. Severe arthritis with osteoporosis blood - 100X.

Plasmacytoid lymphocytes, mott cell, arthritis, osteoporosis, two plasmacytoid lymphocytes; one of which contains numerous vacuoles (mott cell).

Monocyte with RBC, autoimmune hemolytic disease, monocyte with RBC; erythrophagocytosis; many spherocytes.

Bare megakaryocyte nuclei, NRBC, cancer blood, bare lobulated megakaryocyte nuclei, 1 late NRBC and 2 band neutrophils. Dense area.

Late megakaryocyte, buffy coat, lung cancer blood, late megakaryocyte at tail area of buffy coat preparation of blood of a patient with lung cancer.

Megakaryocyte nucleus, lung cancer blood, single megakaryocyte nucleus with scant fragmenting cytoplasm.

Megakaryocytes, lung cancer, buffy coat, left frame: late small mature megakaryocyte - 100X. Right frame: bare lobulated megakaryocyte nuclei. Metastatic cancer of lung.

Mott cell, plasmacytoid lymphocyte, viral infection, one mott cell - plasmacytoid lymphocyte with multiple vacuoles. The round shape and high nuclear/cytoplasmic ratio indicate, it is a plasmacytoid lymphocyte rather than a plasma cell. One band neutrophil is also in the field.

Plasma cell, bacterial infection blood, one plasma cell, oval in shape, with a low nuclear/cytoplasmic ratio, eccentrically located nucleus, deep basophilic blue cytoplasm containing a few vacuoles, dense nuclear chromatin with a few randomly distributed less dense areas. A monocyte is at lower left corner.

Plasmacytoid lymphocyte, renal infection blood, one plasmacytoid lymphocyte.

Plasmacytoid lymphocytes, viral infection, two plasmacytoid lymphocytes.

Conclusions

In this paper, we have presented a method to automatically construct statistical blood cells from segmented microscopical images. Standard models are used to diagnose the disease that can also be used for haematological applications. Corresponding parameters on the surface of the shape are established automatically by the adaptation of back propogation algorithm to segmented volumetric images. To assess the robustness of the method with respect to the parameter settings, the entire statistical model was carried out for a range of parameter values, varying one at a time while keeping the remaining parameters at their optimal values. MS Visual C++ and Matrox Inspector image processing software are used for the experimental work. Nevertheless, this would be an interesting future investigation, since our approach is sufficiently general to be applied to modelling tasks from other two dimensional or three dimensional application domains. The extension of this may open a new area of research in biomechanics (modelling and simulation).

References

1. Abatti, P.J. 1997. Determination of Blood Cell Ability to Traverse Cylindrical Pores. *IEEE Transactions on Biomedical Engineering.* **44(3)**: 209, 211.
2. Sint Jan, S.V., Giurintano, D.J., Thompson, D.E. and Roozie, M. 1997. Joint Kinematics from Medical Imaging Data. *IEEE Transactions on Biomedical Engineering.* **44(12)**: 1175-1184.
3. Kaus, M.R., Pekar, V., Lorenz, C., Truyen, R., Lobregt, S. and Weese, J. 2003. Automated 3D PDM Construction from Segmented Images using Deformable Models. *IEEE Transactions on Medical Imaging.* **22(8)**: 1005-1012.
4. Hoover, A. and Goldbaum, M. 2003. Locating the Optic Nerve in a Retinal Image using the Fuzzy Convergence of the Blood Vessels. *IEEE Transactions on Medical Imaging.* **22(8)**: 951-958.
5. Malhotra, K., Mohan, C.K. and Ranga, S. 1997. *Elements of Artificial Neural Network.* Penram International Publishing Mumbai.

Biomechanics
R.K. Saxena and P. Mishra (Editors)
Copyright © 2005, Anamaya Publishers, New Delhi, India

14. Microbial Bio-fuel Cell: A Potential Application Tool in Biomedical Instrumentation

T.G. Deepak Balaji and A. Siva Kumar
Centre for Biotechnology, Biological Science Group, Birla Institute of Technology and Science, Pilani - 333031, India

Abstract: Various biomedical applications which involve implanted devices like heart pumps require continuous energy flow to activate the same. Though various energy producing nano-materials are being used, Microbial Bio-fuel Cells can be considered as a best alternative strategy as they depict biological system. The principle behind microbial bio-fuel cells is the direct conversion of sugar to electrical power. Bio-fuel cells potentially offer nature's solution to energy generation by immobilizing microorganisms. Since they produce concentrated source of energy, they can be taken from microbes. These microbes generate the fuel substrates by transformation and can participate in the electron transfer chain between the fuel substrates and electrode surfaces. With the advent of sophisticated extraction, purification and electron-coupling techniques, the emphasis has shifted towards enzymatic fuel cells. By tapping complete multi-enzyme metabolic pathways inside living cells, microbial fuel cells live long and can be utilized for complex and nano-type application. When considering the development of possible consumer products based on compact bio-fuel, it is apparent that there is a hybrid approach of utilizing both metabolic and enzymatic properties of microbial cells. Special analyses are being done to check the longevity and stability of microbial fuel cells. Specific types of microbes are immobilized in fuel cells depending on the need of application. With this concept of microbes based energy, the extractable electrical power could then be used to activate implanted devices such as pacemakers, pumps (e.g. insulin pumps) and sensors or activate self fueled prosthetic units.

Introduction

Potential strategies for deriving useful forms of energy from carbohydrates face technical and economical hurdles. An alternative strategy is direct conversion of sugar to electrical power. Existing fuel cells cannot be used to generate electric power from carbohydrates. For decades, microbes that produce electricity were a biological curiosity. Now, researchers foresee a use for them in watches and cameras as power sources for the third world, and for bioreactors to turn industrial waste into electricity [1]. Bio-fuel cells potentially offer solutions to all these problems, by taking nature's solutions to energy generation and tailoring them to our own needs. They take readily available substrates from renewable sources and convert them into benign by-products with the generation of electricity.

Since they use concentrated sources of chemical energy, they can be small and light, and the fuel can even be taken from a living organism (e.g. glucose from the blood stream). The biological fuel cell generates a small electrical current by diverting electrons from the electron transport chain. Bio-fuel cells, in which whole cells or isolated redox enzymes catalyze the oxidation of the sugar, have been developed. Microorganisms have the ability to produce electrochemically active substances that may be metabolic intermediates or final product of anaerobic respiration [2]. For the purpose of energy generation, these fuel-substances can be produced in one place and transported to a bio-fuel cell to be used as fuel. In this case, the bio-catalytic microbial reactor is producing the bio-fuel and the biological part of the device is not directly integrated with the electrochemical part. The two parts can even be separated in time, operating completely individually.

Materials and Methods

Bio-fuel cells use bio-catalysts for the conversion of chemical energy to energy, to generate the fuel substrates by transformations or by participating in the electron transfer chain between the fuel substrates and electrode surfaces. The extractable power of a fuel cell P_{cell} [3] is

$$P_{cell} = V_{cell} \times I_{cell}$$
$$V_{cell} = \left(E_{ox}^{o'} - E_{fuel}^{o'} \right) - h$$

where h is the voltage; $E_{ox}^{o'}$ the potentials of oxidizer compounds; and $E_{fuel}^{o'}$ is the potentials of fuel compounds.

Bio-fuel cells cause electricity generation from organic substrates by the following approaches:

1. It includes the use of microorganisms as biological reactors for the fermentation of raw materials to fuel.
2. It utilizes microbes in the assembly of bio-fuel cells, which includes the *in situ* electrical coupling of metabolites generated in the microbial cells.
3. It involves the application of redox-enzymes for the targeted oxidation and reduction of specific fuel and oxidizer substrates, and generation of electrical power output.

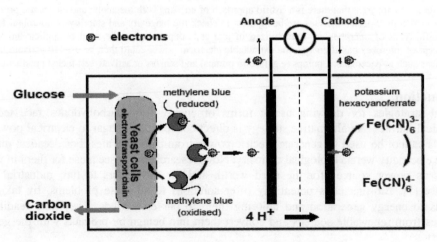

Fig. 1. Typical biofuel cell reaction model.

The most widely used fuel in this scheme is hydrogen gas, allowing well-developed and highly efficient H_2/O_2-fuel cells to be conjugated with a bioreactor. In another approach, the microbial fermentation process proceeds directly in the anodic compartment of a fuel cell, supplying the anode with the *in situ* produced fermentation products [4]. In this case, the operational conditions in the anodic compartment are dictated by the biological system, so they are significantly different from those in the conventional fuel cells.

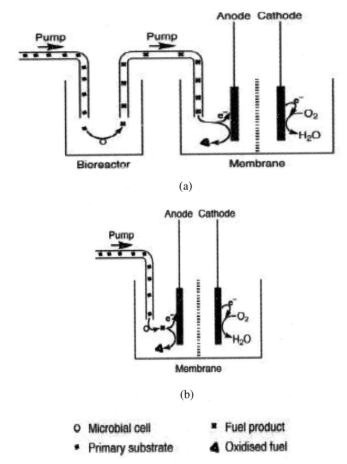

Fig. 2. A microbial bioreactor providing fuel directly in the anodic compartment of electrochemical cell.

Bearing in mind that the organism itself needs some of the glucose for its own biomass production, corresponding type of microbial fuel cell is being selected. In the case of *R. ferrireducens* type 1 microbial biofuel cell is being selected. In addition, electrons generated by *R. ferrireducens* are easily transferred to the anode without the assistance of electron-shuttling mediators and the cells grow at a steady rate, which guarantees a steady supply of electrons, and therefore, a consistent current density [5].

Results

Development and Design of Enzymatic Bio-fuel Cells
Enzymatic bio-fuel cells can utilize a diverse and unlimited supply of fuel sources, which has prompted researchers to focus on miniature, implantable powering devices in living systems, with the

primary fuels being di-oxygen and glucose. Unlike conventional fuel cells that need periodic refueling, the implanted micro enzymatic bio-fuel cells continue to produce electricity as long as the host is alive. Researchers have focused on gas-diffusion laccase-catalyzed cathodes, operating under "air breathing" conditions, which can utilize oxygen directly from atmosphere [6]. There are reports on the integration of such gas-diffusion laccase-catalyzed cathode with glucose electro-oxidizing, glucose oxidase anode. Development of the designs allowing for direct utilization of various fuels harvested from biological sources is addressed. This will include employment of coupled enzymatic reactions between glucose oxidase and invertase, e.g. to provide for the possible use of sucrose as a fuel.

One of such basic anode design concepts relies on mediated enzymatic oxidation of glucose. Being the first enzyme to be used in glucose biosensors, Glucose oxidase (Gox) has been the most commonly used for glucose electro oxidation at the anode of enzymatic bio-fuel cells. Among the numerous mediators and redox polymers that have been used, scientists have exploited nickelocene, which has more favorable quasi-redox potential compared to ferrocene derivatives, and other redox mediators [7]. The composite electrode was prepared using a mixture of teflonized carbon and carbon black, with the mediator and Gox immobilized by physical adsorption, and was designed to serve as anode of the bio-fuel cell. This electrode demonstrates high activity in glucose conversion. The laccase from *Rhus veinicifera* was immobilized at the gas-diffusion cathode. The electrolyte used was 0.1 M phosphate at pH 7, which was found to be optimal for both enzyme electrodes. Optimization of the anode with respect to enzyme and mediator, immobilization methods and use of appropriate membranes is in progress.

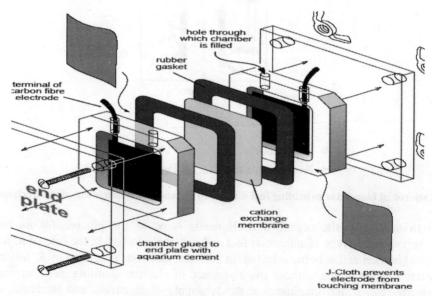

Fig. 4. Typical bio-fuel cell model.

Hybrid Bio-fuel Cell Approach

In the pioneering days of the technology, erstwhile researchers were focused on MFCs and EFCs. When considering the development of possible consumer products based on compact bio-fuel cell technology, such as battery chargers or replacements, it soon becomes apparent that some form of hybrid approach is called for, since individually neither MFCs nor EFCs seem to possess all of the desired traits. The researchers have introduced such a hybrid bio-fuel cell concept that also permits a high degree of integration and miniaturization [8]. *Brewer's Yeast* was chosen as the bio-catalyst since it is readily available, non-pathogenic and easy to culture. However, the reducing centers of eukaryotes are located within the mitochondria deep inside the cytoplasm. Therefore, Lipophilic mediators act lethargically when facilitating yeast electron transport. The cell wall has also frustrated attempts at direct "wiring" yeast to electrodes. Lysis is proposed herein, as a way of overcoming these problems, thereby making yeast a viable bio-catalyst [9].

Yeast is initially grown aerobically in a micro-chamber or "breeder". Under such favorable conditions the yeast reproduces rapidly and develops a respiratory apparatus capable of the complete oxidation of sugar. Next, most of the yeast cells are subjected to a cell disruption treatment that splits them open, spilling out their intracellular contents, but without damaging electro active protein constituents. A variety of dissolutive methods have been demonstrated in miniaturized cell lysis devices, i.e. electroporation, sonication and optics. Then, debris passes into the anaerobic anode compartment of a bio-fuel cell, along with a sugar fuel. Liberated from the constraints imposed by membranes and cell wall, the enzymes are free to interact with the fuel, each other and the anode, thereby exchanging electrons much like a non-wired EFC. This hybrid system potentially possesses the longevity and bio-fuel utilization traits of a MFC, since living yeast cells are continually being cultured as the bio-catalyst. In addition, because electron exchange with the anode is not encumbered by a cell wall, no lipophilic electron transport mediator in solution is required [10]. This characteristically EFC trait facilitates easy waste removal, since no mediator chemicals are lost or need be replaced.

Discussion

Any technology has its own strength, weakness and application. The same is the case for bio-fuel technology. To attain efficient yield of this technology, the researches are doing various strategies and novel approaches. The identification of rate-limiting steps allows the development of strategies to improve and enhance the cell output [11]. The chemical modification of redox-enzymes with synthetic units that improve the electrical contact with the electrodes provides a general means to enhance the electrical output of bio-fuel cells. The site-specific modification of redox enzymes and the surface-reconstitution of enzymes represent novel and attractive means to align and orient biocatalysts on electrode surfaces.

The effective electrical contacting of aligned proteins with electrodes suggests that future efforts might be directed towards the development of mutants of redox-proteins to enhance their electrical communication with electrodes. The stepwise nano-engineering of the electrode surfaces with relay-cofactor-bio-catalyst units by organic synthesis principles allows us to control the electron transfer cascades in the assemblies. By tuning the potentials of the synthetic relays or bio-catalytic mutants, enhanced power outputs from the bio-fuel cells may be envisaged [12].

A bio-fuel cell was made from two 7 μm dia. electrocatalyst-coated carbon fiber electrodes placed in 1 mm grooves machined into a polycarbonate support. One electrode is a glucose-oxidizing anode, and the other is an oxygen-reducing cathode. The electrode areas are $1/60^{th}$ the size of the smallest reported methanol-oxidizing fuel cell and $1/180^{th}$ the size of the smallest previously known bio-fuel cell [13]. The power density of the device is five times greater than that of the previous best bio-fuel cell. It has a power output of 600 nW at 37°C, enough to power small silicon-based microelectronics. An interesting example, which this report will be giving an outline of, is about a robot powered by a bio-fuel cell named *"Gastronome"* [13].

Conclusion

Technology in bio-fuel cells is still in its infancy. Technology developed till now shows that there is tremendous application. Various configurations designed by the scientists are vividly explained here. The configuration of the bio-fuel cells discussed can theoretically be extended to other redox enzymes and fuel substrates from biomass substrates using bio-fuels could complement energy sources from chemical fuel cells.

If current trend of bio-fuel cells towards the development of biologically inspired robotic systems continues, it seems inevitable that a power source based on natural food consumption will play an important role [14]. An important potential use of bio-fuel cells is their *in situ* assembly in human body fluids, e.g. blood. The extractable electrical power could then be used to activate implanted devices such as pacemakers, pumps (e.g. insulin pumps) and sensors or activate self-fueled prosthetic units. It is hard to guess now what the economy of such systems would be, but the development of efficient microbial fuel cells is the first step towards this goal.

References

1. Benntto, P. 1987. Microbes Come to Power. *New Scientist*. April 16, 1987.
2. Chaudhary, S.K. and Lovley, D.R. 2003. Electricity Generation by Direct Oxidation of Glucose in Mediatorless Microbial Fuel Cells. *Nature Biotechnology*. **21**: 1229-1232.
3. *Chem.ch.huji.ac.il/~eugeniik/biofuel/biofuel_cells2_2.html*
4. Wilkinson, S. 2000. Gastrobots: Benefits and challenges of microbial fuel cells in food powered robot applications. *Autonomous Robots*. **9**: 99-111.
5. Davis et al. 1962. Preliminary Experiments on a Microbial Fuel Cell. *Science*. **137 (3530)**: 615.
6. Fritz et al. 2003. Electricity Generation by Bacteria. *Nature Biotechnology*. **21**: 1151-1152.
7. Zhang et al. Modelling of a Microbial Fuel Cell Process. *Biotechnology Letters*. **17(8)**: 809-814.
8. Deepak Balaji, T.G. 2003. *Electricity Generation by Microorganism*. Bio-horizon, I.I.T. Delhi, New Delhi.
9. *www.bioscience-explained.org/EN1.1/PDF/fulcelEN.pdf* - 21 Nov 2003.
10. *www.eng.usf.edu/~wilkinso/*
11. *www.gastrobots.com*
12. *www.me.berkeley.edu/~kblam/*
13. *www2.electrochem.org/cgi-bin/*
14. *www.sciencedaily.com/releases/*

Biomechanics
R.K. Saxena and P. Mishra (Editors)
Copyright © 2005, Anamaya Publishers, New Delhi, India

15. Biomechanical Study of Orthopaedic Implants

R.K. Saxena
Centre for Biomedical Engineering, Indian Institute of Technology Delhi,
New Delhi - 110016, India

Abstract: Despite all the precautions, failure of orthopaedic implants keep on occurring causing loss and inconvenience to the patients and frustration to the surgeons. Failed implants, if properly analysed, can give valuable information which can help the manufacturer of implants, practicing orthopaedic surgeons and even patients to minimize the occurrence of implant failure. Failure/retrieval of orthopaedic implants/protheses is very large in India. In orthopaedic surgery, metallic implants and fixation devices such as bone plates, screws, intramedullary pins, circlage wire, s-p nails, k-nails are commonly used by the orthopaedic surgeons for the repair of bone fractures. Repair of fracture and use of implants depends upon configuration of fracture site, viability of regional soft tissues and surgeon's choice (preference) etc. Unfortunately, still no metallic implant material is completely inert and biocompatible for biological environment, and all metallic material implants can produce corrosive action leading to pathological symptoms and fail biomechanically during use. In order to understand the causes of failure and to suggest cures for it, an attempt has been made to measure and investigate the biomechanical, chemical and structural properties of failed plates, s-p nails, k-nails and rods etc. Biomechanical properties of failed implants were studied by Zwick/Roel Material Testing Machine (ZRTM-Z250/SN5A, USA). Compressive, tensile, bending, shear and torsion mechanical testing of failed implants have been measured and modulus of elasticity, buckling, shearing and fatigue failure properties was calculated. Resilience (Strain energy/ weight: J/N) and endurance strength properties of orthopaedic implants were measured. It is an important property of a material and shows the capacity of the material to bear impacts and shocks during dynamic activities.

Introduction
Orthopaedic implants and prostheses are widely used for internal fixation. Various combinations of alloy complexes are used for enhancing the biocompatibility, strength and endurance of these devices. Both stainless steel and titanium alloys are extensively used for the manufacturing of these implants by orthopaedic industries, because of their wear resistance, mechanical properties, corrosive resistance and biocompatibility properties. Even though metals corrode *in vivo*, a relationship between the tissue reaction and the released metallic products of corrosion has been reported in several retrieved and failed implants studies. In India, stainless steel is the commonly used implantable material for internal fixation because of its low cost and mechanical strength. Major disadvantages of stainless steel are well-known, i.e. surface corrosion phenomena, high rate of locally and systemically release of corrosion products. That is why search for better alloy materials is in progress. But, unfortunately, the failed implants are often not analyzed in strict scientific and technical terms to know the causes of failure, and to take suitable measures for avoiding the recurrence of such failures. Failure of

orthopaedic implants is reported to be high in India. Various factors have been attributed to explain this inadequacy of metallic implants manufactured in India viz. erosion and electrolytic corrosion, improper design and manufacturing defects including use of unsuitable or defective materials. Corrosion fatigue because of imperfect surface treatment of the implants is often found to be one of the common causes of fatigue failure of the implants. Concentration of stresses due to over tightening of the fasteners is also a potent cause of the implant failure. There is an urgent need to study the magnitude and causes of high incidence of fatigue failure of locally manufactured implants because the phenomenon is injurious as well as expensive for the patients. This will require mechanical testing (Hardness and Toughness test, Tension, Compression, Torsion and Fatigue tests etc.) and chemical analysis of the metallic alloys. Simultaneously, specimen of unused local implants procured from the market will also be tested to evaluate their suitability. This study will also provide insight to enable us to suggest ways to improve the quality of implants manufactured in India. Feasibility of manufacturing implants from alternative materials such as Titanium and its alloy will also be examined for prevention of implant failure. Although the Bureau of Indian Standards (BIS) is doing a commendable job in improving the design and manufacture of orthopaedic implants being made in India by formulating suitable Indian standards for the implants etc. and providing certification to the desirous manufacturers. But, in spite of these commendable efforts of BIS and the best efforts of manufacturers and surgeons, implant failures do take place. However, at present there is no proper dedicated research/testing laboratory in India for analysis of the causes of implant failure and for providing relevant findings to dedicated manufacturers, surgeons and also to the public for their benefit. There is indeed an urgent need to create a suitable dedicated laboratory in the country to undertake the analysis of failed implants to minimize their failure. In this investigation, an attempt have been made to study biomechanical and metallurgical properties of indigenous failed and new orthopaedic implants/prostheses, and suggest some remedial measures for the improvement of mechanical and biocompatible properties of Indian manufactured stainless steel implants.

Experimental Methodology
Biomechanical properties of failed/retrieved orthopaedic, and new implants and prostheses have been measured and load (stress) vs. change in dimensions (strain) graphs were obtained during loading. Shear, yield stress (plastic deformation) and fatigue failure properties were studied.

Analysis of Failed/Retrieved Orthopaedic Implants/Prostheses
Failed orthopaedic implants (screws, nails, plates and hip prostheses) have been obtained from different hospitals. Stainless steel implant's biomechanical and metallurgical studies were analyzed macroscopically for the causes of failure and retrieval. New orthopaedics implants (rods, nails, screws and plates) have been purchased from different orthopaedic implant manufacturers/suppliers. The failed orthopaedic implants have been analyzed physically and classified according to their mode of failure. The physical characteristics of failed implants/prostheses (length (mm), diameter (mm), cross sectional area (mm^2), moment of inertia (mm^4) and weight (g)) were measured by digital vernier caliper, screw gauge and electronic weighing machine. The values of different dimensions are summarized in Table 1.

Table 1. Details of failed implants and classification according to the mode of failure

Type of Implants	No. of Implants	Bend Implants	Broken Implants	Corroded Implants	Other
Screws	38	5	3	20	10
Different Plates	7	2	1	2	2
Reconst. Nail	7	2	3	1	1
Hip Prosthesis	4	-	1	2	1

Analysis of Physical Properties of New Orthopaedic Implants

The physical properties of new orthopaedic implants like length, diameter, cross sectional area, weight, moment of inertia were precisely measured and calculated. The results are given in Table 2.

Table 2. Physical properties of new orthopaedic implants

Name of Implant	Part No.	Length (mm)	Diameter (mm)	Cross Sectional Area (mm^2)	Moment of Inertia (mm^4)	Weight (g)
Harrington Rod	1	238	6.5	33.18	87.57	52
	2	240	6.5	33.18	87.57	52
	Average	239	6.5	33.18	87.57	52
Recon. Nail (Right)	1	423	10.10	80.12	510.82	217
	2	417	9.96	77.91	482.82	217
	Average	420	10.03	79.01	496.82	217
Screw	1	80.44	5.62	24.81	48.94	10
	2	79.32	5.70	25.5	51.80	10
	Average	80	5.66	25.16	50.37	10
DCP Borrow Plate	1	151	NA	91.4	238.20	75
	2	154	NA	91.00	238.20	75
	Average	152.5	NA	91.20	238.20	75

Analysis of Mechanical and Metallurgical Properties of New Orthopaedic Implants

The following mechanical properties of new orthopaedic implants have been tested:

Mechanical Testing

1. Compression tests
2. Tensile Tests

Through obtained compression and tensile tests data, the following parameters have been calculated for new orthopaedics implants:

1. Buckling load
2. Compression Stress
3. Deflection at F_{max}

The results of various biomechanical analyses of new orthopaedic implants have been given in Table 3.

Table 3. Biomechanical properties of new orthopaedic implants (A)

Name of Implants	Part No.	Buckling Load (N)	Displacement at F_{max} (mm)	Yield Stress (N/mm^2)
Harrington Rod	1	2521.43	1.97	76.00
	2	2512	1.95	75.70
	Average	**2516.7**	**1.96**	**75.85**
Recon Nail (Right)	1	3784.02	5.3	47.23
	2	3790.12	5.22	48.65
	Average	**3787.01**	**5.26**	**47.94**
Screw	1	6097.01	2.36	245.75
	2	6070	2.40	238.1
	Average	**6083.5**	**2.38**	**241.92**
DCP Borrow Plate	1	7034	3.02	76.95
	2	6890	2.90	75.71
	Average	**6962**	**2.96**	**76.33**

Metallurgical Properties of New and Failed Implants and Prostheses
Failed/retrieved orthopaedic implants obtained from different hospitals metallurgical properties (Hardness (Vicker's value), chemical and microstructural properties etc.) were studied. Also, new orthopaedic implants which have been purchased from different manufacturing companies and suppliers mechanical properties and metallurgical properties (hardness (Vicker's value), chemical and microstructural properties etc.) were measured.

Hardness Test
Failed/retrieved and new orthopaedic implants/prostheses metal hardness characteristics were measured by Digital Vickers Hardness Tester MP-V1 Rockwell testing machine (Micro Photonics, Inc., Irvine, CA, USA). The Vickers hardness test method consists of indenting the test material with a diamond indenter, in the form of a right pyramid with a square base and an angle of 136° between opposite faces subjected to a load of 1-100 kgf. The full load is normally applied for 10-15 s. The two diagonals of the indentation left in the surface of the material after removal of the load are measured using automatically optical micrometer, and their average calculated and displayed on the LCD system. The Vickers hardness is the quotient obtained by dividing the load (kgf) by the area (mm^2) of indentation. Vickers hardness also can be calculated from the formula and convenient conversion tables. If Vickers hardness reported is 800 HV/10, it means a Vickers hardness of 800, was obtained using 10 kgf force. Several different loading settings give practically identical hardness numbers on uniform material, which is much better than the arbitrary changing of scale with the other hardness testing methods. The advantages of the Vickers hardness test are that extremely accurate readings can be obtained by using one type of indenter for all types of metals and surface treatments. Vickers machine is thoroughly adaptable, and very precise for testing the softest and hardest of materials, under varying loads.

Chemical Analysis of New and Failed Implants and Prostheses
New and failed implants and prostheses elemental chemical analysis has been done by Vacuum x-ray (XRF) fluorescence spectrometer (CAMET Research, Inc., Goleta, CA 93117, USA). It interprets the

characteristic radiation emitted by the elements of the sample upon excitation. The analysis is element sensitive. It provides information about the elemental composition of the sample. Vacuum XRF spectrometer with flow proportional detector and an RAP analyzing crystal for lighter elements and above; scintillation counter in combination with highly reflective LiF analyzing crystal for elements of Ti (22) and above; designed for nonstandard, customized experiments and analyses in unusual matrix; alternative calibration approaches, i.e. multiple standard addition; scattered x-ray background-ratio method for nondestructive as-is analysis of objects with rough or uneven surface.

Microstructural Evaluation of New and Failed Implants and Prostheses
Microstructural features of new and failed implants alloy materials have been studied by Nikon OPTIPHOT Transmitted Fluorescence Stereoscopic Optical Microscope SM2 (USA). In this method, simple light imaging, both basic fluorescence and transmitted, including darkfield and phase contrast can be studied. A good selection of objective lenses is available in the microscope. It is capable of magnifying the images to 50×, 100×, 200× and 500×. The system is equipped with a Nikon DXM1200 digital capture camera for image acquisition. The software automatically adds the micron bar on the image and is also capable of time-lapse photography. It has a CoolSnap Pro color camera, automated XY stage and Image Pro Plus software. This microscope is ideal for image photographs.

Microstructural Samples Preparation
Small cut pieces of failed and new implants were fixed, and mounted in resin material plastic blocks for making thin uniform surface for studying the crystalline properties of alloy material. The resin mounted cut implants pieces were subjected to rough and fine grindings, polishing and etching (chemical treatment) etc. for making the surface of the material clear and uniform. For every sample both longitudinal and transverse sections have been made and studied in order to understand details of microstructural characteristics of implants of alloy material. Microstructural photographs were taken at various magnifications for detailed study.

Biomechanical Study of New Orthopaedic Implants
Orthopaedic implant's mechanical properties have been measured on Zwick/Roel Material Testing Machine (ZRTM-Z250/SN5A, USA). The measuring system has capacity range of 1 ton and is controlled by computer program for the measurement of compressive, tensile and bending properties on rectilinear chart recorder for obtaining load deflection graph. Resilience strength, compliance of elasticity and yield stress have been calculated. Also, shear stresses at loading conditions have been measured. Implants in ZRTM system were mounted and gripped properly as per experimental requirement for studying the biomechanical properties. The following parameters were fixed in ZRTM system for the measurement:

1. Speed of loading was set at 1mm/min
2. Load measurement to 1000 kg
3. Deflection equivalent to 50mm

Results and Discussion
Biomechanical properties of orthopaedic implants have been measured to understand material strength and endurance characteristics. In the present study, we have measured load-bearing properties of orthopaedic implants under compression. In living beings, the main role of the skeleton is to provide the support to the body to maintain integrity of tissues and organs, and to provide support for

withstanding the load-associated changes during development and varying activities in adult life. However, fracture of bone causes strain on normal pattern of activity and loading alters the magnitude, rate and distribution of stress and strain behaviour. Orthopaedic implant devices have been developed to provide help in additional mechanical support to the fracture site. But, very close tightening and fixing of orthopaedic implants develops higher shear stiffness and axial stiffness of the affected region, and leads to the decrease in fracture healing. In this study, mechanical properties of local manufactured/indigenous stainless steel implants have been evaluated. Stainless steel (316 L) alloy material orthopaedic implants have been purchased because stainless steel shows good yield and corrosion resistance but marginally shows corrosive properties and is not completely biocompatible. The load deflection graphs have been obtained from Zwick/Roel Material Testing Machine (ZRTM-Z250/SN5A, USA) through their computer software. Loading were given to be 2-3 times of the body weight and maximum up to 1000 kg, till the time the implant starts showing buckling properties. Stress vs. strain data was recorded and buckling load (N), displacement at F_{max} (mm) and yield stress (N/mm^2) were measured and calculated in one set of new orthopaedic implants. In Harrington rod, we found the values as 2516.7, 1.96 and 75.85, respectively. For reconstruction nails (R) they were, 3787.01, 5.26 and 47.94 and for the screw, they were 6083.5, 2.38 and 241.92, respectively. In DCP Borrow Plate, they were 6962, 2.96 and 76.33, respectively. Analyses of data of biomechanical and metallurgical studies are in progress.

Biomechanics
R.K. Saxena and P. Mishra (Editors)
Copyright © 2005, Anamaya Publishers, New Delhi, India

16. Two Layered Model for Experimental and Analytical Study of Drag Reduction in Glass Model of Stenotic Glass Tube

V.K. Katiyar and Manoj Kumar
Department of Mathematics, Indian Institute of Technology Roorkee,
Roorkee - 247667, India

Abstract: Drag reducing agents (DRAs) are high molecular weight polymers that are suspended in a solvent. They are known for the ability to increase the throughput through oil and natural gas pipelines without increasing the horsepower and are vastly used by petroleum companies. Cardiologists have found an increment in blood flow rate through stenotic arteries by injecting drag reducing agents in the blood stream. In this paper, a two layered model for blood flow through stenotic artery is proposed to study the effects of drag reducing agents in increasing the blood flow rate across the diseased artery. All the governing equations are formulated in the cylindrical polar coordinates and analytical solutions obtained. The experiment was performed in the biomechanics laboratory situated in the department. A glass model for the diseased stenotic artery was used for the experiment and Peristaltic pump for generating the pulsatile flow. Physiological relevance, various flow parameters before and after injecting the drag reducing agents, and comparison between experimental and analytical results is discussed.

Introduction

Circulatory disorders are known to be responsible in many cases of death and stenosis or arteriosclerosis is one such cause. Stenosis, a medical term which means narrowing of any body passage, tube or orifice, is the abnormal and unnatural growth in arterial wall thickness that develops at various locations of the cardiovascular system under diseased conditions. This can cause circulatory disorders by reducing or occluding the blood supply which may result in serious consequences (cerebra strokes, myocardial infarction).

Drag reducing agents are known for their ability to increase the flow rate. They are typically long chain high molecular weight polymers that are suspended in a solvent. Drag reducing agents are widely used in the oil and natural gas pipelines to increase the flow rate so that a lot of energy is conserved. Attempts have been made to use drag reducing agents in the stenosed artery to increase the blood flow rate so that highly risky by-pass surgery may be avoided [9]. Some of the most commonly used drag reducing agents are:

1. Polyethylene Oxide: It is known by the trade name Polyox WSR301. It is extremely effective water soluble, drag reduction polymer with molecular weight of 5×10^6. Friction reduction of 40% has been noted in pipe flow with a polymer concentration of 1/2 parts per million (ppm).

2. Polyacrylamide: It is known by the trade name Separan AP30 and has molecular weight 3×10^6 This polymer is an effective drag-reducer in turbulent shear flows and forms a weakly anionic solution.
3. Poly-Iso-Butylene: It is an important drag reducer mainly used in crude oil pipelines and for kerosene like solvents. Many other polymers have been found to give drag-reducing properties to their respective solvents. For water pipelines Hydroxyl-ethyl cellulose and sodium polystylenesulfonate are used as drag reducing agents.
4. Conco CDR: It is a highly effective drag reducing polymer for crude oil pipelines and is being used in the Trans Alaska Pipeline system.

Formulation

For the present study, the laminar flow of viscous incompressible fluid without body forces of a plasma layer and a core layer with viscosity coefficients μ_p and μ_c having density ρ_p and ρ_c, respectively, through a stenosed tube with axial symmetry has been considered (Fig. 1). All the basic equations are considered in cylindrical coordinates (r, θ, z), where z is the direction of flow along the axis of the tube. The radial and tangential components of the velocity are zero. Due to axial symmetry of the flow v_z will be independent of θ. The expression for an axially symmetric stenosis surface is

$$\frac{R}{R_0} = 1 - \frac{\delta}{2R_0}\left(1 + \frac{\pi z}{z_0}\right) \tag{1}$$

and geometry for the stenosed two layered flow is given as follows:

We have $\qquad v_r = 0, \; v_\theta = 0, \; v_z = v(r, z, t), \; p = p(r, z, t)$ \hfill (2)

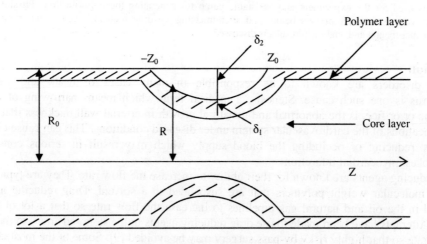

Fig. 1. Stenosed two layered flow.

We also assume that the plasma layer is of uniform thickness δ_1 therefore, the equation of continuity and equation of motion in plasma layer and core layer transformed to

$$\frac{\partial v}{\partial z} = 0 \tag{3}$$

$$\frac{\partial p}{\partial r} = 0, \quad \frac{1}{r}\frac{\partial p}{\partial \theta} = 0 \tag{4}$$

$$\frac{\partial v}{\partial t} + v\frac{\partial v}{\partial z} = -\frac{1}{\rho}\frac{\partial p}{\partial z} + v\left(\frac{\partial^2 v}{\partial r^2} + \frac{1}{r}\frac{\partial v}{\partial r} + \frac{\partial^2 v}{\partial z^2}\right) \tag{5}$$

where v is the velocity of fluid, p the pressure, t the time, r the radial distance and ρ is the density.

From Eq. (3), v is a function of r and t only, while it is clear from Eq. (4) that p is a function of z and t only. Therefore, Eq. (5) reduces to

$$\frac{\partial v}{\partial t} = -\frac{1}{\rho}\frac{\partial p}{\partial z} + \frac{v}{r}\frac{\partial}{\partial r}\left(r\frac{\partial v}{\partial r}\right) \tag{6}$$

For a pulsatile flow, the expressions for the pressure and velocity are

$$\frac{\partial p}{\partial z} = -P(z)e^{i\lambda t}, \quad v(r,t) = V(r)e^{i\lambda t} \tag{7}$$

Boundary Conditions
In the polymer layer

$$\begin{aligned} v_p &= 0 \quad \text{at} \quad r = R(z) \quad -z_0 \leq z \leq z_0 \\ v_c &= 0 \quad \text{at} \quad r = R_0 \quad |z| \geq z_0 \end{aligned} \tag{8a}$$

In the Core Layer (δ_1 is uniform thickness)

$$\begin{aligned} v_c &= v_p \quad \text{at} \quad r = R(z) - \delta_1 \quad -z_0 \leq z \leq z_0 \\ v_c &= v_p \quad \text{at} \quad r = R_0 - \delta_1 \quad |z| \geq z_0 \end{aligned} \tag{8b}$$

Applying boundary conditions and using Eq. (7) in (6) we get expression for velocity in the polymer and core layer as

$$v_p(r,t) = -\frac{iPR^2}{\alpha^2 \mu_P}\left[1 - \frac{J_0\left[i^{3/2}\alpha s\right]}{J_0\left[i^{3/2}\alpha\right]}\right]e^{i\lambda t} \tag{9}$$

$$V = -\frac{iPR^2}{\alpha^2 \mu_P}\left[1 - \frac{J_0\left[i^{3/2}\alpha s\right]}{J_0\left[i^{3/2}\alpha\right]}\right] \tag{10}$$

where $\alpha^2 = \frac{\lambda \rho_P}{\mu_P}R^2$ and $s = \frac{r}{R}$.

$$v_c(r,t) = \left[\frac{J_0\left[i^{3/2}\alpha s\right]}{J_0\left[i^{3/2}\alpha\left(1-\frac{\delta_1}{R}\right)\right]} \left(V_1 + \frac{iP}{\lambda\rho}\right) - \frac{iP}{\lambda\rho_c} \right] e^{i\lambda t} \qquad (11)$$

where $\alpha^2 = \frac{\lambda\rho_C}{\mu_C}R^2$, $s = \frac{r}{R}$ and V_1 is the velocity at the intersection of the two layers which is obtained from Eq. (10) by putting $r = R - \delta_1$. The expression from the flow rate is obtained as

$$Q = \frac{\pi P R^4}{8}\left\{\frac{1}{\mu_c}\left(1-\frac{\delta_1}{R}\right)^3 + \frac{2}{\mu_p}\left(1-\frac{\delta_1}{R}\right)^2\left[1-\left(1-\frac{\delta_1}{R}\right)^2\right] + \left[\frac{1}{2} - \left(1-\frac{\delta_1}{R}\right)^2\right]\left\{1 - \frac{1}{2}\left(1-\frac{\delta_1}{R}\right)^2\right\}\right\} \qquad (12)$$

Experiment

In the experiment, peristaltic pump PP20 has been used to generate the pulsatile flow in the model of the stenosed glass tube. A stenosed glass model was used as the test model to study the flow rate increase due to the addition of drag reducing agents in the flow. The fluid used for the experiment is plain water. Polyacrylamide and polyethylene oxide were used as drag reducing agents. These drag reducing agents (DRAs) are injected in the flow by using a syringe. The viscosity of the DRAs was measured by Ostwald viscometer. Two pressure transducers were inserted at both the ends of glass model to measure pressure at both ends.

Results and Discussion

Figures 2 and 3 depict the relation between flow rate and position in the stenosed glass tube when palyacrylamide and polyethyleneoxide, respectively, were used as a drag reducing agent. From Figs. 2 and 3, it is clear that as the stenosed radius increases, the flow rate also increases and it attains a maximum value when $z = z_0$, which is quite obvious from the geometry of the stenosis. Afterwards, the flow rate attains a constant value.

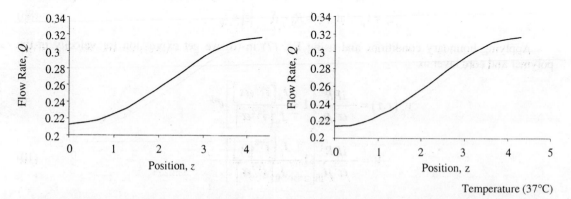

Fig. 2. Flow rate vs. position for polyacrylamide.   Fig. 3. Flow rate vs. position for polyethyleneoxide.

Temperature (37°C)

For the experiment we used plain water. Polyacrylamide and polyethyleneoxide were used as drag reducing agents. Two solutions were prepared by dissolving 10.2 g of polyacrylamide in 100 ml of water and 0.2 g of polyethyleneoxide in 75 ml of water. These two solutions of DRAs were continuously injected in the silicon tube at the entrance using a syringe. The viscosity of these two solutions was measured with Ostwald viscometer. To determine the density of the two solutions, solutions were taken in a 10 ml beaker. Then, the beaker was weighted empty as well as with the solutions on an electronic weighing machine in the Chemistry Department. The density of the polyacrylamide (PAA) solution was measured as 1.049529 g/cm^3 and that of polyethyleneoxide (PEO) as 1.00267 g/cm^3.

Table 1. Experimental values

S. No.	ρ_w (gm/cm^3)	μ_w (cp)	R_0 (cm)	R (cm)	Stenosis Length (cm)	δ_1 (cm)	δ_2 (cm)	μ_{PAA} (cp)	μ_{PEO} (cp)
1	1.03	1.005	0.75	0.68	9.6	0.001	0.07	1.904483	1.8840

With repeated number of trials in the Ostwald viscometer, the viscosity of PAA and PEO was calculated to be 1.904483 and 1.8840 cp, respectively.

The diameter of various glass and silicon tubes involved in the experiment was measured with the help of Vernier Caliper. The radius of unstenosed tube was found to be 0.75 cm and that of stenosed glass model was 0.68 cm. Therefore, the stenosed height is 0.07 cm. The radius of the silicon tube used in the experiment was measured to be 0.3 cm while the length of the stenosed glass model was measured to be 9.6 cm.

We have taken photographs with the help of Digital Camera (Sony Handy Cam) to show the changes in the flow before and after injecting the DRAs. The DRA solution will form a layer adjacent to the wall of the tube. The formation of the two layered flow has been clearly shown in Figs. 4 and 5.

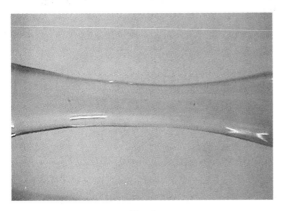
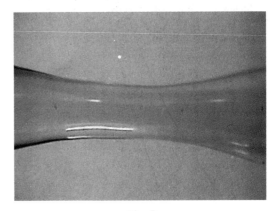

Fig. 4   Fig. 5

Calculation of the flow rate shows that an increase in the flow rate occurs when the DRAs are injected in the flow. When PAA was injected in the flow, flow rate increased by 0.1767% analytically, while experimentally an increment of 0.167507% was observed.

When PEO was injected in the flow, it was found, analytically, that flow rate increased by 0.18693%, while the experimental observations reveal a flow rate increment of 0.19584%. In both the cases, it has been found that analytical and experimental values are in close agreement.

The pressure difference has been measured by pressure transducers before and after injecting the polymers. The value for plain water is found to be 125.9621 dyne/cm^2, while it was 123.2379 and

124.5312 dyne/cm² for polyacrylamide and polyethyleneoxide, respectively. The local drag reduction has been calculated by using

$$\frac{\Delta p_w - \Delta p_p}{\Delta p_w} \times 100$$

where Δp_w and Δp_p are the pressure drops without and with polymer, respectively, and have been found as 2.162 and 1.1398% for PAA and PEO, respectively.

References

1. Cavalcanti, S. 1995. Hemodynamics of an Artery with Mild Stenosis. *J. Biomechanics*. **28**: 387.
2. De Angelis, E., Casiciola, C.M., L'vov, V.S., Piva, R. and Procaccia, I. 2003. Drag Reduction by Polymers in Turbulent Channel Flows: Energy redistribution between invariant empirical mode. *Phy. Rev.* **E 67**: 056312.
3. Haldar, K. 1985. Effects of the Shape of Stenosis on the Resistance to Blood Flow through an Artery. *Bull. Math. Biol.* **47**: 545.
4. Kim, J. and Lim, J. 2000. A Linear Process in Wall-bounded Turbulent Shear Flows. *Phy. of Fluids*. **12(8)**.
5. Kapur, J.N. 1992. *Mathematical Models in Biology and Medicine*. East West Press. p. 326.
6. Long, Q., Xu, X.Y., Ramnarine, K.V. and Hoskins, P. 2001. Numerical Investigation of Physiologically Realistic Pulsatile Flow through Arterial Stenosis. *J. Biomechanics*. **34**: 1229.
7. Sawchunk, A.P., Unthank, J.L. and Dalsing, M.C. 1999. Drag Reducing Polymers May Decrease Atherosclerosis by Increasing Shear in Areas Normally Exposed to Low Shear Stress. *J. Vasc. Surg*. **30 (4)**: 761-764.
8. Berger, T.W., Kim, J., Lee, C. and Lim, J. 2000. Turbulent Boundary Layer Control Utilizing the Lorentz Force. *Phy. of Fluids*. **12 (3)**.
9. Unthank, J.L., Lalka, S.G.J.M., Nixon, J.C. and Sawchunk, A.P. 1992. Improvement of Flow through Arterial Stenoses by Drag Reducing Agents. *J. Surg. Res*. **53 (6)**: 625-630.

17. Thermodynamic Properties of Bone

Rinku Singh, Reeva Gupta, L.M. Aggarwal, A. Koul and D.V. Rai
Department of Biophysics, Panjab University, Chandigarh - 160014, India

Abstract: Thermophysical properties of bone and its composites have been investigated in the present study. Goat femurs were obtained from local abattoirs and machined into desired dimensions. The samples were divided into control, deproteinized (apatite) and demineralized (collagen) groups. Thermal gravimetry analysis (TGA) and differential scanning calorimetry (DSC) were performed on these samples using TGA-851E, Metallar Toledo and DSC-821E, Metallar Toledo respectively. Results showed a decrease in weight in the bone and its composites in both test groups, i.e. 27.09%, 49.56% and 26.65% at 500°C respectively. Similarly, the data on DSC showed significant change in exogenous behavior of these samples. The changes in enthalpy pattern were found to be greater in apatite than in collagen i.e. −38.25 mJ and −618.25 mJ in range of 600-700°C. The values for entropy and free energy changes were also calculated in these samples. A possible mechanism underlying the said thermal phenomenon is discussed further in the article.

Introduction

Considerable attention has been given to the development of demineralized bone matrix (DBM) as osteogenic material. It has potential of osteoinductive features and is considered as a good substitute in bone related therapy. It is biocompatible, biodegradable, osteoinductive, cost effective, readily available from approved tissue banks and suitable for biomedical application. Bone composites, i.e. collagen and hydroxyapatite have been recognized as natural biocompatible materials (Liu, de Wign and Van Blitterewijck, 1998). Autogenous bone is the ultimate choice for repair of defective bones. Bone material show interactions of different components with change in temperature (Liu, de Wign and Van Blitterewijck, 1998). The demineralization process destroys antigenic substances present in the bone tissue that make the DBM less immunogenic than calcified allograft. Grossly visible but imprecise signs of thermal damage in tissues include apparent drying, shrinkage, whitening, disruption, charring, combustion and ablation.

Collagen fiber coatings are being used to increase the osteoinductive and osteointegerative properties of material coated with sintered apatites (Danielsen, 1990; Danielsen and Andreassen, 1998; Vogel, 1998). It has been reported that calcification increases the thermal stability of collagen fibers (Kronick and Cooke, 1996). The increased crystallinity reduces the solubility of bone mineral and thus affects the remodelling behavior in bone when used as bone substitute (Doi et al., 1993). Using sintered apatite coating performed at high temperature increases the osteointegrating properties of bone implant. The increased crystallinity can affect the osteointegration of hydroxyapatite coated implants (Bagambisa et al., 1993). Thermal gravimetry has been used in the kinetic analysis of polymer stability, compositional analysis of multicomponent materials, atmospheric analysis and corrosion studies, moisture and volatile determinations and accelerated tests of aging. DSC is quite useful in determining the compositional phase changes in bone samples. It also gives quantitative interpretation of entropy and free energy behavior of the samples.

The thermal properties are important because demineralized bone grafts are often stabilized with bone cements, which sets up exothermic reactions. The present study has investigated the multiphase behavior of bone composite material in terms of thermal characteristic. The thermal stability of bone, collagen and apatite was determined in terms of weight loss with temperature under static heat treatment and thermal decomposition behavior of bone and composites in terms of enthalpy, entropy and free energy under dynamic heating were also calculated. An attempt has been made to explain thermokinetic behavior of bone and its composites.

Material and Methods

Five pairs of goat femur bone were obtained from local abattoirs. The soft tissues were excised and cleaned with standard samples. The samples were machined for dimensions 10×6×5 cm and divided into control and experimental groups. The experimental groups were categorized into collagen and apatite samples. All samples were made into fine powder using mortar and pestle. The powdered form of control and experimental i.e. demineralized and deprotenized bone were placed in crucibles and their initial weight recorded. The weight loss was continuously recorded as a function of temperature with TG-assemble (851E Mettler Toledo). Temperature was increased at a constant rate of 10°C/min and upto 1000°C. A moving light point indicating an increase of temperature and corresponding weight loss in the material was recorded on paper to develop TG-analysis of the sample. Differential scanning calorimetry of the same assembly was used to monitor the enthalpy changes in these bone samples using model (821E Mettler Toledo). In this technique, the sample and reference materials were subjected to a precisely programmed temperature change. When a thermal transition (a chemical or physical change that results in the emission or absorption of heat) occurred in the sample, thermal energy was added to either the sample or the reference containers in order to maintain the sample and the reference at the same temperature. Linear heating rates from 5-10°C were maintained. The moving light point demonstrated different thermal peaks indicating enthalpy changes in bone samples. Computer software was involved in kinetic interpretations and processes in the samples.

Results and Discussions

Thermal gravimetric analysis (TGA) provided a quantitative measurement of any weight changes associated with thermally induced transitions (Figs. 1 to 3). Thermal gravimeter recorded directly loss in weight as a function of temperature or time (when operating under isothermal conditions) for transitions that involved dehydration or decomposition. Thermogravimetric curves presented a unique sequence of physical transitions and chemical reactions occurring over different temperature ranges, i.e. 50-500°C.

Table 1 shows percentage rate loss with continuous rise in temperature in the range of 50-150°C, 270-370°C and 400-500°C, respectively. The data suggested that control bone showed appreciable loss of organic component in the range 50-150°C as compared to collagen and apatite, which was 10.98 and 8.30% with in the same range of temperature. The result may be attributable to the presence of free water in the hydroxyapatite channels, which evaporated initially at this range, whereas the rate of loss measured for collagen and apatite was 38.75 and 17.5% in the range of 270-370°C as compared to 9.13% observed in control bone within this range. This could be explained on the basis of bound water and certain oxides present in the hydroxyapatite channels. In the range of 400-500°C, rate of loss for control bone was 96.5% as compared to collagen and apatite samples, which were 19.23% and 8.73%.

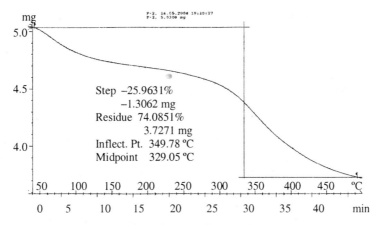

Fig. 1. Plot of weight loss in bone with temperature.

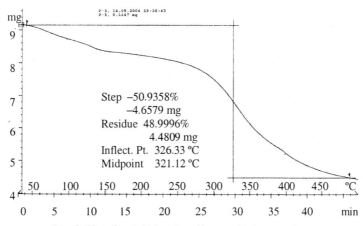

Fig. 2. Plot of weight loss in collagen with temperature.

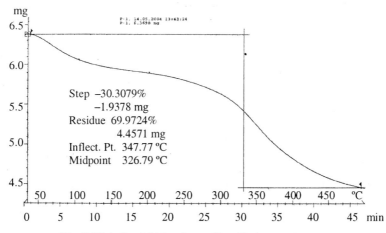

Fig. 3. Plot of weight loss in apatite with temperature.

Table 1. Gravimetric study of bone and its composite

Sample	Percentage Rate Loss		
	50-150°C	270-370°C	400-500°C
Control	20	9.13	96.5
Collagen	10.98	38.75	19.23
Apatite	8.30	17.5	8.73

This may be explained on the basis of abrupt volatility of oxides, phosphates and carbonates of these minerals at these temperature ranges. The extent of demineralization showed significant weight loss with progressive rise in temperature (Kronick and Cooke, 1996). The progressive loss of bone strength due to erosion of minerals may not be able to maintain bone elasticity and degradation at different temperatures. This may be due to loss of water molecules (Doi et al., 1996) from its network and loss of phosphorous in the form of phosphoric acid, which increases its degradation at higher temperature.

The bone crystalline network was maintained upto 250-300°C, but got disturbed at higher temperatures as shown in Table 2, which suggests the loss of crystalline pattern at more than 400-500°C range. TGA analysis confirms that the thermal stability is more pronounced in control bone and it continues to decrease with the extent of demineralization, which may be explained on the basis of reduction in the content of its hydroxyapatite pattern. The hydroxyapatite provides rigidity to bone composition and structure, which together with the protein content provides thermal stability to bone overall structure. This explains reduced thermal stability with increase in demineralization of bone. The erosion of the hydroxide channels in which apatite crystals are embedded decreases its stability at higher temperature ranges. When the same data is compared to deproteinized bone specimens show decreased stability with excessive loss of protein component. However, at higher temperature above optimum range, starts denaturation of protein. The overall result may have influenced over the ultra structure of bone which ultimately somehow brought change in the loss of weight with increase in temperature. This is in accordance to earlier studies done by Doi et al. (1996).

Table 2. Measurement of Enthalpy changes in bone and its composites

Sample	Temp. (K)	Time (s)	Enthalpy Change ΔH (mJ)	Entropy ΔS (mJ/K)	Free Energy ΔG (mJ)
Control Bone	315	358	−360	−2309.33	−0.02
Apatite	343	319	−2.009.24	−5.857	−0.789
	623	1950	−618.35	−0.9925	−0.03
Collagen	348	330	−873.41	−2.509	−0.278
	418	720	−425.84	−1.018	−0.316
	683	38	−38.256	−0.056	−0.02

DSC curves of bone, apatite and collagen shown in Figs. 4, 5 and 6 can measure both the temperature and the enthalpy of a transition or the heat of a reaction. It is often substituted for differential thermal analysis as a means of determining these quantities except at higher temperatures. The differential increase in number of thermal peaks was observed in bone, apatite and collagen.

Thermodynamic Properties of Bone   129

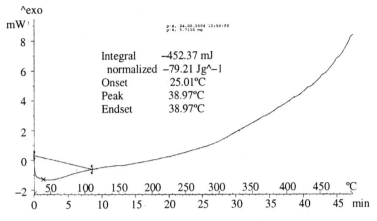

Fig. 4. DSC thermogram of control bone.

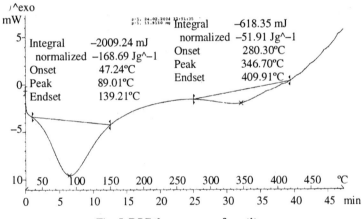

Fig. 5. DSC thermogram of apatite.

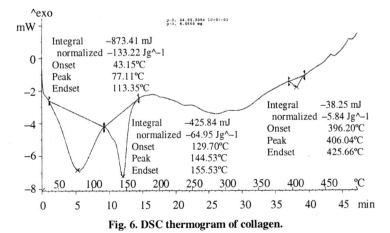

Fig. 6. DSC thermogram of collagen.

From the results shown in Table 2, the value of enthalpy change is found to be greater in case of collagen. The value of ΔH is very less for control bone as well as for apatite due to their structural framework. In case of collagen the mineral loss facilitates more enthalpy change, which is characterized by three peaks as compared to one peak in control and two peaks in apatite. This study relates similar conclusion as earlier studies in alloys by Li et al. (1999). DSC results confirm that biological processes are involved in heat change relationship based on the spontaneity of the process. In biological system such as bone enthalpy change is least for control bone. As decomposition of bonds of hydroxy channels in its apatite structure is followed by exothermic process where heat change is negative due to evolution of heat. As the process of demineralization occurs decomposition of various bonds influence the exothermic process. Similarly, the change in entropy in these samples follows the same trend as enthalpy pattern. However, not much significant changes had been observed in free energy of these samples. The analysis of thermal properties of bone and its composite using thermal gravimetric analysis and differential scanning calorimetry could be helpful in explaining the possible physiochemical mechanism in bone and its composites along with compositional phase changes during thermal induced transitions.

Acknowledgements

We express our sincere thanks to Dr. R.S. Aggarwal, for extending the facility of TGA assembly. The technical help of Mr. Vikas and Mr. N.P. Singh is also recognized.

References

1. Bagambisa, F.B., Joos, U. and Schilli, W. 1993. Mechanisms and Structure of Bond between Bone and Hydroxy Apatite Ceramics. *J. Biomed. Mater. Res.* **27**: 1047-1055.
2. Danielsen, C.C., Li Mosckilde, Bollerslew, J. and Le Mosekilde. 1994. Thermal Stability of Cortical Bone Collagen in Relation to Age is Normal Individuals with Oseoporosis. *Bone.* **15(1)**: 91-96.
3. Danielson. 1990. Age Related Thermal Stability and Susceptibility to Proteolysis of Rat Bone Collagen. *Biochem. J.* **272**: 697-701.
4. Danielson, C.C. and Andreassen, T.T. 1988. Mechanical Properties of Rat Tail Tendon in Relation to Proximal Distal Sampling Positions and Age. *J. Biomech.* **21**: 207-212.
5. Doi, Y., Koda, T., Wakamat Sun, Goto, T., Kainemizuh, Moriworki, Y., Adachi, M. and Suwa, Y. 1993. Influence of Carbonates on Sintering of Apatites. *J. Dent. Res.* **72**: 1279-1284.
6. Kronick, P.L. and Cooke, P. 1996. Thermal Stabilization of Collagen Fibrils by Calcification. *Connect Tiss. Res.* **33(4)**: 275-282.
7. Li, B.Y., Rong, L.J., Luo, X.H. and Yi-Li. 1999. Transformation Behaviour of Sintered Porous Ni-Ti Alloys. *Metallurgical and Material Transactions.* **A Vol. 30 A**: 2753-2755.
8. Li, B.Y., Rong, L.J. and Li, Y.Y. 1998. *J Mater. Res.* **Vol. 13**: 2847-2851.
9. Liu, Q., de Wign, J.R. and Van Blitterewijck, C.A. 1998. A Study on Grafting Reaction of Isocynates with HA Particles. *J. Biomed. Mater. Res.* **40(3)**: 358-364.
10. Murugan, R. and Ramakrishna, S. 2004. Modification of Demineralized Bone Matrix by a Chemical Route. *I Mater. Chem.* **14**: 2040-2045.
11. Rai, D.V. and Singh, K.V. 2000. Effect of Mineral Loss on Elemental Composition and Thermostability of Bone Collagen. *J. PAS.* **2(1)**: 15-18.
12. Reddi, A.H. 1998. Role of Morphogenetic Protcins in Skelctal Tissue Engineering and Regeneration. *Nature Biotechnal.* **16**: 247-52.
13. Saha, S. 1986. *Biomedical Engineering vs Recent Developments.* Pergamon Press, New York.
14. Vogel, H.G. 1998. Influence of Maturation and Age on the Mechanical and Biochemical Parameters of Connective Tissue of Various Organs in the Rat. *Connect Tiss. Res.* **6**: 161-166.

Biomechanics
R.K. Saxena and P. Mishra (Editors)
Copyright © 2005, Anamaya Publishers, New Delhi, India

18. Stress Analysis of Human Hip Joint Using CT-Scan Data

Dibyendu Chakraborty, Bidyut Pal and Subrata Pal
School of Bio-Science and Engineering, Jadavpur University, Kolkata - 700032, India

Abstract: Stress analysis of any human joint is the basic requirement for any prosthesis design. Today, whatever prostheses are available mostly are based on the stress analysis of 2-D system using conventional design procedure and on the western lifestyle and osteological parameters. Their lifestyle, working posture etc. are different from the Asian people in general. Asians sit cross-legged, squat for household scores, harvesting and villagers do lot of bicycling regularly. Again, their socioeconomic status is also different. All these demand prostheses of different design. To fulfill that demand we have to undertake a detailed stress analysis of natural joints in a different approach than the existing, considering the above factors. So, 3-D stress analysis of normal human hip joint is done using CT scan data. The CT slices were reconstructed using software and then transferred to ANSYS, where the detailed stress analysis was conducted with exact force application at right areas. In this method, exact muscle and joint force positioning is possible, as it provides accurate biomechanical model. Also, as it is taken from direct CT images, accurate anatomy of the joint can be obtained and load can be distributed on the appropriate quadrant. Hence, the prosthesis designed with the help of this method would be tailor made for the Asian people. The prosthesis can also be prepared using Rapid Prototype Technology.

Introduction

The need for better understanding of the complicated mechanical aspect of joints and joint replacements has given rise to collaboration between orthopedic surgeons and engineers in the field of biomechanics and biomaterials. The purpose of this investigation is to develop a better understanding of the mechanical aspects of the musculo-skeletal system and provide criteria on the mechanical aspects of diagnostic methods and surgical interventions.

In the musculo-skeletal system, hip joint is one of the most vital components. The lower extremities are connected to the upper portion of the body through pelvis and hip joint. The lower extremities not only support the central mass or the upper extremities but are also subjected to high variable forces, generated via repetitive contacts the foot makes with the ground. Also, at the same time, it offers the mobility to human beings by increasing the range of motion in the lower extremities. The pelvic girdle and hip joints are the part of a closed kinetic chain system, whereby, forces travel up through the hip and pelvis to the trunk, or down from the trunk through the pelvis and the hip to the knee and foot and ground. Thus, hip joint is subjected to all these loads, which make it as the most frequently affected human joint.

Due to such nature, artificial hip joint study has gained a lot of attention. It has a long history of more than 100 years. To design the most effective and user-friendly artificial hip joint prosthesis the designers has done rigorous stress analysis of hip joint. Still, whatever prostheses are available are based on two dimensional (2-D) stress analyses with conventional design procedure, which were designed for the western population, keeping western lifestyle in view and taking western osteological parameters in consideration. But all these are quite different from the Asian people. Their lifestyles, socio-economic condition, living requirements, working postures are all of different nature from the western people. Asians sit cross-legged, squat for household scores, daily living, harvesting and villagers have to do lot of bicycling regularly. All these demand prostheses of different design.

Now, to fulfill that demand a detailed stress analysis of normal human hip joint is undertaken considering all the aforesaid factors. So, a complete three-dimensional (3-D) stress analysis is performed using Computerized Tomography (CT) scan data. The rapid development of spiral (helical) CT has resulted in exciting new applications for CT. One of these applications is 3-D volume rendering from the CT scan data. 3-D volume rendering is proving to be far more than just a solution in search of a problem. One of the greatest advantages of spiral CT with 3-D volume rendering is that it provides all the necessary information in a single radiological study in cases that previously required two or more studies. 3-D volume rendering generates anatomically accurate and immediately available images from the full CT data set without extensive editing, giving all the parameters, joint axes, mechanical axes retroversion etc.

The purpose of this study is to develop a 3-D volume of the femur, which is the long shaft like bone in hip joint that is exposed to all the forces, then, to create a finite element solid model of that volume and after that performing stress analysis. Also, it is tallied with the existing data. This analysis not only offers exact 3-D volume of the femur, and exact force and constraint application, but also offers flexibility to apply different mechanical properties at required positions, i.e. cortical and cancellous bone properties, including anisotropy. It will offer far more accurate result than the previous ones, taking all aspects in consideration.

Steps Followed for Solution

1. CT scan data of one subject's hip joint was collected in dicom (*.dcm) format in compact disc (CD).
2. The data was saved into the computer.
3. Then, it was imported to image processing software MIMICS (demo version 8.11).
4. In MIMICS, the dataset was stacked according to the order of slices.
5. The orientation of the slices was to be given manually.
6. MIMICS automatically stack the slices in order of selection to give a complete picture of the portion of the CT scan.
7. Now, according to the requirement of the analysis, thresholding is to be done. In this case, as the bony parts were only required, the thresholding was done accordingly (Hounsfield unit were to be selected between 1200 and 2800) [7].
8. Then, the required portion, i.e. femur was selected from each slice from each of the three views (coronal, sagital and transverse plan).
9. For the selected portion, a 3-D volume was calculated.
10. That selected portion was then exported as DXF (*.dxf) format.
11. The DXF file was then opened in MECHANICAL DESKTOP software, from where it was exported as IGES (*.igs or *.iges) format.

12. The IGES file was then imported in ANSYS, where all the stress analysis using finite element method was to be done.
13. The file exported in ANSYS often cannot generate volume itself, so the volume was generated from all the areas presented there, as connected loop.
14. Material property was assigned in the form of Elastic Modulus and Poisson's Ratio. Elastic Modulus was taken as 17,000 MPa and Poisson's Ratio was taken as 0.33 [2, 5].
15. Element type was selected as "structural solid of ten nodded tetrahedron" [2].
16. The volume was then meshed in finite element with defined mesh size of 15.
17. Boundary constraint was positioned at the bottom of the femur.
18. Two types of forces were taken to be acting on the femur. First one as Joint Force F_J acting on the femur due to its articulation with the acetabulum, and the other one as hip abductor group of muscle pull F_M. F_J works over almost 70% of the surface of the femur head, whereas, F_M works on femoral neck [3]. Here, F_J was taken as 4 times of Body Weight (BW) and F_M as 2.5 times of BW. Body Weight was taken as 50 kg. F_J works at 20° to the vertical axis and F_M at 17° to the vertical axis [3].
19. In the ANSYS file, required portion was selected to know the number of nodes present. From there number of node was taken and the forces were divided equally at each node for both F_J and F_M [1].
20. Then, the finite element model was given command to solve, which solves the problem and generate results in raster and vector mode and can be displayed as per the need.

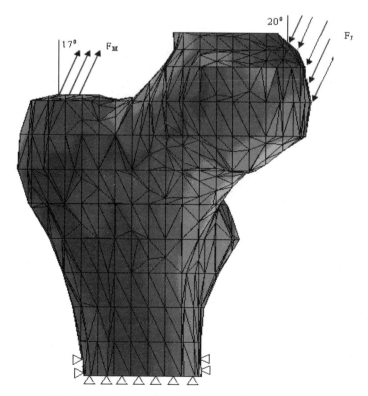

Fig. 1. Finite element model of human femur in 3-D.

Results and Discussion

The result obtained from the aforesaid solution procedure can be tallied with the results obtained from 2-D analysis. The 2-D analysis was done using the femur image of same dimension. Mesh size was the same, and as with 2-D analysis the mesh type was selected as "8 nodded quadrilateral". Also, the force values were also same, but for sake of 2-D analysis the whole force was proportionally decreased with the articulating surface area. The total force was then divided by number of node point, from selected area of force application, to get the force per node point which was then applied to each node.

In the 3-D model analysis total 9249 nos. of elements are connected through 14,026 nos. of nodes, whereas, in 2-D model analysis total 1480 nos. of elements are connected through 4643 nos. of nodes. The whole finite element stress calculations were carried out using ANSYS software. Here, emphasis is given on Von-Misses stress, shear stress in frontal, sagital and transverse plane and maximum principal stress. The stress model as well as path plot of various stresses are discussed for both 3-D and 2-D model.

As Fig. 2 describes, stress generated in the 2-D analysis is of higher order than 3-D analysis. For 3-D analysis its maximum value is 44.943 MPa and for 2-D it is 88.642 MPa. Here, the maximum value of stress is ignored as it was generated near the applied constraint, which in actual case should not be there itself. So, the next higher range is selected as highest value of generated stresses. As Fig. 2 implies, stress variation in 3-D model (Fig. 2 (a)) is much clear as well as it is more accurate due

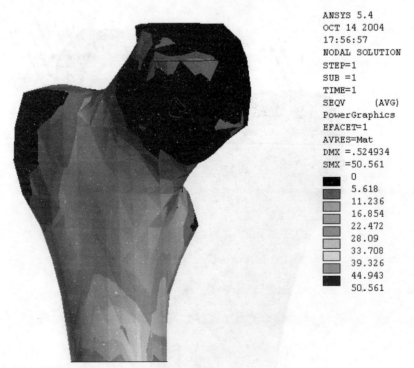

left femar of rohina molla(stress value in MPa)

(a)

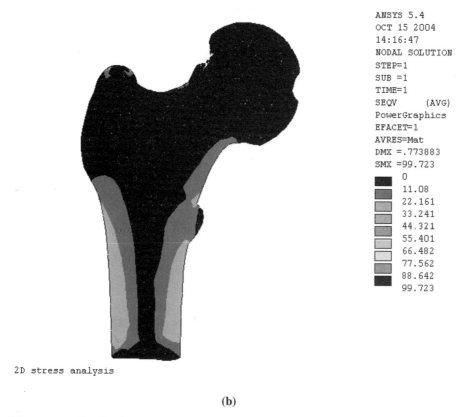

(b)

Fig. 2. Von Misses stress distribution pattern in the femur for (a) 3-D, and (b) 2-D finite element stress analysis.

to its originality. 2-D analysis is based on projected view, so added simplification comes in to the analysis automatically. It also can be seen as the joint force is acting on the femur head it is of much more mass than the other areas. Again wherever the stress concentration happens the femur is heavier in that region. It confers with the fact that nature itself generates path to overcome its problem. Fig. 2a also reveals that there is no sharp stress variation i.e. sudden stress variation is very much avoided, by the nature itself.

Here, in Figs. 3, 4 and 5, the path plots are presented in different directions for both analysis, 3-D and 2-D. It displays the stress profile along a particular line, inside the femur.

The change in the stress level can be easily seen in the be-siding figures; also the stress profiles are different though the dimension of the femur or magnitudes of the forces are same. It is very obvious due to different type of geometry of the femur in 3-D and 2-D.

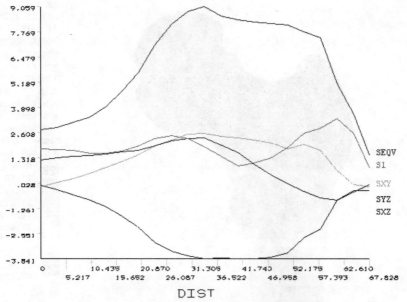

(a) Left femur of rohina molla (stress value in MPa)

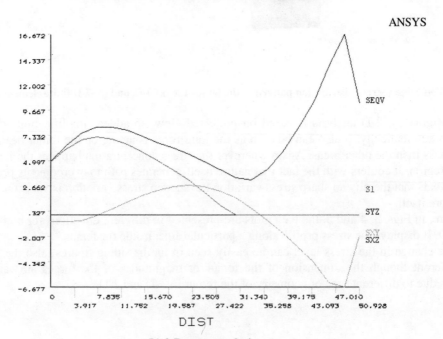

(b) 2-D stress analysis

Fig. 3. Path plot along transverse direction at femur neck (a) 3-D and (b) 2-D finite element stress analysis. SEQV = Von Misses stress in MPa; SXY = shear stress in transverse plane in MPa; SYZ = shear stress in sagital plane in MPa; SXZ = shear stress in coronal plane in MPa; S1 = maximum principal stress in MPa (here in transverse plane).

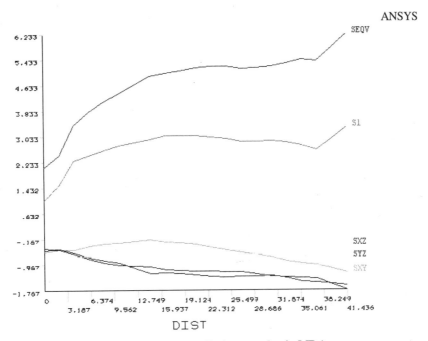

(a) Left femur of rohina molla (stress value in MPa)

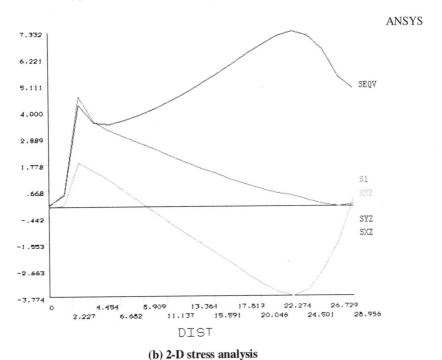

(b) 2-D stress analysis

Fig. 4. Path plot along oblique direction at femur neck (a) 3-D and (b) 2-D finite element stress analysis.

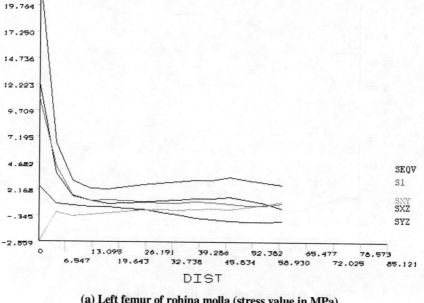

(a) Left femur of rohina molla (stress value in MPa)

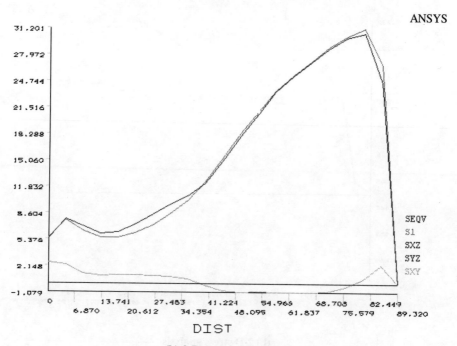

(b) 2-D stress analysis

Fig. 5. Path plot along sagital direction from the greater trochenter region (a) 3-D and (b) 2-D finite element stress analysis.

Conclusion

The discussion definitely indicates that 3-D stress analysis using CT scan data is a very powerful tool for analyzing human joints' stress condition. Consequently, it helps in prosthesis design more accurately, so that the subject to whom the implant will be fitted can live a normal life with capability to fulfill all his/her requirements. As everything is based on CT data, the whole analysis will be tailor made for the person intended. So this is just a footstep forward to improve quality of life of a patient.

Acknowledgement

The first author (DC) duly acknowledges the UGC fellowship.

References

1. Senapati, S.K. and Pal, S. 2002. Finite Element Analysis: An effective tool for prosthesis design. *Trends in Biomaterials and Artificial Organs.*
2. Majumdar, S., Roychowdhury, A. and Pal, S. 2002. Variations of Stress in Pelvic Bone During Normal Walking, Considering all Active Muscles. *Trends in Biomaterials and Artificial Organs.*
3. Harmill, J. and Knutzen, K. *Biomechanical Basis of Human Movement*, Ch. 6. p. 207.
4. Calhoun, P.S., Kuszyk, B.S., G, Jennifer, David, G., Carley, J.C. and Fishman, E.K. *Three-dimensional Volumes Rendering of Spiral CT Data: Theory and method.*
5. Dalstra, M. and Huiskes, R. 1995. Load Transfer across the Pelvic Bone. *Journal of Biomechanics.* **28**: 715-724.
6. Cook, R.D., Malkus, D.S. and Plesha, M.E. 1989. *Concept and Application of Finite Element Analysis.* John Willey & Sons, New York.
7. Mimics 8.0 Tutorial. Materialize software (demo version).

Biomechanics
R.K. Saxena and P. Mishra (Editors)
Copyright © 2005, Anamaya Publishers, New Delhi, India

19. Visualization of Brain Deformation for Neurosurgery

Akash Kumar Singh
IBM India, Bangalore, India
Computation Intelligence Leeds Metropolitan University, United Kingdom

Abstract: Monitoring the change of anatomical structures in the brain provides information that can be used in both medical research and patient management, e.g. brain change can be an indication for the progression of diseases such as Alzheimer's disease, multiple sclerosis and prion disease. Detecting these changes can improve the accuracy of early diagnosis. Additionally, the ability to track changes in tumors and the surrounding brain tissue is important for identifying the patient's response to therapies and to assess the therapeutic efficacy. In neuroscience, monitoring, for instance, the change in the growing brain of children provides insight into the process of brain development. This research focuses on analyzing brain deformation during neuro-surgical procedures. Image-guided surgery (IGS) systems are more and more routinely used in standard neurosurgical procedures. The registration of preoperative image volumes with the physical space occupied by the patient during the neurosurgery is a fundamental step when using such neuro-navigation systems. The resulting transformation or mapping is used to display the position and orientation of tracked surgical instruments on reformatted image slices and renderings of the brain and anatomical structures of interest. The overall clinical accuracy of IGS systems depends on technical accuracy of the system itself, the quality of imaging and the registration process (Maciunas et al., 1994). Furthermore, accidental movement of the head in relation to the reference frame can cause a misalignment of the physical space and the image space. Current commercial IGS systems make the assumption that a patient's head and brain form a rigid body and that the image space and physical space can be aligned by a combination of translation and rotation. While this assumption is likely to be true for the initial steps of the procedure, it has long been recognized that as the surgery progresses, there is a potential loss of validity due to brain deformation (e.g. Kelly et al., 1986). Several "ad-hoc" strategies which use the common sense of trained neurosurgeon have been suggested to reduce the error of IGS caused by brain shift, e.g. positioning of the craniotomy in the horizontal plane, avoidance of intra-operative osmotics, or attention to sites most vulnerable to displacement early in the procedure (Roberts et al., 1998). Recent studies report on significant brain deformation (brain shift) after the skull is opened and before the interventional procedure is started (e.g. Dickhaus et al., 1997; Hill et al., 1998b; Dorward et al., 1998; Roberts et al., 1998; Maurer et al., 1998; Rubino et al., 1999). If the tissue deformation is large relative to the level of surgical accuracy required, then the overall accuracy of the IGS system will be substantially reduced. Because of the widespread use of IGS systems, neurosurgeons have been increasingly interested in the nature of brain shift and the limits of IGS caused by brain deformation. For example, Prof. David G.T. Thomas from the National Hospital for Neurology and Neurosurgery in London considers deformation of brain during craniotomy as:

"[...] One of the most important topics in operative neurosurgery today. [...]. This is vital to the assessment of how accurate sophisticated neuronavigation systems may be." (Thomas, 1998).

Several approaches have been developed to address the brain deformation problem. A MR scanner that has been modified for intra-operative surgery can be used to scan the patient multiple times during the intervention and monitor the brain deformation (e.g. Schwartz et al., 1999). Alternatively, real-time ultrasound systems have been proposed to correct for the brain deformation (Iseki et al., 1994; Bucholz et al., 1997). Although these devices can provide surgeons with updated images during the intervention, the use of pre-operative data remains of clinical interest. For instance, it may be desirable to have images from other modalities (e.g., PET, FMRI) and pre-operatively prepared data (e.g., segmentations of clinically important anatomical structures) displayed in an IGS system. In particular, the fusion of these data with intra-operatively acquired images provides surgeons with additional information. Furthermore, since interventional MR scanners are expensive, complicate access to the patient and prevent the use of standard metallic surgical instruments because of their high magnetic fields, their usage is restricted and for some cases it may be preferable to use pre-operatively acquired images. Recent studies developed biomechanical models which estimate displacements in order to update the pre-operative images (Paulsen et al., 1999; Miga et al., 1999; Miga et al., 2000a and b; Miga et al., 2001).

Introduction

Castellano-Smith et al. (2001) models are based on physical brain deformation and require intra-operative measurements to constrain their model. In general, these studies assume that the brain deformation can be estimated by introducing simple physical models for the cause of deformation, e.g. direction of gravity or size of the resection (Miga et al., 2001; Castellano-Smith et al., 2001). Investigating brain deformation might indicate if their assumptions are feasible for such models and might provide adequate constrains. Altogether, investigating brain deformation during neuro-surgery is important because it:

1. Provides information for surgeon about possible brain deformation pattern during neuro-surgical procedures,
2. Yields measurements for the error of IGS systems caused by deformations,
3. Improves understanding of the cause of brain shift,
4. Potentially provides simple rules to predict or describe brain shift that could be incorporated in a IGS system in order to update, for instance, pre-operative image data, and
5. Indicates constraints for biomechanical models.

This research aims to devise computational tools to quantify and visualize brain deformation, to investigate brain deformation during neurosurgery using these computational tools, and to consider the implications of the measured brain deformation on the practice of image guided surgery.

Approach

Firstly, three-dimensional serial MR images are registered rigidly to correct for any global movement between the images. Secondly, on the basis of the calculated rigid transformation registering non-rigidly the MR images captures local changes in the brain. Finally, the deformation field defined by the resulting non-rigid transformation is analyzed both qualitatively by displaying the deformation field and quantitatively, for instance, by applying vector field operators on the deformation field or by analyzing the displacement directions. This approach is based on the assumption that the global motion between the images can be captured by rigid registration leaving only local deformation between the images. If the global motion is not sufficiently corrected, it is potentially captured by the non-rigid registration and, therefore, added to the brain shift measurements. Thus, the non-rigid registration needs a good rigid or affine starting estimate. No one has yet thoroughly investigated the way rigid or affine registration error propagates into non-rigid registration solutions. In this work we apply a well-validated rigid registration algorithm (e.g. West et al., 1997) and visually inspect individually the alignment after rigid registration for each case. Sophisticated visualization tools are necessary to inspect the registration results in order to verify their quality. Since the deformation measurements are calculated on the basis of the transformation obtained by the non-rigid registration, the investigation depends on the registration quality. Ideally, non-rigid registration is based on a one-to-one correspondence between the images, i.e. the assumption that each voxel in the source image can be mapped unambiguously to one point in the target image. As pointed out (Crum et al., 2003) the one-to-one correspondence assumption breaks down in practice, since:

1. Vowels that represent different biological structures can have the same intensity values,
2. Distinct tissue boundaries can be blurred by a partial volume effect,
3. Image noise can corrupt the true signal, and
4. Biological features might be present in one image but not in the other. This can be caused, for instance, by human intervention like resection or by the growth of a lesion.

Either matching different anatomical structures in the image or not matching of homologous structures, therefore, typically characterizes mis-registration. In our investigation, morphological changes can be divided into (a) direct interventional changes like tissue resection or craniotomy, and (b) brain deformation which can be considered as a "side-effect" of the intervention. Since there are no one-to-one correspondences between biological features at the location of the intervention, e.g. between the tissue before the resection and the space after the resection, registration algorithms are likely to result in a mis-alignment of anatomical boundaries at this location. This behavior is desirable in our investigation because it corresponds to the fact that no information about deformation is available at this point. In such a case, correspondence of surrounding tissue specifies the image transformation at this location according to the regularization scheme of the registration algorithm. In contrast, one-to-one correspondences can be established in the case of brain shift, because the biological features are only deformed but are still present in the image. Theoretically, registration algorithms can, therefore, match these deformed structures. The validation of non-rigid registration is highly important when quantitative measurements are derived from its results. Schnabel et al. (2003) suggest that non-rigid registration can be validated:

1. By simple visual inspection which can be supported by robustness and consistency checks,
2. By visual assessment techniques judged by experts, e.g. inspection of subtraction images, contour or segmentation overlays, alternate pixel displays, or
3. By quantitative measures assessing the accuracy of the registration method.

However, accuracy is often difficult or impossible to measure because of the absence of a gold standard. The non-rigid registration algorithm used in our study has been previously quantitatively validated for the registration of 3D deformed brain MR images (Hill et al., 1999) in which the displacement vectors calculated by the non-rigid algorithm were found to be consistent with the displacements of manually detected point landmarks within the precision of the manual detection. Denton et al. (1999) evaluated the registration algorithm for 3D breast MR images and Schnabel et al. (2003) compared it and validated with deformations calculated with a finite element model of the female breast. Furthermore, the algorithm works well on serial MR images of the brain (Holden et al., 2000) and serial MR images of the heart (McKleish et al., 2002). Finally, in this study, we will visually inspect the registration results for each individual case using assessment techniques as suggested by Schnabel et al. (2003).

This research will present a novel software package to investigate, quantitatively and qualitatively, brain change in 3D MR images and presents its application to the analysis of brain deformation during neurosurgery. This study aims to investigate the pattern of brain deformation with respect to location, magnitude and direction, and to consider the implications of this pattern on models for correcting brain deformation or on the errors in IGS systems due to brain deformation. Three questions are in the focus of this investigation:

1. Where is the deformation, e.g. does it occurs only at the brain surface or also in deeper brain structures?
2. What is the magnitude of the deformation?
3. Can simple rules describe or predict the deformation, e.g. loss of cerebrospinal fluid (CSF) causes the brain to sink in direction of gravity?

The novel software package provides the tools for such investigations using the approach described earlier, i.e. it combines established affine and non-rigid registration algorithms with an interactive image viewer to assess their registration results and user interfaces to visualize and analyze 3D deformation fields. Additionally, we propose a novel semi-automatic registration approach that incorporates feature information such as corresponding surfaces or points in the original non-rigid registration algorithm. We hope to demonstrate that our approach will yield considerably better registration results than the original algorithm in cases with substantial change between the images and in cases with a lack of voxel intensity correspondences.

Biomechanics
R.K. Saxena and P. Mishra (Editors)
Copyright © 2005, Anamaya Publishers, New Delhi, India

20. Effect of Physical Exercise on the Relationship between Selected Kinanthropometric Variables and Percentage Height of Centre of Gravity of Male

Dhananjoy Shaw[1] and Seema Kaushik[2]

[1]Biomechanics Laboratory, Department of Natural/Medical Sciences,
I.G.I.P.E.S.S., University of Delhi, New Delhi - 110018, India

[2]Department of Physical Education, Lakshmibai College,
University of Delhi, Delhi - 110007, India

Abstract: The purpose of the investigation was to study the relationship between selected kinanthropometric variables and percentage height of centre of gravity as an effect of exercise on male students of University of Delhi. 135 male students of University of Delhi, age ranging between 11 and 25 yrs were selected randomly for the purpose of the study. The sample was classified into three groups viz. conditioning male ($n_1 = 43$), non-conditioning male ($n_2 = 62$) and sedentary male ($n_3 = 30$). The conditioning program consisted of three meso cycles viz. M-1, M-2 and M-3 of six weeks duration each with the target training intensity (HR) of 130 ± 10, 150 ± 10 and 170 ± 10 beats/min, respectively for a session of 1 h duration (45 min for general conditioning program and 15 min for warming up and cooling down) regularly for five sessions per week. The subjects were tested four times viz. T-1 (at zero weeks of training), T-2 (after 6 weeks of training), T-3 (after 12 weeks of training) and T-4 (after 18 weeks of training). Body weight, height, sitting height, leg length, length of arm, length of foot, biacromian breadth, bicristal breadth, bitrochanterion breadth, wrist diameter, knee diameter, ankle diameter, elbow diameter, chest circumference, upper-arm circumference, fore-arm circumference, thigh circumference, calf circumference, biceps skinfold, triceps skinfold, forearm skinfold, subscapular skinfold, suprailiac skinfold, thigh skinfold, calf skinfold and percentage height of centre of gravity were the selected variables for the purpose of the study. Equipments included lever based weighing scale, a wooden board of $183 \times 100 \times 2$ cm fitted within a steel frame and fixed fulcrum, anthropometer, steel tape and skinfold caliper. The data was statistically analyzed while computing mean, standard deviation and percentage etc. The study concluded that physical exercise affects the relationship of selected kinanthropometric variables and percentage height of centre of gravity of male.

Introduction

Many research findings have shown that physical inactivity and negative lifestyle habits are a serious threat to an individual's health. Movement and activity are basic functions needed by the human organism to grow, develop and maintain health (Hoeger and Hoeger, 1990). However, vigorous physical activity is no longer a natural part of our existence. We live in an automated world where, most of the activities that used to require strenuous physical exertion can be accomplished by machines with the simple pull of a handle or push of a button. In fact, the modern life-style fosters a lack of physical fitness (Hoeger and Hoeger, 1990). Technological advances such as computers, automobiles, elevators, escalators, telephones, intercoms, remote controls, electric garage door openers etc. contribute to a sedentary life-style while minimizing the amount of movement, effort and physical exertion required by the human body (Prentice and Bucher, 1988).

One of the most significant detrimental effects of modern day technology has been an increase in chronic conditions, which are related to a lack of physical activity (e.g. hypertension, heart disease, chronic low back pain and obesity etc.).

Moreover, it is predicted that by 2020, 30% population of the world will be in the age of above 65 years, i.e. old age. Aging is associated with degenerations and is detrimental to physical and physiological well-being (due to reduced maximum aerobic power and muscle strength, i.e. reduced physical fitness). As a consequence of diminished exercise tolerance, a large and increasing number of elderly persons will be living below, at, or just above "thresholds" of physical ability, needing only a minor inter-current illness to render them completely dependent.

Among the number of sufferings/problems, persons by and large suffer from posture, stability and balance related problems. The balance, posture and stability in the whole spectrum of life are important for work, working environment and sports performances. For such circumstances, the location of centre of gravity (CG) is always found to be the most attributing factor. The height of CG and location of CG plays a vital role in controlling the posture though maintaining the natural alignments, which may positively be effected by physical training/conditioning. Hence, the role of location of centre of gravity in solving the problems of the society, job conditions, working environment as well as sports performance can't be overlooked.

For achieving excellence, one must be habitual of vigorous and strenuous training to meet the demands of hardships of life. For instance, in sports, this aim is fulfilled by means of physical training/conditioning that leads to various physical, physiological and psychological changes (Blomquist, 1983; Shephord, 1983; Monahan, 1987; Harris et al., 1989; and McHenery et al., 1990). Though the rate of change may vary from one body part to another or from one variable to another, for instance, the lengths and skeletal diameters are likely to be affected to a less extent especially at college level of students, where growth doesn't have much influence whereas, skinfold variables, circumferences, fat percentage, water percentage and lean body mass in the body etc. are likely to change in different ratios, may be at a faster or a slower pace due to the nature of activity involved or other related factors, thereby affecting the somatotypes, body composition and different indices. Such changes may lead to certain biomechanical changes also (Jensen, Schultz and Bamgartner, 1984).

Any such momentarily change is likely to deviate the location of CG at any of the three planes and axes. Due to the little deviation of CG, the unstable equilibrium of the body is disturbed, leading to the postural disturbances and may further cause postural deformities or functional impairment.

Thus, the present investigation was conducted to study the effect of physical exercise on the relationship between selected kinanthropometric variables and percentage height of CG of male.

Material and Methods

Sample
One hundred and thirty five (n = 135) male students of University of Delhi, age ranging between 17 and 25 years were selected randomly for the purpose of the study. The sample was classified into three groups namely, conditioning ($n_1 = 43$), non-conditioning ($n_2 = 62$) and sedentary ($n_3 = 30$) male.

Selection of Variables
Body weight, height, sitting height, leg length, length of arm, length of foot, biacromian breadth, bicristal breadth, bitrochanterion breadth, wrist diameter, knee diameter, ankle diameter, elbow diameter, chest circumference, upper-arm circumference, fore-arm circumference, thigh circumference, calf circumference, biceps skinfold, triceps skinfold, forearm skinfold, subscapular skinfold, suprailiac skinfold, thigh skinfold, calf skinfold and percentage height of CG were the selected variables for the purpose of the study.

Experimental Protocol
The conditioning programme consisted of three meso cycles namely, M-1, M-2 and M-3 of six weeks duration each with the target training intensity (HR) of 130 ± 10, 150 ± 10 and 170 ± 10 beats/min, respectively for a session of 1 h duration (45 min for general conditioning programme and 15 min for warming up and cooling down) regularly for 5 sessions per week where, conditioning group participated in the described conditioning programme of physical exercises along with the regular physical education programme of Indira Gandhi Institute of Physical Education and Sports Sciences (IGIPESS) curriculum, however, non-conditioning group participated in the regular physical education programme of IGIPESS curriculum only, whereas, sedentary group did not participate in any kind of physical exercise for the period of experimentation. The detailed experimental protocol is presented in Table 1.

Table 1. Experimental protocol

S. No.	Meso Cycles	Target HR/Intensity (beats/min)	Target Components
1	Meso cycle one (M-1)	130 ± 10	Flexibility, cardio-respiratory endurance
2	Meso cycle two (M-2)	150 ± 10	Muscular endurance, strength
3	Meso cycle three (M-3)	170 ± 10	Speed, power (explosive strength)

Testing Protocol
Each subject was tested four times during 4½ month of physical exercise programme viz. T-1 (at zero weeks of training), T-2 (after 6 weeks of training), T-3 (after 12 weeks of training) and T-4 (after 18 weeks of training). The detailed testing protocol is exhibited in Table 2.

Table 2. Testing Protocol

S. No.	Tests	Test Code	Time of Testing
1	Testing one (Pre-test)	T-1	Before the start of physical training/conditioning at zero weeks of training, i.e. 3rd and 4th week of July.
2	Testing two (First post test)	T-2	At the end of first meso-cycle (M-1) after 6 weeks of training, i.e. 1st and 2nd week of September.
3	Testing three (Second post test)	T-3	At the end of second meso-cycle (M-2) after 12 weeks of training, i.e. 4th week of October.
4	Testing four (Third post test)	T-4	At the end of 3rd meso cycle (M-3) after 18 weeks of training, i.e. 2nd and 3rd week of December.

Collection of Data

For the purpose of collection of data, the method explained by Sen and Ray (1983), Das and Ganguli (1982), scientifically authenticated by Shaw, Kaushik and Kaushik (1998) and validated by Shaw et. al. (1998) was strictly administered. A lever based weighing machine (Avery) with a range from 0 to 100 kg with the balancing accuracy of ±10 grams, a wooden board of 183 cm in length, 100 cm in width and 2 cm in thickness were among the equipment used to collect the data for determination of centre of gravity.

The partial body weight of the wooden board was determined (S_0). The two knife edged wooden blocks, one on the platform of the balance and other on the wooden box (placed opposite end to the balancing scale) below which the footrest was fabricated. Thereafter, the scale reading was recorded (S_0). The subjects were instructed to be flat in supine position on the board in such a way that the imaginary vertical plane passed through the longitudinal axis of symmetry of the board and the sagittal axis of the body are coincident. The feet were kept flush at the end of the board (foot rest), opposite to the direction of the weighing scale was placed. The scale reading in this position of the body was recorded (S_1).

The following equation was used to calculate the height of centre of gravity:

$$\text{Height of CG} = \frac{(S_1 - S_0)}{M} \times L$$

where S_0 = scale reading of the system; S_1 = scale reading when subject is lying flat on the system; L = length of the reaction board and M = mass of the body.

The height of centre of gravity was converted to percentage height of centre of gravity by the following formula:

$$\text{Percentage Height of CG} = \frac{\text{Height of CG (m)}}{\text{Stature/Height (m)}} \times 100$$

All the kinanthropometric measurements were taken from the left side of the individual. Standard landmarks and measurement protocols were followed to measure the selected fundamental variables (Martin and Carter, S. P. Singh, H. S. Sodhi and D. K. Kansal).

Analysis of Data

The collected data was analyzed while computing mean, standard deviation and percentage. The analysis of data pertaining to the effect of physical exercise on the relationship between selected kinanthropometric variables and percentage height of centre of gravity of male have been presented in Table 3 and illustrated (Figs. 1 to 4).

Further, a seven-point scale was prepared i.e. extremely high, very high, above average, average, below average, very poor and extremely poor to understand the relationship among the selected variables. The three seven-point scales developed for each category of the sample namely, conditioning male (CM), non-conditioning male (NCM) and sedentary male (SM) are presented in Tables 4, 5 and 6, respectively. The analysis of data in Table 3 pertaining to the effect of physical exercise on selected variables of male students of University of Delhi reveals that the variables namely weight, chest circumference, upper-arm circumference, forearm circumference, thigh circumference, calf circumference, biceps skinfold, triceps skinfold, forearm skinfold, subscapular skinfold, suprailiac skinfold, thigh skinfold, calf skinfold and percentage height of centre of gravity were observed to be decreasing in the conditioning male (CM) whereas, the variables namely height, sitting height, leg length, length of arm, length of foot, biacromial breadth, bicristal breadth, bitrochanterion breadth, wrist diameter, knee diameter, ankle diameter, elbow diameter and ankle circumference were observed to be increasing.

In regard to non-conditioning male (NCM), the variables namely biceps skinfold, triceps skinfold, forearm skinfold, subscapular skinfold, suprailiac skinfold, thigh skinfold, calf skinfold and percentage height of centre of gravity were observed to be decreasing whereas, the variables namely weight, height, sitting height, leg length, length of arm, length of foot, biacromial breadth, bicristal breadth, bitrochanterion breadth, wrist diameter, knee diameter, ankle diameter, elbow diameter, chest circumference, upper-arm circumference, forearm circumference, thigh circumference, calf circumference and ankle circumference were found to be increasing.

Table 3. Effect of physical exercise on selected variables of male students of University of Delhi

S. No.	Variable	Percentage Change		
		Conditioning Group	Non-Conditioning Group	Sedentary Group
1	Body Weight	1.13% (Decrease)	0.65% (Increase)	2.83% (Increase)
2	Height	0.22% (Increase)	0.14% (Increase)	0.17% (Increase)
3	Sitting Height	0.11% (Increase)	0.04% (Increase)	0.26% (Increase)
4	Leg Length	0.32% (Increase)	0.23% (Increase)	0.07% (Increase)
5	Length of Arm	0.25% (Increase)	0.18% (Increase)	0.33% (Increase)
6	Length of Foot	0.71% (Increase)	0.52% (Increase)	0.33% (Increase)
7	Biacromian Breadth	0.81% (Increase)	0.79% (Increase)	0.65% (Increase)
8	Bicristal Breadth	1.14% (Increase)	0.72% (Increase)	0.48% (Increase)
9	Bitrochanterion Breadth	0.83% (Increase)	0.62% (Increase)	0.63% (Increase)
10	Wrist Diameter	1.86% (Increase)	1.49% (Increase)	1.90% (Increase)
11	Knee Diameter	0.88% (Increase)	0.88% (Increase)	1.55% (Increase)
12	Ankle Diameter	1.45% (Increase)	1.31% (Increase)	1.62% (Increase)
13	Elbow Diameter	1.41% (Increase)	1.13% (Increase)	1.44% (Increase)
14	Chest Circumference	0.36% (Decrease)	0.25% (Increase)	0.97% (Increase)
15	Upper-arm Circumference	0.73% (Decrease)	1.38% (Increase)	1.32% (Increase)
16	Forearm Circumference	0.64% (Decrease)	0.73% (Increase)	1.34% (Increase)
17	Thigh Circumference	0.99% (Decrease)	1.25% (Increase)	1.72% (Increase)
18	Calf Circumference	0.60% (Decrease)	0.42% (Increase)	1.13% (Increase)
19	Ankle Circumference	1.33% (Increase)	1.90% (Increase)	1.07% (Increase)
20	Biceps Skinfold	5.26% (Decrease)	5.13% (Decrease)	11.06% (Increase)
21	Triceps Skinfold	13.92% (Decrease)	2.69% (Decrease)	11.76% (Increase)
22	Forearm Skinfold	11.13% (Decrease)	6.26% (Decrease)	4.78% (Increase)
23	Subscapular Skinfold	4.98% (Decrease)	2.99% (Decrease)	9.92% (Increase)
24	Suprailiac Skinfold	6.10% (Decrease)	2.03% (Decrease)	3.34% (Increase)
25	Thigh Skinfold	11.24% (Decrease)	1.38% (Decrease)	7.81% (Increase)
26	Calf Skinfold	9.08% (Decrease)	1.34% (Decrease)	3.59% (Increase)
27	Percentage Height of CG	4.26% (Decrease)	2.00% (Decrease)	1.30% (Increase)
	Mean (%)	3.03	1.42	2.72
	S.D. (%)	3.94	1.45	3.39

Effect of Physical Exercise on the Relationship between Selected Kinanthropometric Variables 149

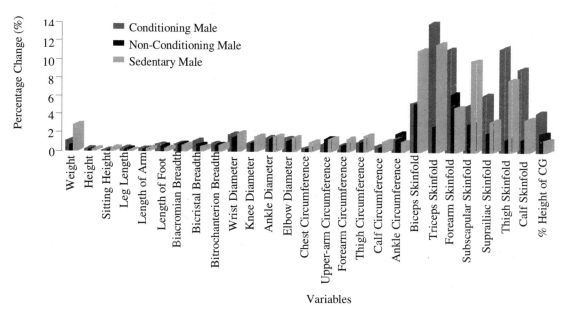

Fig. 1. Effect of physical exercises on selected variables of male students of University of Delhi.

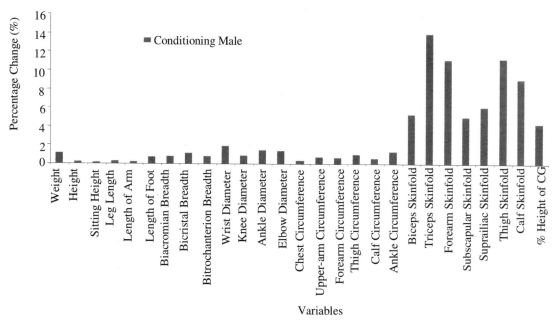

Fig. 2. Effect of physical exercises on selected kinanthropometric variables of male students (conditioning group) of University of Delhi.

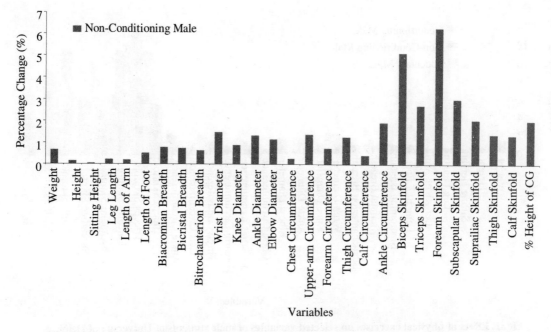

Fig. 3. Effect of physical exercises on selected kinanthropometric variables of male students (non-conditioning group) of University of Delhi.

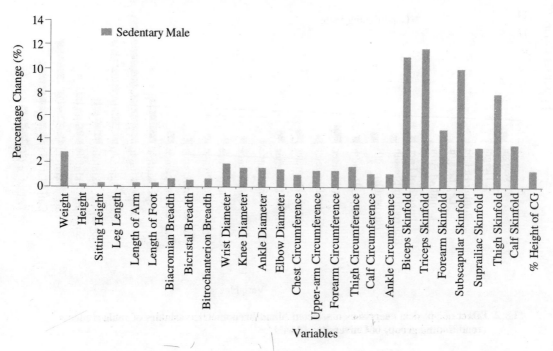

Fig. 4. Effect of physical exercises on selected kinanthropometric variables of sedentary male students of University of Delhi.

Table 4. Seven-point scale for conditioning male

S. No.	Group for Relationship	Range (Min. to Max. Value)	Variables
1	Extremely high	12.89 onwards	Triceps skinfold, forearm skinfold
2	Very high	8.95 to 12.88	Thigh skinfold, calf skinfold
3	Above average	5.01 to 8.94	Biceps skinfold, suprailiac skinfold
4	Average	1.06 to 5.00	Weight, subscapular skinfold, % height of centre of gravity
5	Below average	4.99 to −2.88	Height, sitting height, leg length, length of arm, length of foot, biacromial diameter, bicristal diameter, bitro-chanterion diameter, wrist diameter, knee diameter, ankle diameter, elbow diameter, chest circumference, upper-arm circumference, forearm circumference, thigh circumference, calf circumference, ankle circumference
6	Very poor	−2.89 to −6.82	
7	Extremely poor	−6.83 downwards	

Table 5. Seven-point scale for non-conditioning male

S. No.	Group for Relationship	Range (Min. to Max. Value)	Variables
1	Extremely high	5.06 onwards	Biceps skinfold, forearm skinfold
2	Very high	3.61 to 5.05	
3	Above average	2.16 to 3.60	Triceps skinfold, subscapular skinfold
4	Average	0.70 to 2.15	Suprailiac skinfold, thigh skinfold, calf skinfold, % height of centre of gravity
5	Below average	2.14 to −0.76	Weight, height, sitting height, leg length, length of arm, length of foot, bicristal diameter, bitrochanterion diameter, knee diameter, chest circumference, forearm circumference, calf circumference
6	Very poor	−0.77 to −2.21	Biacromial diameter, wrist diameter, ankle diameter, elbow diameter, upper-arm circumference, thigh circumference, ankle circumference
7	Extremely poor	−2.22 downwards	

Table 6. Seven-point scale for sedentary male

S. No.	Group for Relationship	Range (Min. to Max. Value)	Variables
1	Extremely high	11.21 onwards	
2	Very high	7.82 to 11.20	
3	Above average	4.43 to 7.81	
4	Average	1.03 to 4.42	
5	Below average	4.41 to −2.37	Height, sitting height, leg length, length of arm, length of foot, biacromial diameter, bicristal diameter, bitrochanterion diameter, wrist diameter, knee diameter, ankle diameter, elbow diameter, chest circumference, upper-arm circumference, forearm circumference, thigh circumference, calf circumference, ankle circumference, % height of centre of gravity
6	Very poor	−2.38 to −5.76	Weight, forearm skinfold, suprailiac skinfold, thigh skinfold, calf skinfold
7	Extremely poor	−5.77 downwards	Biceps skinfold, triceps skinfold, subscapular skinfold

In regard to sedentary male (SM), all the selected variables namely weight, height, sitting height, leg length, length of arm, length of foot, biacromial breadth, bicristal breadth, bitrochanterion breadth, wrist diameter, knee diameter, ankle diameter, elbow diameter, chest circumference, upper-arm circumference, forearm circumference, thigh circumference, calf circumference and ankle circumference, biceps skinfold, triceps skinfold, forearm skinfold, subscapular skinfold, suprailiac skinfold, thigh skinfold, calf skinfold and percentage height of centre of gravity were observed to be increasing.

Conclusions

On the basis of developed scales, the following conclusions have been drawn:

1. The percentage change in the variables namely weight and subscapular skinfold are corroborating with percentage height of centre of gravity in conditioning male (CM) and falling in the average category.
2. The percentage change in the variables namely triceps skinfold and forearm skinfold (falling in extremely high category); thigh skinfold and calf skinfold (falling in very high category); biceps skinfold and suprailiac skinfold (falling in above average category) indicating a higher change than the percentage height of centre of gravity (falling in average category) in conditioning male.
3. The percentage change in the variables namely height, sitting height, leg length, length of arm, length of foot, biacromial diameter, bicristal diameter, bitrochanterion diameter, wrist diameter, knee diameter, ankle diameter, elbow diameter, chest circumference, upper-arm circumference, forearm circumference, thigh circumference, calf circumference and ankle circumference (falling in all variables below average category) was lower than the change in percentage height of centre of gravity (average category) in conditioning male.
4. The percentage height of centre of gravity is observed to be decreasing in conditioning male. Similar trends have been observed by the variables namely weight, chest circumference, upper-arm circumference, forearm circumference, thigh circumference, calf circumference, biceps skinfold, triceps skinfold, forearm skinfold, subscapular skinfold, suprailiac skinfold, thigh skinfold and calf skinfold, whereas, reverse trends have been observed by the variables namely, height, sitting height, leg length, length of arm, length of foot, biacromial diameter, bicristal diameter, bitrochanterion diameter, wrist diameter, knee diameter, ankle diameter, elbow diameter and ankle circumference.
5. The percentage change in the variables namely suprailiac skinfold, thigh skinfold and calf skinfold are corroborating with percentage height of centre of gravity in non-conditioning male, falling in the average category.
6. The percentage change in the variables namely biceps skinfold and forearm skinfold (falling in extremely high category) triceps skinfold and subscapular skinfold (falling in above average category) demonstrates a higher change than the percentage height of centre of gravity in non-conditioning male.
7. Percentage change in the variables namely weight, height, sitting height, leg length, length of arm, length of foot, bicristal diameter, bitrochanterion diameter, knee diameter, chest circumference, forearm circumference, calf circumference (falling in below average category); biacromial diameter, wrist diameter, ankle diameter, elbow diameter, upper-arm circumference, thigh circumference and ankle circumference was lower (falling in very poor category) than the change in percentage height of centre of gravity (falling in average category) in non-conditioning male.

8. The percentage height of centre of gravity was observed to be decreasing in non-conditioning male. Similar trends have been observed by the variables namely biceps skinfold, triceps skinfold, forearm skinfold, subscapular skinfold, suprailiac skinfold, thigh skinfold and calf skinfold, whereas, reverse trends have been observed by the variables namely, weight, height, sitting height, leg length, length of arm, length of foot, biacromial diameter, bicristal diameter, bitrochanterion diameter, wrist diameter, knee diameter, ankle diameter, elbow diameter, chest circumference, upper-arm circumference, forearm circumference, thigh circumference, calf circumference and ankle circumference.
9. The percentage change in the variables namely height, sitting height, leg length, length of arm, length of foot, biacromial diameter, bicristal diameter, bitrochanterion diameter, wrist diameter, knee diameter, ankle diameter, elbow diameter, chest circumference, upper-arm circumference, forearm circumference, thigh circumference, calf circumference and ankle circumference are corroborating with percentage height of centre of gravity in sedentary male, falling in the below average category.
10. No other variable is found to be higher in percentage change than the percentage height of centre of in sedentary male (falling in extremely high, very high, above average or average category).
11. The percentage change in the variables namely weight, forearm skinfold, suprailiac skinfold, thigh skinfold and calf skinfold (falling in very poor category), biceps skinfold, triceps skinfold and subscapular skinfold is lower (falling in extremely poor category) than the change in percentage height of centre of gravity in sedentary male.
12. The percentage height of centre of gravity was observed to be decreasing in sedentary male. Similar trends have been observed by the variables namely biceps skinfold, triceps skinfold, forearm skinfold, subscapular skinfold, suprailiac skinfold, thigh skinfold and calf skinfold, whereas, reverse trends have been observed by the variables namely, weight, height, sitting height, leg length, length of arm, length of foot, biacromial diameter, bicristal diameter, bitrochanterion diameter, wrist diameter, knee diameter, ankle diameter, elbow diameter, chest circumference, upper-arm circumference, forearm circumference, thigh circumference, calf circumference and ankle circumference.

References

1. Blomquist, C.G. 1983. CV Adaptation to Physical Training. *Annual Review of Physiology.* **45**: 169.
2. Bunn, J.W. 1957. *Basket Methods.* The McMillan Company, New York.
3. Bunn, J.W. 1972. *Scientific Principles of Coaching* 2nd *ed.* Prentice Hall, Inc., Englewood Cliffs, NJ.
4. Burleigh, L., Horak, F.B. and Malquin, F. 1994. Modification of Postural Responses and Step Initiation: Evidence for goal-directed postural interactions. *Journal of Neuro Physiology.* **72(6)**: 2892-2901.
5. Charles, S. 1976. *Fundamentals of Sports Biomechanics.* Kendall/Hunt Publishing Company, Dubuque, Ia.
6. Das, R.N. and Ganguli, S. 1982. Mass and Centre of Gravity of Human Body and Body Segments. *Journal of the Institution of Engineers.* **62**: IDGE-3.
7. Das, R.N. and Ganguli, S. 1982. Moment of Inertia of Living Body and Body Segments Geometrically Shaped Timber and Sleep Specimens. *Journal of the Institution of Engineers.* **62**: IDGE-3.
8. Harris, S.S., Caspersen, C.J., Defriese, G.H. and Estes, E.H. 1989. Physical Activity Counseling for Health Adults as a Primary Preventive Intervention in the Clinical Setting. **JAMA 261**: 3590-98.
9. Hoeger and Hoeger. 1990. *Fitness and Wellness.* p. 3.
10. Jensen, C.R., Schultz, G.W. and Bangerter, B.L. 1984. *Applied Kinesiology* 3rd *ed.* Mc Graw Hill, Singapore.
11. Johnson, P. and Strolberg, D. 1971. Conditioning. Prentice Hall, Englewood Cliffs, N.J.

12. Kaushik, R., Kaushik, S. and Shaw, D. 1995. Changes in the Location of Three-Dimensional Centre of Gravity as a Result of Six-weeks Conditioning Programme on Male Athletes. *Souvenir: 5th National Conference of NAPESS and GANSF* (Delhi: October 28-29, 1995). p. 33.
13. Kawamura, T. et al. 1984. An Analysis of Somatotypes and Postures of Judoist. *Bulletin of the Scientific Studies on Judo: Report VI* (Tokyo: Kodokan, 1984): 107-116.
14. Logan, G.A. 1976. *Adaptation of Physical Activities*. Prentice Hall, Englewood Cliffs, N.J.
15. McHenery, P.L. et al. 1990. Statement of Exercise. *Special Report Circulation*. **51**: 1.
16. Monahan, T. 1987. Is Activity As Good As Exercise. *The Physician and Sports Medicine*. **15(10)**: 181.
17. Ray, G.G. and Sen, R.N. 1983. Determination of Whole Body Centre of Gravity in Indians. *Journal of Human Ergo*. **12**: 3-4.
18. Shephord, J.R. 1983. Employee Health and Fitness-State of the Art. *Preventive Medicine*. **12**: 644-653.

21. Effect of Cyclic Impact Mechanical Stresses on Biochemical Properties of Weight Bearing Synovial Joint Articular Cartilage Chondrocytes in Alginate Matrix

Garima Sharma[1], R.K. Saxena[1] and P. Mishra[2]
[1]Centre for Biomedical Engineering, [2]Department of Biochemical Engineering and Biotechnology, Indian Institute of Technology Delhi, New Delhi - 110016, India

Abstract: Physico-chemical stimulations of articular cartilage have been found to cause degeneration of tissue material and osteoarthritis. Degenerative joint disease or osteoarthritis (OA) is the most common form of joint disability, which affects population of all ages in all parts of the world and no permanent remedy is available for the cure of the disease. An attempt has been made to elucidate the cellular/molecular correlations between mechanical stimuli and cartilage physiology *in vitro* environment. In this study, the effect of moderate cyclic pressure on biochemical and synthetic behavior of articular chondrocytes cells isolated from femaropatellar joint of young goat and encapsulated in alginate matrix has been studied. Alginate matrix allows cells to maintain their normal differentiated phenotype and can synthesize cartilage specific macromolecules such as proteoglycan, collagen, hyaluronic acid etc. in culture media. The viability of chondrocytes in alginate beads has been studied for up to 7 days. The cellular biosynthetic properties have been measured and found to be normal. In this study, chondrocytes cells entrapped in alginate beads have been subjected to moderate cyclic loading (1.5 Mpa for 4 h at 0.66 Hz) by 'cyclic loading electronically controlled machine' designed and developed by the authors. The chondrocytes in alginate beads have been developed in matrix having two compartments: the cell associated matrix (CM) and further removed matrix (FRM) compartments. After loading experiments, the CM and FRM of chondrocytes were analyzed for proteoglycan, total protein, proteases and DNA content of the cells. The results of the study found that the total proteoglycan content of chondrocytes increased in all matrix compartments, whereas, a decreasing pattern has been observed in total protein and proteases level under the moderate cyclic loading. There were no significant changes observed in the rate of cell proliferation as demonstrated by DNA content of the chondrocytes under mechanical stress conditions. The results of this study indicate that moderate cyclic loading modulates the cartilage metabolism and helps in regeneration of cartilage matrix.

Introduction
Human synovial joints are responsible for the frictionless and free movement of the skeleton and locomotion. The potential functioning of synovial joints without resistance is possible by virtue of a smooth, soft and glistening connective tissue material articular cartilage, which covers the articulating surfaces of the condyles of bones of weight bearing synovial joints, as knee and hip. The main function of articular cartilage is shear less articulation and dissipation of mechanical loads by acting as

a shock absorber during mechanical stress conditions. *In vivo* cartilage is continuously exposed to moderate to high cyclic mechanical pressure during various physiological activities. The human synovial joints are subjected to an average of 1-4 million load cycles per year [1] and the stress level may rise up to 10-20 Mpa (100-200 times the atmospheric pressure) [2]. It has been realized that this level of repeated exposure to stress may cause injuries to the articular cartilage and affect the health and functional properties of physiological synovial joint. Recent epidemiological and clinical studies have found that articular cartilage fatigue resistance properties get decreased due to abnormal and high mechanical stresses and postures, and in turn can lead to a very common disorder, osteoarthritis (OA) or degenerative joint disease [3]. Various *in vitro* studies have been carried out on cartilage explants to monitor the effect of mechanical stimulations on structural and functional properties of articular cartilage [4-7]. It has been suggested that physiological or moderate level of intermittent hydrostatic pressure enhance the matrix properties of cartilage [8]. However, these studies do not elucidate the underlying mechanism of cellular response of chondrocytes to moderate intermittent mechanical loadings. It is presumed that chondrocytes may respond to mechanical signals and remodel according to their physical and biochemical demands by altering their synthetic and structural properties [9, 10]. But, how the chondrocytes perceive the mechanical signals, and utilize this feedback mechanism to alter its metabolic properties, is not yet clear. The molecular events, which are taking place during joint loading conditions at the cellular and sub-cellular level have to be understood in details, in order to identify the factors responsible for degeneration and regeneration phenomenon in articular cartilage. In our experimental investigation, we studied the effect of moderate cyclic loading on biochemical properties of lamb articular chondrocytes isolated and cultured in alginate beads. The alginate culture system provides a suitable matrix for growth and differentiation of chondrocytes and they form a matrix, which is similar to the native environment of cartilage. The matrix accumulation and turnover occurs in alginate in two distinct compartments: Cell Associated Matrix (CM) and Further Removed Matrix (FRM) [11, 12]. The changes in biochemical contents of chondrocytes (proteoglycan, total protein, DNA content and level of proteases) as a result of moderate cyclic mechanical stimulations have been measured in both the matrix compartments and results have been presented.

Materials and Methods

Isolation and Culturing of Chondrocytes from Goat Articular Cartilage

Freshly slaughtered lamb (age: 2-3 months) knee joint specimen were obtained from a local slaughterhouse within 3-4 h of slaughter, washed with 70% ethanol before removing the articular cartilage. The sterile full thickness pieces of cartilage tissue were removed and cut into very small pieces (1 mm^3). Cartilage cells were released from tissue by sequential enzymatic treatment of the cartilage matrix with papain, hyaluronidase and type II collagenase (Sigma) [13] in Dulbecco's Modified Eagle Medium (DMEM) containing glucose (4.5 g/l), glutamine (0.584 g/l), HEPES (12.5 mmol/l), proline (0.4 mmol/l), ascorbic acid (50 mg/l), streptomycin (100 µg/ml) and fetal bovine serum (Sigma) 10% at pH 7.4. [14]. The cartilage tissue pieces were incubated sequentially with 1 mg/ml of hyaluronidase and 0.5 mg/ml of papain at 37°C for 1 h each with mild shaking. Chondrocytes cells were released by overnight treatment of cartilage with 0.15 mg/ml of type II collagenase. The digestate is centrifuged at 4000 rpm for 10 min at 4°C. The supernatant was discarded and pellet was washed twice with 0.02% EDTA. The cells were finally washed with 0.9% NaCl and resuspended in DMEM containing 10% FBS and 100 µg/ml streptomycin. The viable chondrocyte cells count were determined under phase contrast microscope by haemocytometer after staining with trypan blue. The cell count was 3×10^6 cells/ml and cell viability was found to be approximately 95% after isolation.

Alginate Culture of Chondrocyte

The isolated chondrocytes were encapsulated in alginate as described by Hauselmann et al. [15]. The isolated chondrocyte suspension in 0.9% NaCl was mixed with low viscosity alginate solution (3% w/v) in 1:1. The alginate-cell suspension was slowly mixed by adding drop wise in 102 mM ice-cold $CaCl_2$ solution, through a 22-gauge needle with mild stirring. The beads were further allowed to polymerize for 10 min in $CaCl_2$ solution. After washing the beads twice in 10 volume of 0.9% NaCl, they were kept in complete DMEM containing 10% FBS at 37°C at 5% CO_2 incubator. 5 ml of medium containing approximately 10 alginate beads was used as experimental for mechanical loading studies and 5 ml of medium with 10 alginate beads containing chondrocytes was used as control for comparison, under similar conditions.

Mechanical Stimulations of Chondrocyte Cells in Alginate Matrix

The lamb chondrocytes cultures in alginate beads were cyclically loaded with a peak load of 1.5 MPa for 4 h with loading frequency of 0.6 Hz. The hydrostatic mechanical load was applied via a 'Cam regulated cyclic loading electronically controlled machine', designed and developed by the authors [16].

Brief Details of Cam Based Cyclic Loading Electronically Controlled Machine

Cam based cyclic loading electronically controlled machine is a rotary motion impact cyclic loading system for generating repeated hydrostatic pressure on the chondrocyte cells entrapped in alginate beads via an air column above the medium containing the beads. The cyclic loading is given by cam, which is attached to the shaft of a 2-phase stepper motor, which provides the driving mechanism to the system. The frequency of the load cycle is regulated by the control system of stepper motor (Fig. 1).

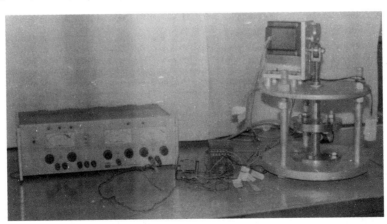

Fig. 1. The complete assembly of 'Cyclic loading electronically controlled machine'.

Details of Mechanical Loading Experiments

The sample tube of the cyclic loading tube was detached from the assembly and sterilized by autoclaving. 10 alginate beads containing articular chondrocytes were taken in 5 ml of DMEM culture medium and kept in sampling tube under aseptic conditions. On top of the medium, a layer of 2 ml of non-reactive, non-miscible, high-density paraffin oil was put to avoid direct stress effect of impact of loading on the cells. The loading experiment was performed at a thermo-neutral temperature of $37 \pm 1°C$. The control experiment was carried out by taking 10 alginate beads in 5 ml medium and keeping it in a sample tube under similar conditions without applying mechanical stimulations. The stepper motor driver card was given 23 ± 0.5 V power via DC power supply and stepper motor control card is attached to an AC power supply. A cyclic load of 1.5 MPa was applied on cells for 4 h with a frequency of 0.6 Hz. The changes in load magnitude were monitored via pressure gauge. The duration for peak load during one complete cycle was 0.5 s.

Biochemical Analysis of Chondrocytes after Loading Experiments
After loading experiment, the sample tube was detached from the assembly and the medium containing alginate beads was removed from it. The medium was stored for analysis of biochemical contents. The alginate beads were disrupted and chondrocytes were recovered.

Release of Chondrocytes by Disruption of Alginate Beads
Alginate beads containing the chondrocytes were dissolved in Ca^{2+} free medium by the protocol given by Hauselmann et al. [12]. Alginate beads (approximately 25) were added to 1 ml of 55 mM sodium citrate, 50 mM EDTA, 0.15 M NaCl and pH 6.05 at 4°C for 20 min. The resulting suspension was centrifuged at 2000 rpm for 10 min at 4°C. The pellet consisting of cell associated matrix was termed as CAM and the supernatant consists of further removed matrix (FRM). The pellet was washed twice with PBS pH 7.4.

Solubilization of Matrix Molecules and Biochemical Analysis of Matrix Compartments
The pellet consisting of chondrocytes and pericellular matrix (CM), supernatant consisting of FRM and medium, were all digested with papain at 20 µg/ml in 0.1 M sodium acetate, 50 mM Sodium EDTA, 5 mM Cysteine-hydrochloride, pH 5.53 and kept overnight at 60°C [17]. The samples were boiled for 10 min to deactivate the enzyme and stored at −20°C for biochemical analysis. The CAM, FRM and medium were analyzed for measuring the content of proteoglycan, total protein, proteases and DNA in loaded and unloaded chondrocytes.

Measurement of Total Proteoglycan Content
The content of sulfated proteoglycan in the papain digest of CM, FRM and medium for loaded and unloaded chondrocytes were measured by glycosaminoglycan (GAG)-alcian blue binding method, described by Whiteman 1972 [18]. The absorbance of the blue colored complex was measured at 620 nm. Chondrotin sulfate (1 mg/ml) was used as standard for determining the concentration.

Measurement of Total Protein Content
The total protein content in determined by Lowry's method [19] in FRM, CM and medium. The papain-digested samples of CM, FRM and medium from loaded and unloaded cells were treated with alkaline reagent and Folin ciocelateu reagent was added. The optical density of the samples was measured at 660 nm by spectrophotometer using tyrosine as standard.

Measurement of Protease Activity
The Protease activity was assayed in all compartments by the modified method of Anson [20] using casein as substrate. The reaction mixture, consisting of 2 ml of 1% casein in 20 mM borate buffer, pH

8.0 and 0.5 ml of suitably diluted enzyme from CM, FRM and medium was incubated at 45°C for 30 min. The reaction was terminated by the addition of an equal volume of 10% trichloroacetic acid and filtered through Whatman no.1 filter paper. To 1 ml of filtrate, 5 ml of 0.5 M Na_2CO_3 solution and 0.5 ml of 3 fold diluted FC reagent were added and mixed thoroughly. The color developed after 30 min of incubation at 30°C was measured at 660 nm.

One unit of enzyme activity was defined as the amount of enzyme required to liberate 1 μmol of tyrosine in 30 min at 45°C. The specific activity is expressed in units of enzyme activity/mg of protein.

Measurement of DNA content
The DNA content of the chondrocytes was measured by the method based on measurement of absorbance of colored products at 595 nm, formed by reaction of DNA with diphenylamine (DPA) in acidic conditions [21]. The calf thymus DNA in buffered saline (0.15 M NaCl and 0.015 M sodium citrate, pH 7.0) was used as standard.

Statistical Analysis of Data
Both control and experimental data were statistically analyzed and have been presented as mean ± standard deviation of the five sets of experiments. The proteoglycan and total protein contents have been expressed as per μg of DNA.

Results

DNA Content and Rate of Chondrocytes Proliferation
To assess the chondrocyte proliferation under the influence of loading, the DNA content of chondrocytes was measured in unloaded (control) and loaded (experimental) alginate beads containing chondrocytes encapsulated into them. The DNA content of chondrocytes in CAM, FRM and medium was evaluated. After loading of 1.5 Mpa for 4 h no significant change occurred in the mean value of DNA content μg/ml of the chondrocytes in all the three matrix compartments ($P > 0.1$) (Fig. 2).

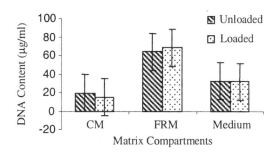

Fig. 2. Changes in the DNA content due to mechanical loading in different matrix compartments of alginate culture (mean value ± standard deviation).

The mean value of DNA content is highest in FRM in comparison to CM and medium. The chondrocytes does not show proliferation under moderate cyclic mechanical loading. The DNA content of chondrocytes has been normalized as 1 μg/0.25×10^6 cell.

Total Proteoglycan Content

The amount of total sulfated proteoglycan released from chondrocytes cultured in alginate beads varies in the various compartments of alginate beads and measured by alcian blue-GAG binding method. The PG content has been expressed as per μg of DNA. The amount of sulfated proteoglycan is approximately 6 times higher in FRM in comparison to CM and medium. In experimental, cyclically loaded (1.5 MPa for 4 h) chondrocytes, the total proteoglycan increased in all compartments, i.e. CM, FRM and medium (Fig. 3). However, the differences between the net content of proteoglycan in loaded and unloaded chondrocytes were statistically significant in CM ($P < 0.02$) and FRM ($P < 0.01$) but there occurs no significant increase in PG content of medium under cyclic loading ($P < 0.1$).

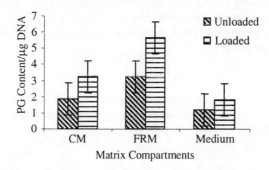

Fig. 3. Changes in proteoglycan content of chondrocytes (per μg DNA) in various matrix compartments of alginate in loaded and unloaded cells (mean value ± standard deviation).

Total Protein Content

The total protein content of chondrocytes in various matrix compartments was measured in unloaded and loaded chondrocytes. The amount of total proteins has been presented as per μg DNA. It was found that the mean value of total protein content released by chondrocytes gets increased in CM after cyclic loading at level of ($P < 0.1$), whereas, it gets decreased in the medium ($P < 0.05$). However, no significant change occurs in the protein content of FRM compartments due to mechanical stimulations in comparison to unloaded cells (Fig. 4).

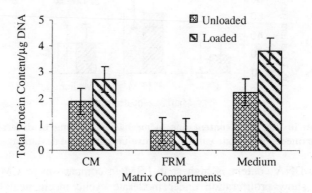

Fig. 4. Changes in total protein content of chondrocytes (per μg DNA) in various matrix compartments of alginate in loaded and unloaded cells (mean value ± standard deviation).

Specific Activity of Proteases
The specific activity of proteases released after mechanical stimulations of chondrocytes has been measured in the alginate matrix compartments and presented as µmol/mg of protein. It was found that the specific activities of proteases decrease in all matrix compartments, i.e. CM, FRM and medium in comparison to unloaded cells. The mean value of specific activity of proteases decreases significantly in CM ($P < 0.05$), whereas, the mean value of FRM and medium does not show any significant decrease after cyclic mechanical stimulations (Fig. 5).

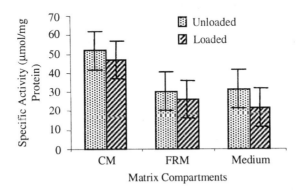

Fig. 5. Changes in specific activity of protease in chondrocytes (per μg DNA) in various matrix compartments of alginate in loaded and unloaded cells (Mean value ± standard deviation).

Discussion

The alginate matrix culture offers a suitable model for characterization and identification of metabolism of various matrix molecules under varying culture conditions. In the present study, we have demonstrated the effect of moderate cyclic pressure on the chondrocyte metabolic properties. The results presented in this study suggested that cyclic mechanical loading significantly alters the net content of proteoglycan and total proteins in various matrix compartments. Earlier studies have suggested the beneficial role of moderate cyclic loading on the health and physiology of articular cartilage of weight bearing regions [22, 23]. The results of this study suggest that under the moderated level of cyclic loading, the synthesis of proteoglycan molecule and their incorporation increases in pericellular matrix as compared to further removed matrix and medium. This may have a beneficial effect on health and functioning of the tissue and the increase in cartilage thickness. Previous *in vivo* and *in vitro* studies have demonstrated that the joint regions under moderate loading activity have higher proteoglycan content and more resilient load bearing properties [24, 25].

However, in our study, it was found that the moderate cyclic loading has less significant effect on the net content of total protein synthesized by chondrocytes in CM and FRM as compared to the medium. Also, the analysis of specific activity of proteases demonstrated that under the influence of mechanical cyclic loading, as the total protein content decreases subsequently, the level of proteases in all matrix compartments also decreases.

The results of this study suggested that the approach of using alginate matrix for chondrocyte culture holds promise as a method to study the matrix metabolism and turn over under mechanical stimulations. Further work is in progress to study the effect of high mechanical cyclic loading on the properties of articular cartilage chondrocytes so as to study the interplay of mechanical and biochemical factors in cartilage degeneration and regeneration. Our study clearly demonstrates that articular cartilage chondrocytes cells grown and cultured in alginate beads respond and remodel according to their mechanical environment and their functional and synthetic properties are governed by the mechanical stimulations of the cells.

References

1. Seedhom, B.B. and Wallbridge, N.C. 1985. Walking Activities and Wear of Prostheses. *Ann. Rheum. Dis.* **44**: 838-843.
2. Hodge, W.A., Fijan, R.S., Carlson, K.L., Burgess, R.G., Harris, W.H. and Mann, R.W. 1986. Contact Pressure in the Human Hip Joint Measured *in vivo*. *Proc. Natl. Acad. Sci. USA*. **83**: 2879-2883.
3. Dieppe, P. and Kirwan, J. 1994. The Localization of Osteoarthritis. *Br. J. Rheumatol.* **33**: 201-204.
4. Torzilli, P.A., Grigiene, R., Huang, C., Friedman, S.M., Doty, S.B., Boskey, A.L. and Lust, G. 1997. Characterization of Cartilage Metabolic Response to Static and Dynamic Stress Using a Mechanical Explant Test System. *Journal of Biomechanics.* **30**: 1-9.
5. Saxena, R.K., Sahay, K.B. and Guha, S.K. 1991. Morphological Changes in the Bovine Articular Cartilage Subjected to Moderate and High Loadings. *Acta. Anat. (Basel).* **142 (2)**: 152-157.
6. Steinmeyer, J. and Knue, S. 1997. The Proteoglycan Metabolism of Mature Bovine Articular Cartilage Explants Superimposed to Continuously Applied Cyclic Mechanical Loading. *Biochem. Biophy. Res. Commun.* **240**: 216-221.
7. Clements, K.M., Bee, Z.C., Crossingham, G.V., Adams, M.A. and Sharif, M. 2001. How Severe Repetitive Loading is to Kill Chondrocytes in Articular Cartilage. *Osteoarthritis Cartilage.* **9**: 499-507.
8. Smith, R.L., Rusk, S.F., Ellison, B.E., Wessells, P., Tsuchiya, K., Carter, D.R. et al. 1996. *In vitro* Stimulation of Articular Cartilage mRNA and Extracellular Matrix Synthesis by Hydrostatic Pressure. *J. Orthop. Res.* **14**: 53-60.
9. Handa, T., Ishihara, H., Ohshima, H., Osada, R., Tsuju, H. and Obata, K. 1997. Effects of Hydrostatic Pressure on Matrix Synthesis and Matrix Metalloproteinase Production in the Human Lumbar Intervertebral Disc. *Spine.* **22**: 1085-1091.
10. Wong, M., Wuethrich, P., Buschmann, M.D., Eggli, P. and Hunziker, E. 1997. Chondrocyte Biosynthesis Correlates with Local Tissue Strain in Statically Compressed Articular Cartilage. *J. Orthop. Res.* **15**: 189-196.
11. Mok, S.S., Masuda, K., Hauselmann, H.J., Aydelotte, M.B. and Thonar, E.J. 1994. Aggrecan Synthesized by Mature Bovine Chondrocytes Suspended in Alginate. Identification of Two Distinct Metabolic Matrix Pools. *J. Biol. Chem.* **269**: 33021-33027.
12. Hauselmann, H.J., Aydelotte, M.B., Schumacher, B.L., Kuettner, K.E., Gitelis, S.H. and Thonar, E.J.M.A. 1992. Synthesis and Turn Over of Proteoglycan by Human and Bovine Adult Articular Chondrocytes Cultured in Alginate Beads. *Matrix.* **12**: 116-129.
13. Hernotin, Y.E., Labasse, R.G. and Simonis, P.E. 1999. Effect of Nimesulide and Sodium diclofenac on IL-6, IL-8, Proteoglycan and Prostaglandin E2 Production by Human Articular Chondrocyte *in vitro*. *Clin. Exp. Rheumatol.* **17**: 151-160.
14. Freed, L.E., Hollander, A.P., Marquis, J.C. and Nohria, A. 1993. Neocartilage Formation *in vitro* Using Cell Cultured on Synthetic Biodegradable Polymers. *J. Biomed. Mater. Res.* **27**: 11-23.
15. Hauselmann, H.J., Fernandes, R.J., Mok, S.S., Schmid, T.M., Block, M.B., Kuettner, K.E. and Thonar, E.J.M.A. 1994. Phenotypic Stability of Bovine Articular Chondrocytes after Long-term Culture in Alginate Beads. *J. Cell. Sci.* **107**: 17-27.
16. Sharma, G., Saxena, R.K. and Karol, S. A Cam based Cyclically Loading Electronically Controlled Machine for Three Dimensional Culture of Chondrocytes. (*Communicated*).
17. Chiba, K., Anderson, G.B.J., Masuda, K. and Thonar, E.J.M.A. 1997. Metabolism of the Extracellular Matrix formed by Intervertebral Disc Cells Cultured in Alginate. *Spine.* **22**: 2885-2893.
18. Whiteman, P. 1973. The Quantitative Measurement of Alcian Blue Glycosaminoglycan Complexes. *Biochemical Journal.* **131**: 343-345.
19. Lowry, O.H., Rosenbrough, N.J., Farr, A.L. and Randal, J.R. 1951. Protein Measurement with Folin Phenol Reagent. *J. Biol. Chem.* **193**: 265-275.
20. Anson, M.L. 1938. *J. Gen. Phy.* **22**: 79-89.

21. Setaro, F. and Morley, C.D.G. 1977. Anal. Biochem. 81: 467-471.
22. Smith, R.L., Lin, J., Trindade, M.C., Shida, J., Kajiyama, G., Vu, T., Hoffman, A.R., Van der Meulen, M.C., Goodman, S.B., Schurman, D.J. and Carter, D.R. 2000. Time-dependent Effects of Intermittent Hydrostatic Pressure on Articular Chondrocyte Type II Collagen and Aggrecan mRNA Expression. *J. Rehabil. Res. Dev.* **Mar.-Apr. 37(2)**: 153-61.
23. Hung, C.T., Mauck, R.L., Christpher, C.B.W., Lima, E.G. and Athesian, G.A. 2004. A Paradigm for Functional Tissue Engineering of Articular Cartilage via Applied Physiologic Deformational Loading. *Annals Biomedical Engg.* **Jan. 32 (1)**: 35-49.
24. Beagle, A.D. 1975. Content and Composition of Glycosaminoglycan in Human Knee Joint Cartilage Variation with Site and Age in Adults. *Conn. Tiss. Res.* **3**: 141-147.
25. Slowmann, S.D. and Brandt, K.D. 1986. Composition and Glycosaminoglycan Metabolism of Articular Cartilage from Habitually Loaded and Unloaded Sites. *Arthritis Rheum.* **29**: 88-94.

Biomechanics
R.K. Saxena and P. Mishra (Editors)
Copyright © 2005, Anamaya Publishers, New Delhi, India

22. Neural Network Based Inverse Kinematics Analysis for Design and Control of Artificial Leg

R.P. Tewari and Sachin Chaudhry
Instrumentation and Control Engineering Division, Netaji Subhas Institute of Technology,
Dwarka, New Delhi - 110075, India

Abstract: Human gait comprises of various phases in the walking cycle of a person. The various phases in human gait are distinct to each other in terms of relative spatial orientation and position of segments of human leg (i.e. thigh, shank and foot). Each stride contains 8 relevant phases. Stance is comprised of 5 gait phases (i.e., initial contact, loading response, mid stance, terminal stance, pre swing), with the remaining 3 phases occurring during swing. An artificial neural network was used for inverse kinematics analysis of human gait. The neural networks were trained using back propagation and radial basis function algorithms. Artificial neural network systems for human gait analysis are computationally simple and fast. This would allow further dynamic analysis of human gait and real time control of multi-link leg motion of prosthetic leg.

Kinematics of Human Gait Analysis

Kinematics is the study of the motion of bodies without reference to the forces that cause the motion. Fundamental objective is to capture a record of the dynamic range of motion of each of the lower extremity joints as well as the spatial orientation of the pelvis, and perhaps the trunk and head. Human walking cycle involves a frame-by-frame analysis of the three-dimensional joint motion and limb kinematics.

Lower limb kinematics deals with the analytical study of the geometry of limb locomotion of a lower limb with respect to a base coordinate system. Kinematical analysis gives the relations between joint angles and position and orientation of the end foot of lower limb. There are two techniques of kinematical analysis:

(a) Direct kinematical analysis
(b) Inverse kinematical analysis

Three Dimensional Kinematic Model of the Lower Limb

The lower limb can be modeled as a sequence of four rigid links connected by three universal rotary joints representing the hip, knee and ankle joints. Each joint is modeled as a sequence of three single axis rotational joint. Each link in the lower limb model represents one of the segments of the leg (pelvis, thigh, shank, foot). The joints of the lower limb are modeled as a sequence of single axis rotational joints. A specific set of rules has been developed by Paul (1981) in which coordinate frames are assigned to each link n and the transformation matrices relating the nth coordinate frame to the $(n-1)$th coordinate frame can be obtained. Using this rule and taking link parameters of the generalized rotary joint, the transformation matrix for a rotary joint is given by

$$A_{n-1}^{n} = \begin{bmatrix} \cos\theta_n & -\sin\theta_n \cos\alpha_n & \sin\theta_n \sin\alpha_n & a_n \cos\theta_n \\ \sin\theta_n & \cos\theta_n \cos\alpha_n & -\cos\theta_n \sin\alpha_n & a_n \sin\theta_n \\ 0 & \sin\alpha_n & \cos\alpha_n & d_n \\ 0 & 0 & 0 & 1 \end{bmatrix}$$

where θ_n is the joint variable, d_n the joint offset distance, α_n the link twist and a_n the link length.

The kinematics chain model of lower limb may be modeled as an open kinematical mechanism consisting of a sequence of 10 rigid links connected by 9 rotary joints, where each joint has one degree of freedom. The sequence of rotation for each joint is abduction or adduction followed by internal rotation or external rotation and then flexion or extension.

For the present model of leg, the values of the parameters are:

Link	θ	α	d	a
1	$\theta + 90$	90.0	0.0	0.0
2	$\theta - 90$	-90.0	0.0	0.0
3	$\theta - 90$	90.0	0.0	L_3 (thigh)
4	$\theta + 90$	90.0	0.0	0.0
5	$\theta - 90$	-90.0	0.0	0.0
6	$\theta - 90$	90.0	0.0	L_6 (shank)
7	$\theta + 90$	90.0	0.0	0.0
8	$\theta - 90$	-90.0	0.0	0.0
9	$\theta + 180$	0.0	0.0	L_9 (foot)

The angles are defined positive for:

θ_1 = Adduction of hip (+)
θ_2 = External rotation of hip (+)
θ_3 = Extension of hip (+)
θ_4 = Varus or Adduction of knee (+) (inside towards mid line)
θ_5 = External Rotation of knee (+)
θ_6 = Flexion of knee (+)
θ_7 = Varus or Adduction of ankle (+)
θ_8 = External rotation of the ankle
θ_9 = Plantar flexion of the ankle

All these movements are for right leg.
The total motion of the lower limb (ankle joint, knee joint and hip joint at a time) is given by

$$T_{\text{Hip-Ankle}} = \begin{bmatrix} T_P^{Th} \end{bmatrix} \begin{bmatrix} T_{Th}^{S} \end{bmatrix} \begin{bmatrix} T_S^{F} \end{bmatrix}$$

This gives final orientation and position of leg with respect to base coordinate

$$= \begin{bmatrix} A_0^1 A_1^2 A_2^3 \end{bmatrix} \begin{bmatrix} A_3^4 A_4^5 A_5^6 \end{bmatrix} \begin{bmatrix} A_6^7 A_7^8 A_8^9 \end{bmatrix}$$

The transformation Matrix from shank to the foot [at ankle joint]

$$T_S^F = A_6^7 \times A_7^8 \times A_8^9$$

The transformation matrix from the thigh to the shank [at knee joint]

$$T_{Th}^S = A_3^4 \times A_4^5 \times A_5^6$$

The transformation matrix from the pelvis to the thigh [at hip joint]

$$T_P^{Th} = A_0^1 \times A_1^2 \times A_2^3$$

Considering from hip joint to ankle joint for kinematical analysis, the total motion below knee is comprised of the thigh motion, shank motion and the foot motion.
Total motion of below knee leg = [hip motion] [shank motion] [foot motion]

$$T_{\text{Hip-Ankle}} = T_P^{Th} \cdot T_{Th}^S \cdot T_S^F$$

Here T describes the end effector (final motion of foot) position and orientation with respect to base coordinate frame.

Neural Network Modeling of Inverse Kinematics for Leg

Direct kinematics analysis for lower limb modeled as robotic problem is done as explained previously. Position and orientation of foot with respect to pelvis are calculated for the given angles i.e. adduction/abduction, external/internal rotations and flexion/extension at hip, knee and ankle joints.

Now, a multilayer feed forward neural network model is trained for inverse kinematics using the back propagation algorithm. The position and orientation of foot with respect to pelvis is presented as input and the corresponding joint angles data set is presented as output.

This inverse kinematics model of lower limb is used as the trajectory planner unit for the control of the motion of the human walking cycle. This model easily maps the trajectory of foot given in the cartesian coordinates with respect to pelvis to the joint variable space at hip, knee and ankle joints. The trajectory provided is free from any path constraints and has been taken for the person walking at normal speed.

Conclusion

The inverse kinematics neural network model for human locomotion was simulated. Error was compared for the feed forward neural network trained using back propagation algorithm and radial basis function neural network as shown in simulation results (Figs. 1 to 8).

Simulation Results

Joint Angles of Legs

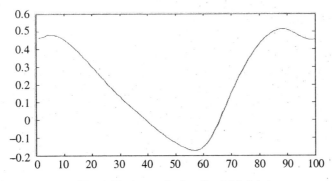

Fig. 1. Flexion angle data for thigh joint.

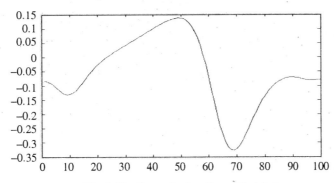

Fig. 2. Flexion angle data for ankle joint.

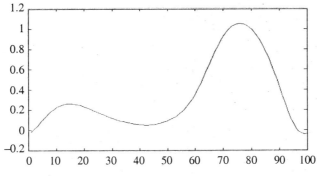

Fig. 3. Flexion angle data for knee joint.

Simulation Results for Inverse Kinematic Model of Leg

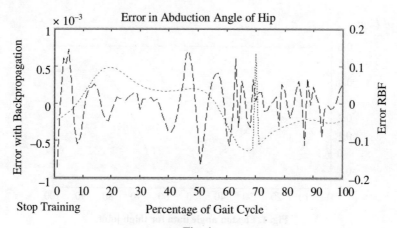

Fig. 4

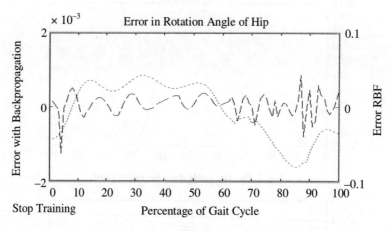

Fig. 5

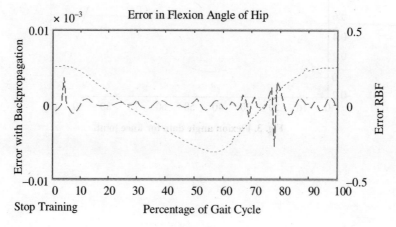

Fig. 6

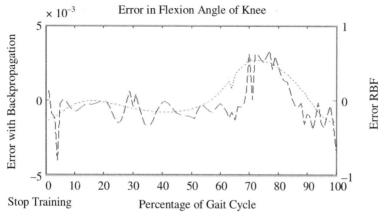

Fig. 7

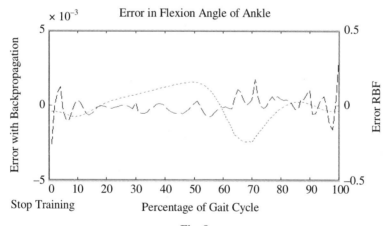

Fig. 8

References

1. Koopman, B.F. and de Jongh, J.H. 1995. An Inverse Dynamics Model for the Analysis, Reconstruction and Prediction of Bipedal Walking. *J. Biomech.* **Vol. 28**: 1369-1376.
2. Kumpati, S.N. and Kannan, P. 1990. Identification and Control of Dynamics System using Neural Network. *IEEE Trans. On Neural Networks.* **Vol. 1(1)**: 4-27.
3. Lee, F. 1987. *Robotics: Control, sensing, vision and intelligence.* McGraw-Hill Book Company.
4. Liu, Li, Wright, A.B. and Anderson, G.T. Trajectory Planning and Control for a Human-Like Robot Leg with Coupled Neural-Oscillators. *University of Arkansas at Little Rock Little Rock, AR 72204*
5. Schilling, R.J. 1996. *Fundamentals of Robotics, Analysis and Control.* Prentice-Hall of India.
6. Vojislav, K.D., Dejan, P. and Sakug, N.T. 2000. Feedback Error Learning Neural Network for Trans-Femoral Prosthesis. *IEEE Transactions on Rehabilitation Engineering.* **Vol. 8**.
7. Wright, A.B. 1999. Derivation of Equations of Motion for a Four-link Robotic Leg for Walking Vehicle. *Journal of Arkansas Academy of Science.* **Vol. 53**: 137-142.
8. Zurada, J.M. 1997. *Introduction to Artificial Neural Systems.* Jaico Publishing House.

Biomechanics
R.K. Saxena and P. Mishra (Editors)
Copyright © 2005, Anamaya Publishers, New Delhi, India

23. Biomechanics of Foot, Its Deformity and Conservative Treatment

M.D. Burman and Ranjan Das
Department of Prosthetics and Orthotics, Swami Vivekanand National Institute of Rehabilitation
Training and Research, Cuttack - 754010, India

Abstract: The biomechanics of the foot describes the mechanics by which the foot is converted from a flexible to a rigid structure. The main function of the foot at initial ground contact is to absorb impact and adopt the ground. When weight is borne by a foot, the bone and ligaments on the dorsal aspect are under compressive stress, while the bones, ligaments and planter aponeurosis on the planter aspect of the foot are under tensile stress. During the push off phase of the walking cycle the toes are dorsi-flexed and they greatly tightened the planter fascia, which produces a higher foot arch and inverted foot, a shorter foot and a more rigid foot by the so called wind lass action to propel the foot forward.

The foot is probably in a state of relatively rapid evaluation consequent upon man's assumptions of the upright posture and is prone to variation in structure. The main deformities of foot are pesplanus, pescavus, CTEV, hallux valgus. In pesplanus medial longitudinal arch of the foot is reduced and foot-ware is custom made to uphold longitudinal arch to relieve ligamentous strain by forming medial heel and sole wedge along medial longitudinal arch support. In pescavus lateral heel and sole wedge distribute weight over entire foot. In CTEV we use normal alignment by shifting weight medially and posteriorly with the help of a splint and modified foot-ware. In Hallux Valgus modified foot-ware along with long vamp and broad toe to prevent aggravation of valgus inclination is used.

Introduction

The opportunity to be independently mobile is a key factor in the quality of life of every one. For a person with lower limb disorder, the ambulation with gait is not at all possible without good quality foot support, which not only helps for ambulation but also protects the foot from external barrier. Those who have got misfortune of foot deformities due to either congenital or acquired reasons could not do their basic activities of daily living. Timely management treatment could prevent these deformities further. With a mobility aid the person can be forced in to a life style that is both sedentary and dependent on others.

Biomechanics of Foot

The subject of biomechanics of foot and ankle during gait is a complex one. The foot is a unique structure, which produces balance and propulsion of the body. It acts not only as the foundation of the rest of the body but also its function affects every motion of the body. It must initially be a loose 'bag of bones' capable of being placed down on any surface, flat or angled, smooth firm and yet maintain the equilibrium of the leg and the body above and act as a shock absorber. It must be capable of locking itself into a rigid structure so that it can act as a lever, first to stabilize and lift the body weight and then propel it forward. The healthy foot satisfies the seemingly paradoxical requirements of both shock absorption and thrust through an interaction of interrelated joint connective tissue and muscles.

During normal walking each lower segments of the skeleton (composed of part of the pelvis, femur, tibia, and fibula) rotates in the transverse plane. The degree of rotation progressively increases from the more proximal segments to the more distal. When the person is walking on level ground, the pelvis rotates at averages of 6°, the femur 13°, and the tibia 18°. The lower limb rotates internally during swing phase and first 15% of stance phase. The direction is then reversed and the external rotation begins, reaching its peak just after toe-off, when internal rotation again occurs. This transverse rotation is passed to the talus through its articulation with tibia and fibula.

The axis of the ankle passes just distal to the hip of each malleolus and may be reasonably accurately estimated by placing one figure on each malleolus. The axis of ankle rotation is directed laterally and posteriorly in the transverse plane and laterally and downward in the frontal plane. The angle between it and the long axis of tibia is approximately 80°, ranging from 68 to 88°. In the transverse plane it is externally rotated 20 to 30° with respect to the knee axis. The longitudinal axis of the foot, which passes between 2nd and 3rd toe, is internally rotated 6° to the axis of the ankle joint ranging from 21° of internal rotation to 9° of external rotation.

The medial longitudinal arch is the primary load bearing and shock absorbing structure in the foot. The bones that contribute the medial arch are calcaneus, talus, nevicular, cuneiforms and three medial metatarsal. A secondary arch is the transverse arch covering the distal inter tarsal joint. The planter fascia of the foot provides the primary support of the medial longitudinal arch. The fascia consists of an extensive series of thick, very strong longitudinal and transverse bands of collagen rich tissue. The medial longitudinal arch in the healthy foot is supported by two primary forces (1) active muscles forces and (2) passive force produced by the combine elasticity and tensile strength of the connective tissue and the shape of the bones. When standing at ease, passive forces are generally sufficient to support the arch. Active forces are required, however, during more dynamic and stress full actions such as tiptoes, walking and running.

Distribution of Compression Forces (by %) Across the Foot while Standing
- Rear foot (heel): 60%
- Fore foot: 28%
- Mid foot: 8%

The planter fascia is attached to the calcaneus which extend forward to span all the tarsal and metatarsophalangeal joint and attaches to the planter aspect of the proximal phalanges resulting a truss like structure whose links are the tarsal bones and ligaments of the foot. These are held at their base by a tether, the planter fascia. Since the planter fascia spans the entire longitudinal arch and has relatively little intrinsic ability to lengthen, it acts as a cable between the heel and the toe. A Spanish-windlass mechanism is formed at the metatarsophalangeal attachment of the fascia. As the metatarsophalangeal joints are extended passively when one stands on the ball of the foot the planter fascia is pulled distally across the metatarsophalangeal joints, shortening the distance from the calcaneus to the metatarsal heads. This process makes the base of the truss shorter. As the tether is shortened and the distance between the heel and ball of the foot is reduced, the tarsal joints are locked into a forced flexed position and the height of the longitudinal arch of the foot is increased.

Growth of the Foot

The foot is formed at the time that the limb buds develop during the 8th week of gestation. Following birth the growth of the foot in both male and female child proceeds at a steady rate, although it is recognized that the child as a whole has two periods of fast growth namely the first two years of life and puberty. On an average at the age of one year in female child and 18 months in male child, the length of the foot is one half the length of the respective adult foot.

Walking uses a complex interaction of hip, knee, ankle and foot motion to advance the body in the desired line of progression. The moment of the initial floor contact has been designated as the start of gait cycle. Within each cycle the period of floor contact by any part of the foot is called stance. This is followed by an interval of limb advancement called swing.

Percentage of walking cycle:

- Stance phase: 60%
- Swing phase: 40%

Deformities

As the foot carries the entire load of the body, any disorder changes the individuals gait. The disorders may be congenital or acquired. The main deformities of the foot are:

1. CTEV or club foot
2. Pesplanus
3. Pescavus
4. Calcaneal spur-heel pain
5. Hallux valgus

CTEV or Club Foot

It is a congenital deformity. Both the feet are affected but one foot is uncommon.

Causes: In most causes a defect of foetal development is responsible.

Pathology: Soft tissues at the medial side of the foot are under developed and shorter than normal. The foot is adducted and inverted at the sub-talar, mid-tarsal and held in equinus at the ankle.

Clinical Features: The deformity is much common in male than female child. When the infant is born it is noticed that the foot is turned inward so that the sole is directed medially.

Deformity consists of the following three elements:

(i) Inversion of the foot
(ii) Adduction of the forefoot related to the hindfoot.
(iii) Equinus

Pesplanus or Flat Foot

A pronated and everted foot with depressed medial longitudinal arch in which excessive shoe wear is found on the inner aspect of the sole and heel. Medial border is close to or in contact with ground.

Causes: In many causes it probably has a congenital basis but it may be caused by selective muscles weakness or paralysis.

Pathology: All infants have flat feet for a year or two after they begin to stand. When the deformity persists in adult life, it becomes a permanent structural defect. The tarsal bone being so shaped that when articulated, they tend to form a straight line rather than an arch.

Pescavus or High Arch

A foot with excessively high arch in which excessive shoe wear affects the ball and heel of the shoe. It is sometimes called 'Hollow Foot'.

Causes: In many causes it is congenital. It is sometimes familiar, in other cases they are associated with spina bifida or it may follow poliomyelitis.

Pathology: The metatarsal heads are lowered in relation to the hind part of the foot, with exaggeration of the longitudinal arch. The soft tissue in the sole is abnormally short and eventually the bones themselves alter shape perpetuating the deformity. There is always associated clawing of the toes, which are hyper-extended as the metatarsophalageal joint and flexed at the proximal and distal interphalangeal joints. This clawing seems to be the result from defective action of the intrinsic muscles-lunbricals and interossei. The effect is that the toes are almost functionless and unable to take their normal share in weight bearing. Consequently excessive weight falls upon the metatarsal heads on walking and standing and hard callosities formed in the underlying skin. The mal-alignment of the tarsal joint predisposes to the latter development of osteoarthritis.

Clinical Features: The deformity becomes evident in childhood. It may affect one foot or both. In some cases the symptoms are negligible. It can be of three forms:

- Painful callosities beneath the MT head.
- Tenderness over the deformed toes from pressure against the shoe.
- Pain in the tarsal region from osteoarthritis of the tarsal joint.

On examination deformity can be easily recognized, longitudinal arch is high, the forefoot is thick and the toes are clawed. Callosities beneath the MT head indicate that they take excessive weight.

Calcaneal Spurs or Heel Pain

In this condition, which is believed to be inflammatory, there is a pain beneath the heel on standing or walking.

Pathology: The lesion affects the soft tissues at the site of the attachment of the plantar aponeurosis to the inferior aspect of the tuberosity of the calcaneus.

Clinical Features: The complaint is of pain beneath the heel on standing or walking, the pain extends medially and into the sole. The disability is sometimes severe. On examination there is a marked tenderness over the site of attachment of plantar fascia to the calcaneus. A sharp spur projected forwards from the tuberosity of the calcaneus is sometimes found.

Hallux Valgus

In hallux valgus the great toe is deviated laterally at the metatarsophalangeal joint. It is common in middle-aged women.

Causes: In few cases hereditary factors are responsible. But in most, the deformity is caused by the toes persistently forced laterally by enclosure in narrow pointed shoes. The wearing of high heels favours the development of hallux valgus because the fore foot is forced into the narrow pointed part of the shoe.

Pathology: Outward deviation of the big toe is the most obvious feature of the deformity, but a further, almost constant feature is that the first metatarsal is deviated medially, so that the gap between the heads of the first and second metatarsal is unduly wide. It is a primary defect. After several years two secondary changes occur. One is the formation of a thick walled bunion over the medial prominence of the metatarsal head. This may become inflamed, the second, a later development is osteoarthritis of the MT joint consequent upon its mal-alignment.

Clinical Features: The patient is mostly middle-aged women who seek advice. The early symptoms arise from tenderness over the bunion from pressure against the shoe. Additional symptoms arise from osteoarthritis of the metatarsophalangeal joint and from flattening of the transverse arch. On examination, the skin over the prominent joint is hard, tendered and often a thick-walled bursa can be felt and the movement limited and painful.

Conservative Treatment
In conservative treatment foot orthosis, mechanical devices are used to:

- Align and support the foot
- Prevent, correct or accommodate foot deformities
- Improve the overall function of the foot

Purpose of Shoe Modification
- Support foot to improve balance in standing and walking.

    (i) Support or accommodate deformities.
    (ii) Correct deformities.
    (iii) Protect and maintain plaster and surgical correction.
    (iv) Equalize foot and/or leg lengths.

- Relieve painful area from abnormal pressure.

    (i) Protect painful area.
    (ii) Limit motion or provide extra shoe stability.
    (iii) Transfer weight bearing stresses.

Method of Accomplishing Purposes
- Shoe selection.

    (i) Stock shoe.
    (ii) Custom made shoe.

- Shoe alteration.

    (i) Internal.
    (ii) External.

General Principles
- Accommodate fixed deformity.
- Actively correct flexible deformity.
- Shoe with modification should be lightweight.

In CTEV conservative treatment is done by correction of adduction and inversion first and then equines. The foot is placed in the corrected position by the following manipulations:

(i) Plaster of paris
(ii) Dennis Browne Splint

The plaster must extend to the upper thigh with the knee flexed at 90°. Plaster must be changed every week at first but as the child grows it can be extended to 2-3 weeks.

In Dennis Browne Splint method both feet are fixed to metal splints by strips and splints are clamped rigidly to a cross bar with the feet rotated outwards. To maintain effectiveness the splints must be re-applied every day or at least on alternate days. When the feet come in plantigrade position a pair of corrected shoes with medial straight border with lateral heel and sole raise to distribute the weight lateral to medial and in step straps to prevent equanius must be used. Polypropylene AFO or metallic AFO with lateral 'T' straps can be used.

In pesplanus treatment is required for children under 3 years. But for children over 3 years the method of treatment is to tilt the shoe slightly to the lateral side by inserting a wedge, base medially between the layer of heel with a medial longitudinal arch support.

In pescavus a provision is made to place a sponge rubber pad beneath the MT head, to distribute the weight more widely. In adult cases the high arch can be accommodated by building and distributing weight on the entire foot.

In calcaneal spur or heel pain (a) gauged out heel (b) UCBL shoe insert can be fitted with a custom made shoe to relieve pressure on the heel.

In hallux valgus, mild cases do not require treatment but footwear must be selected carefully. It may also be worthwhile to protect the bunion with pad or felt and wear a wedge of plastic foam between the great and second toe to reduce the deformity.

Conclusion
In our center where such types of lower limb deformities cases are coming and after evaluation and assessment through clinical meeting final prescription are made. These patients are being successfully fitted conventional and modified foot orthosis and are advised to come for periodical check ups. We are getting good feedback and the results are satisfied.

References
1. Elmslie, M. 1989. Prevention of Foot Deformities in Children. *Lancet.* **2**: 1260-1260.
2. John, Adams, C., Homblen, D.L. *Outline of Orthopedics* 7th *Ed.* ELBS.
3. Mann, R.A. 1985. Biomechanics of the Foot. *Atlas of Orthotics.* American Academy of Orthopaedic Surgeons. pp. 112-125.
4. Orthotics Lower. Vol. 8: Artificial Limbs Manufacturing Corporation of India. 1980.
5. Wu, K.K. Foot Orthotics. *Atlas of Orthoses and Assistive Devices.*

Biomechanics
R.K. Saxena and P. Mishra (Editors)
Copyright © 2005, Anamaya Publishers, New Delhi, India

24. Finite Element Analysis of Balloon and Artery Interaction During Stent Deployment

C.S. Ramesh[1] and Prashant K. Marikatti[2]

[1]Department of Mechanical Engineering, Ghousia College of Engineering,
Ramanagaram - 571511, India

[2]Department of Mechanical Engineering, S.D.M. College of Engineering and Technology,
Dharwad - 580002, India

Abstract: Modern angioplasty is alternative to critical bypass heart surgery because of safety and viability. To open arteries or veins where deposits restricts the blood flow cylindrical sleeve type of devices termed as stents are being used. Stent is mounted on balloon catheter in collapsed or crimped state. When the balloon is inflated the stent expands and pushes itself against inner wall of coronary artery with the balloon being deflated and removed. During the inflation of balloon, artery-balloon interaction is one of the causes of artery injury. The extent of injury is determined by the contact stresses between artery and balloon during stent deployment. As the artery is hyper-elastic in nature, FEA is used to simulate the contact between the balloon and the artery. The present paper discusses the variation of contact stresses for various deployment pressures. It is found that as the deployment pressure increases, the contact stresses also increases. The maximum contact stress is observed at the center of stent strut.

Introduction

In recent years, cardiac diseases represent the most common cause of death, which are often related to coronary arteriosclerosis (Freed et al., 1997). Arteriosclerosis is a process wherein there is deposition of cholesterol on the inner surfaces of artery producing a local lumen (occlusion) restricting flow of blood. Such artery pathologies have been traditionally treated through pharmacological therapies or invasive surgeries (such as by passes). Recently, non-invasive approaches are more preferred.

Non-invasive treatments are in general based on the insertion of a guide-wire in the vascular system through a peripheral artery (such as the femoral or the brachial one) and on the subsequent use of a catheter, whose tip is equipped with a mechanical device, is used to remove the occlusion.

Classical examples of non-invasive treatments are the PTCA and the stenting techniques. The PTCA (Percutaneous Transluminal Coronary Angioplasty) uses a balloon as a mechanical device to re-open the occluded artery, while the stenting technique is based on the insertion of a permanent tubular structure termed as stent at the infected artery. In general, the stent has the role of supporting the arterial walls, after the catheter retrieval.

In PTCA the expandable balloon is inserted into the blockage, where it is then inflated to several atmospheres of pressure. As the balloon expands, the deposited cholesterol (plaque) is forced up against the walls of the artery and broken up to allow restoration of blood flow. Typically, the blockage in an artery can be reduced from 90% to about 30% by PTCA technique, which has many shortcomings (Auricchio et al., 2000). To overcome these shortcomings, stenting technique is used.

Compared to the PTCA, the use of stents results in better success rates of opening up of the blocked arteries resulting in more effective cardiac operations. These considerations justify the high interest both commercial and the scientific communities are showing towards the stenting technique for future work.

From the available literature, it is clear that stent-artery and balloon-artery interactions are the major cause of injury during angioplasty (Auricchio et al., 2000). The extent of injury to artery depends on the stent design, material, balloon material and the pressure applied during the process. Hence it is essential to optimize these parameters for better, safe and successful treatment of cardio vascular diseases. Improper selection of these parameters results in not only artery injury but also restenosis (Freed et al., 1997; Auricchio et al., 2000; Schwartz et al., 1992; Trepanier et al., 1997).

Finite element analysis is one of the most powerful and popular tools in biomedical applications in recent years (Borgersen and Sedeghi). It allows modeling and analysis of various biological tissues with a variety of visco, shell and hyper-elastic elements under various loading conditions. FEA software Ansys 5.4 provides necessary tool to perform modeling, meshing as well as analysis. It allows simulating the highly non-linear contact existing between artificial implants and tissues. This study deals with simulation contact between artery and balloon during stenting using Ansys 5.4.

Stenting Methodology

Heart Diseases

In recent years, heart disease has risen to become the number one cause of death due to the alarming rates of obesity, drug dependence, and diabetes. As time passes, the symptoms of coronary heart disease can rapidly develop and will continue until intervention is carried out (*www.medicinenet.com/diseases and conditions.html*). Coronary heart disease may take years to develop but if proper care is not taken it leads to death. These alarming numbers have led to many interventions for coronary artery disease. Since 1977, the percutaneous transluminal coronary angioplasty (PTCA) and recently stenting procedures have become the most widely used form of intervention and has led to endless research.

Atherosclerosis

It has been found that, as the body ages, lipids undergo an oxidation process, which leads them to become harmful to the walls of the arterial vessels. As they pass through the artery vessel, the walls get damaged and the damaged walls are repaired with fatty substances, which leave a scar. Calcium and oxidized cholesterol are incorporated into the resulting scar tissue as they pass. The resulting lesion is called atherosclerotic plaque and this disease process is known as atherosclerosis (Squire, 2000). With atherosclerosis, as the years pass, calcium deposits build up, and calcified atherosclerotic plaque forms, lining the walls of the arterial vessels. This plaque is composed of various lipids, foam cells, scar tissue, and overgrown smooth muscles cells from the artery wall (*www.medicinenet.com/diseases and conditions.html*; Squire, 2000). In many people, this process begins in early childhood and progresses over time. The exact content of the plaque is determined by the individual's diet, exercise regiment, antioxidant intake and duration of the plaque formation process (Squire, 2000). High cholesterol levels will induce fatty deposits into the bloodstream and hasten the plaque formation process. The developed plaque causes increase in the size of the blockage of artery resulting in delivering of lower oxygen to the most important organ of human body, heart.

Coronary Angioplasty with Stent

To open arteries or veins where deposits restricts blood flow, cylindrical sleeve type devices known as stents are used. The stents are intended to keep artery walls from collapsing under spasmodic contractions and leave vessel open for continuous blood flow.

Coronary stent is a stainless tube with slots. It is mounted on balloon catheter in collapsed or crimped state. When balloon is inflated, the stent expands and pushes itself against inner wall of coronary arteries, holds the artery open after the balloon is deflated and removed as shown Fig. 1. Coronary stents are used to overcome some of the shortcomings of PTCA (Borgersen and Sedeghi).

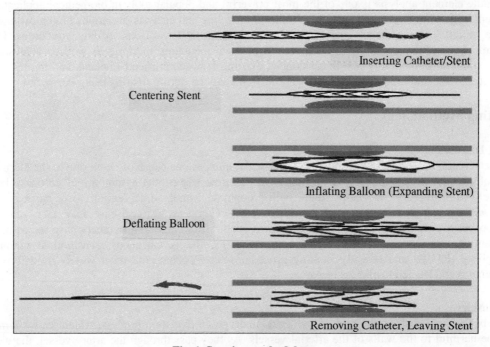

Fig. 1. Stenting methodology.

Biomechanical Studies on Stent

Despite the need for accurate mechanical studies regarding stent and PTCA techniques and their interaction with artery, these have received limited attentions. The developments in finite element analysis have helped researchers to carry out various analyses, as problems related to this field are highly nonlinear. There are several research papers and articles relating to this work.

Rogers et al. (1999) have concentrated on the balloon/artery interaction between stent struts during the stent apposition phase. The aim is to provide a better understanding of effect of balloon and artery interaction during balloon expansion. Both artery and the balloon have been assumed to behave as linearly elastic.

Auricchio et al. (2000) have studied the effect of stent deployment on artery and plaque using finite element analysis.

Holzapfel (2000) has obtained the relationship between different components during PTCA and mechanics of angioplasty wall. They have discussed balloon and stent at length. An attempt for prediction of the stent-induced wall-stresses is made. The strains have been determined by investigation of the morphological and mechanical information acquired through the MRI.

Finite Element Modeling

From the literature, it is clear that stenting technique involves stent, balloon, artery and plaque. Pressure is applied to inner surface of balloon, which expands and pushes itself against the plaque resulting in opening up of the blockage. FEM provides engineers and scientists information about the effects on the artery of applying pressure to the balloon. The pressure must be great enough to expand stent and break up the plaque while minimizing the damage to the artery. The composition of diseased artery is shown in Fig. 2.

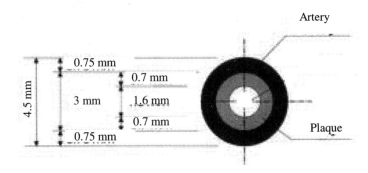

Fig. 2. Pictorial description of artery plaque before angioplasty.

Here the artery and plaque model are considered as hyper-elastic incompressible isotropic. Mooney Rivlin model is chosen for analysis. Balloon is considered as isotropic material.

The slot length in this analysis is considered as 2.88 mm. Plaque and artery are discussed in following section. The problem model is shown in Fig. 3.

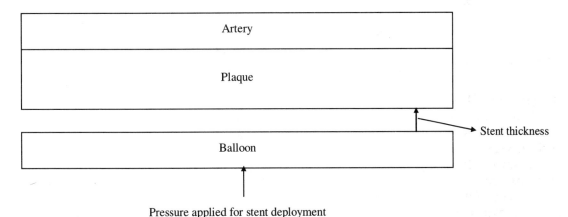

Fig. 3. Contact between balloon and artery during stent deployment.

Modeling and Meshing

The artery and plaque are modeled using 8 noded hyper elastic 183 plane stress elements and balloon with 8 noded 82 plane stress elements.

Creating Contact between Plaque and Balloon

As pressure is applied, balloon in between stent struts pushes against plaque and contact is created between plaque and balloon. As the balloon is stiffer than artery and plaque, it is treated as target surface and plaque as contact surface. Balloon surface is meshed with target 169 element and plaque surface with contact 171 element. Contact created is of surface-to-surface contact. Both artery and balloon are treated as flexible. Penetration tolerance is taken as 0.1 and normal penalty stiffness as 1.0.

Boundary Conditions

The boundary conditions applied are:

- The pressure is applied to the inner surface of the balloon
- All dofs are constrained at both sides of balloon ends as balloon expansion is restricted by stent.
- During the stenting process, outer diameter of the artery is assumed to expand from 3 to 3.1 mm. Hence the top surface nodes of artery are loaded with displacement of 0.05 mm. Fig. 4. shows model applied with all boundary conditions.

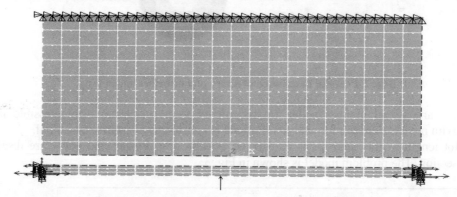

Fig. 4. Meshed model with boundary conditions applied.

The FEA simulation is performed for various pressures ranging from 0.6 to 2 Mpa.

Results and Discussions

The contact stresses during interaction of balloon with artery through the slot of a stent during its deployment are shown in Fig. 5. It is observed that contact stress at the interface of the artery and balloon within the slot of stent is maximum at the center of slot.

A gradual decrease in the contact stress from a maximum value to zero is observed at a distance of 0.1 mm from the center of slot indicating that there is a maximum interaction of balloon with artery at the center of the slot. Variation of contact stresses along the strut is shown in Fig. 6. Further, it is observed that increased balloon pressure resulted in higher contact stresses for the same slot dimension as shown in Fig. 7.

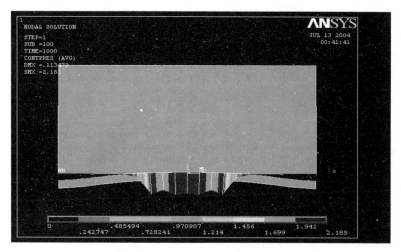

Fig. 5. Contact stress distribution.

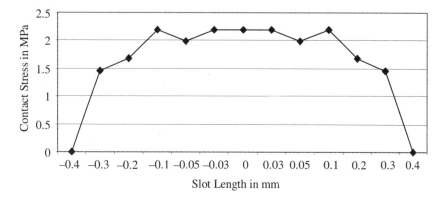

Fig. 6. Contact stress distribution variation along slot length.

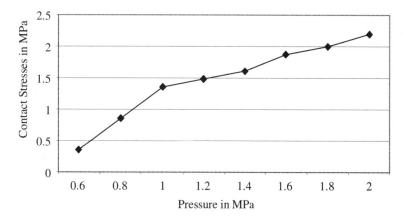

Fig. 7. Variation of contact stresses.

Conclusion

The artery balloon interaction during stent deployment is maximum at the center of slot of the stent. The contact stress is maximum at center of strut and it tapers down to zero along its length. Increased balloon pressure increases contact stresses at the balloon artery interface.

References

1. Auricchio, F. et al. 2000. Finite-element Analysis of a Stenotic Artery Revascularization through a Stent Insertion. *Computer Methods in Biomechanics and Biomedical Engineering, Vol. 01.* pp 1-15.
2. Borgersen, S. and Sedeghi, R. Nonlinear FEA Helps Researchers to Design Biomedical Stent. www.mscsoftware.com.
3. Freed, M., Grines, C. and Safian, R.D. 1997. *The New Manual of Interventional Cardiology.* Physician Press.
4. Holzapfel, G.A. 2000. Biomechanics of Soft Tissue. In *Handbook of Material Behavior-Nonlinear Models and Properties.* J. Lemaitre (ed.).
5. Learn About High Cholesterol. http://www.medicinenet.com/diseases and conditions.html
6. Raghavan, M.L. Ph.D., Lecture Notes on Cardiovascular, Bio-solid Mechanics Section. University of Iowa.
7. Rogers, C., Tseng, D.Y., Squire, J.C. and Edelman, E.R. 1999. *Circulation Research.* **84**: 378-383.
8. Schwartz, R.S., Huber, K.C. et al. 1992. *Journal of the American College of Cardiology.* **19**: 267-274.
9. Squire, J.C. 2000. Dynamic Expansion of Stents. *Ph.D. Thesis.* Massachusetts Institute of Technology.
10. Trepanier, C., Leung, T.K., Tabrizian, M., Yahia, L.H., Bienvenu, J.G., Tanguay, J.F., Piron, D.L. and Bilodeau, L. 1997. *In vivo* Biocompatibility Study of NiTi Stents. *Second International Conference on Shape Memory Superelastic Technologies.* A. Pelton, D. Hodgson, S. Russell and T. Duerig (eds.). pp. 423-428.

25. Evaluation of Various Biomaterials Used in Treating Maxillofacial Injuries

Smriti Sharma, Sandeep Tiwari, Vivek P. Soni and Jayesh Bellare
School of Biosciences and Bioengineering, Indian Institute of Technology Bombay,
Powai, Mumbai - 400076, India

Abstract: Maxillofacial injuries are a common occurrence due to accidents, assaults, falls and industrial injuries. Fractures of the bones of the maxillofacial region are treated by using bone plates. It was the aim of this presentation to evaluate the biomechanical behavior of commercially available bone plates and of those designed by us at IIT Bombay. Different designs of bone plates were made with stainless steel. Replica of the mandible was made with timber wood and fracture was created in these models. Finite element modeling was done to gain an insight into the biomechanics of the jaw. With the background of this information, the bone plates were tested for their mechanical behaviour by an experimental set up designed by us. The results were tabulated and statistically evaluated to get an insight into the biomechanical behaviour of various bone plates. It was found that rectangular bone plates designed by us exhibited the best biomechanical behaviour and strength amongst all the other bone plates.

Introduction

Maxillofacial injuries have increased over the past few years. The change in mechanisms of injury as well as in the type and severity of trauma has caused a change in the treatment modalities for fractures of the facial skull. New developments in the manufacture of corrosion proof metal alloys, introduction of new therapeutical methods, e.g. osteosythesis, which is stable even in early function, improvement in modern anesthesia, improved radiographic diagnosis with 3-D CT scan, and finally development in the antibiotic field as well as in intensive care have made it possible that today we treat complicated cases and perform immediate reconstructions, which earlier were either contraindicated or associated with great risks. Stable osteosynthesis offers the patient a certain amount of comfort in comparison to previous therapeutic methods. Additionally, an osteosynthesis gives a high degree of assurance to the surgeon that can simultaneously and anatomically reduces the fracture and achieve a satisfactory occlusion. When using conservative fracture treatment we cannot expect to achieve this therapeutic goal.

Aims and Objectives

The conventional bone plates and screws can be improved to facilitate better healing. The essential requirements of a good bone plate system are rigidity, good pressure distribution on the fracture zone etc. It is therefore important to examine the effect of different plate geometries on the rigidity of the joint and assess the effect of plate orientation and screw arrangement on the pressure distribution.

Different biomaterials used to prepare bone plates are also evaluated. Both experimental and theoretical studies (2-D and 3-D) were used to establish results mainly the ideal shape and size of the bone plate and its orientation and location with respect to the fracture.

Materials and Methods

Experimental and theoretical studies have been carried out using ANSYS (4.3 version) package. Both 2-D and 3-D finite element analysis are done in the vicinity of the crack. The effect of compression plate location on the stress distribution and the span of contact are studied. The mandible has also been treated as a 3-D object. A 3-D stress analysis is carried out to evaluate effect of muscles and biting forces. Experimental investigations were done to evaluate the merits of different plate geometries and different screw arrangements on the rigidity of the joint. Timber specimens in simple geometries are tested under both in plane and out of plane loading. An experimental setup for testing the timber specimens was designed.

Fig. 1. Plating.

Results and Discussion

The ANSYS FEM package provided facilities to model and solve 2-D and 3-D problems using variety of elements, common as well as special ones. The package works in 3 levels

1. Preprocessing
2. Execution
3. Post processing level

In the preprocessing stage the ANSYS programme was processed to generate mesh and the input data concerning boundary conditions stored. In the execution phase equation solution and stress calculation were carried out. The frontal solution technique was used. The post processor phase stored and presented the output quantities like stresses, displacements, reaction forces etc.

Very relevant results were obtained with the study. The results showed that the plate should be located in the tension zone to neutralize the separation forces generated by the functional loading. Compression plating also gave rise to additional stresses across the crack and opening at the far end. Hence compression plates located in the tensile zones would nullify the separation forces more effectively.

Results of different plate geometries were analyzed. The experimental setup consisted of a vice, loading pan and dial gauge. The specimen was firmly clamped in the vice. Loads were applied in steps of 0.25 kg on the pan from 0.25 to 2 kg. The results of the different arrangements were tabulated. The stiffness k is calculated in each case by considering the increase in load from 0.25 kg to 2 kg and the corresponding increase in deflection.

Conclusion
It was concluded that:

1. Plates should be oriented as perpendicular to the fracture as possible.
2. The proposed triangular and rectangular geometries provide more rigid connection than the standard bone plates.
3. The oblong plates provide better rigidity in case of out of plane loading.
4. Bioresorbable plates and titanium plates are more superior to stainless steel and vitallium.

References

1. Andersen, K.I., Mortensen, H.T., Pedersen, E.H. and Malsen, B. 1991. Determination of Stress Levels and Profiles in the Periodontal Ligament by Means of an Improved Three Dimensional Finite Element Model for Various Types of Orthodontic and Natural Force System. *Journal of Biomedical Engineering*. **13**: 293-303.
2. Askew, M.J., Wirth, R.C. and Campbell, C.J. 1975. Analysis of the Intraosseous Stress Field due to Compression Plating. *Journal of Biomechanics*. **8**: 203-212.
3. Cheal, E.J., Hayes, W.C., White III, A.A. and Perren, S.M. 1984. Three Dimensional Finite Element Analysis of A Simplified Compression Plate Fixation System. *Journal of Biomechanical Engineering*. **106**: 295-301.
4. Knoell, A.C. 1977. A Mathematical Model of an *in vitro* Human Mandible. *Journal of Biomechanics*. **10**: 159-166.
5. Rybicki, E.F. and Simonen, F.A. 1977. Mechanics of Oblique Fracture Fixation Using a Finite Element Model. *Journal of Biomechanics*. **10**: 141-148.

Biomechanics
R.K. Saxena and P. Mishra (Editors)
Copyright © 2005, Anamaya Publishers, New Delhi, India

26. Sports Biomechanics and Human Performance

Dhananjoy Shaw

Biomechanics Laboratory, Department of Natural/Medical Sciences, I.G.I.P.E.S.S.,
University of Delhi, Delhi - 110018, India

Abstract: In recent years for excellent human performance, especially in competitive types of games, sports biomechanics has been found to be playing a crucial role for getting the medal. In general, sports biomechanics is concerned with the enhancement of the understanding and performance in events through modeling, simulations and measurements in order to understand the mechanics of various sports activities, including both the motion of athletes and motion of sports implements and vehicles. The emphasis is on developing mathematical and computational models and experimental protocols for events, which allow increased understanding, performance evaluation and optimization, as well as intelligent implement and vehicle design. Often the goal is the development of methods and systems for providing real time feedback of athlete performance during training. Biomechanics is also a diverse interdisciplinary field in which medical, engineering and basic sciences disciplines such as zoology, botany, physical anthropology, orthopedics, bioengineering and human performance play a crucial role. In all of these disciplines the general purpose of biomechanics is the same: to understand the mechanical cause-effect relationships that determine the motions of living organisms. However, within each discipline, sports biomechanics tries to solve problems specific to that discipline (athletics, swimming, gymnastics, weight lifting, boxing, archery etc.). In sports biomechanics the main goal is to understand the relationships between structure and function. Orthopedics and bioengineering are the main focal areas for understanding the role of structures and functions of body kinesiology, but with a special added emphasis on practical applications, such as the development of prosthetic devices that can help in rehabilitation. In human performance, biomechanics contributes to the description, explanation and prediction of the mechanical aspects of human exercise, sport and play, which has been found necessary for optimization of activity. In this article, interdisciplinary and multifarious approaches for optimization of performances by sports biomechanical techniques are discussed.

Introduction

The scope of sports biomechanics has been overemphasized in the Athens Olympics by the countries representing the European continent (especially Russia and Germany), as well as Australia in all the spheres of biomechanical applications for excellence. The scope of biomechanics is emphasized in the area of discoursing the underlying principle of efficient structure of competitive performance, specific fitness and its application, biomechanics aided teaching and learning, selection of players, formation of teams, coping with the changing rules and regulations, best utilization of facilities and equipment, prevention of injuries, management of sports injuries as well as sports physiotherapy, by which the above mentioned countries dominated the rest of the world. It is the knowledge of biomechanics, which has developed the technique to its excellent standard and unmatched by others, by means of tactical, statistical and strategic applications. Hence, the importance of biomechanics should not be ignored. What India is practicing significantly and resulting to worst, from a population of more than 1 billion people India could manage only a silver medal. There can be hundreds of reasons for this, but one reason is the science and technology, particularly sports biomechanics, at the apex of it.

It has been observed that biomechanics and kinesiology related to human motion and its teaching with specific reference to physical education and sports have been greatly ignored in India. The reason may be the lack of expertise, misinterpretation of the subject, lack of practice and/or lack of literature. No book has been written on the subject till 1998 by an Indian author, specifically catering to the requirements of physical education and sports sciences. Most of the available published books are imported from outside India and are very costly, not easily available to common readers/students. At the same time those literatures are not suitable or not much friendly to the people of physical education profession, whether they are teachers, researchers or students in India.

Biomechanics and kinesiology is one of the most upcoming subjects in India. Before 1980, some practices of biomechanics were observed in India, but they markedly deviated from the actual concept of the subject or sub-situated with some other title or sub-title. Later, since the beginning of the 1980s, biomechanics and kinesiology have been placed in the right perspectives and have been incorporated as an independent subject in various disciplines at the undergraduate, postgraduate levels and doctorate research programs. In 1988, the Indian Society of Biomechanics was founded with the assistance of collaborated experts, researchers from medical, engineering and other subject areas with kind motivation and initiations of the author's Ph.D. supervisor Prof. K.B. Sahay, IIT, Delhi. So far, the society has conducted eight national conferences with great success.

Biomechanics is a term formed by combining the words "Biology" and "Mechanics", and thus, it deals with the principles and methods of mechanics, applied to the study of the structure and function of biological system (Fung, 1966; Hatze, 1974; Lissner, 1967; Warten Weiter, 1973). Biomechanics is shown to be based on, and drawing substances from, a number of fundamental disciplines ranging from classical mechanics to the life sciences. It includes theoretical mechanics, anatomy, anthropology, neuro-muscular physiology, kinesiology and psycho-motorics (Contini and Drillis, 1954, 1966 a).

The area of general biomechanics deals with the basic laws and rules which govern human and animal bodies in motion (biodynamics) and at rest (biostatics). Applied biomechanics is concerned with the more practical problems of improving movements, positions and human well-being in industry, agriculture, medicine, military, sports, the arts and everyday life.

Two Swedish men were among the first in the United States to formally apply the term kinesiology for analysis of muscles and movements in physical education setting. Posse's book on 'The Special Kinesiology of Educational Gymnastics' actually appeared before the turn of the century. Skarstorm's 'Gymnastic Kinesiology' (1909) was a more scientific practice on kinesiology (Cooper and Glassow, 1976; Cooper, Note 2). In 1912, Wilbur Bowen published, 'The Action of Muscles in Bodily Movement and Posture'. In 1917, jointly with R. Tait Mekenzee, this title was applied to anatomy and kinesiology.

The first International Seminar on Biomechanics in Zurich, Switzerland in 1967 was an important event. As indicated by the proceedings (Wartenweiler, Joki and Hebbelinck, 1968), the papers presented were concerned with techniques of motion study, biomechanical principles of human motion and application of biomechanics in work, sports and clinical settings. Two journals were initiated during this period, the first for electromyography and the second for biomechanics.

In 1968, the 'Journal of Biomechanics' was first published with the broadly stated purpose of serving as a focal of application of mechanical principles to biological properties. Among the papers published during 1968, few dealt with biomechanical analysis of sports skills. The mechanical energy transformations in pole vaulting (Dillman and Nelson, 1968) and others presented computer programs used in analyzing a basketball throw (Plagenhoef, 1968).

The highlight of the fourth seminar in 1973 was the founding of the International Society of Biomechanics (ISB). Although, the ISB initially grew out the interest of those in the field of sports and physical education, its program was defined as being broad in the scope of participation and interaction of researchers representing a variety of fields concerned with biomechanics of human movements (Nelson and Morehouse, 1974). Various organizations and publications gave recognition to the increased interest in biomechanics during the 1970s. The Applied Mechanics Division sponsored a symposium on "Mechanics and Sports" (Bleustein, 1973). In 1974, the number of papers presented at the annual meeting of the American College of Sports Medicine was such, to the extent that since 1977, two symposium sessions have been regularly devoted to biomechanics. Medical journals have included papers on biomechanics in the study of sports injuries (e.g. Frankel and Hang, 1975; Whieldon, 1979; Zingg, 1975).

The exponential growth of biomechanics involved professionals from physical education, medicine, ergonomics, biology and engineering to formulate different National Societies of Biomechanics to be affiliated to the ISB to sustain the movement of biomechanics worldwide.

After reviewing all the literature related to the associations and conferences, it may be concluded that during the 1970's there was a marked increase in the number of meetings held for those whose special interest was the science of human movements. The majority of these workshops, symposia, seminars and conferences had titles, which included the term 'biomechanics' rather than kinesiology.

It may be concluded that from 1981 to 2004, the International Congress of Biomechanics and International Conference of Biomechanics are regular features organized annually or in alternate years, in addition to International symposiums, seminars and workshops.

Change of Concept of Biomechanics
During the 1970s, there was marked increase in the number of publication and research articles. All those literature shows the changing concept of kinesiology and biomechanics. The term biomechanics was widely used rather that kinesiology.

Increase in the Horizon of Scope
Kinesiology literature first appeared with the combination of gymnastics and other sports. Then, in 1937, in an article by McCloy the word "Mechanics" appeared for the first time. After that, books on kinesiology and mechanics were published in various combination and scope. The word 'biokinetics' appeared for the first time in 1942 as a text. After the 1960's, the biomechanics literature came in different combinations and scope. In 1968, the name 'biomechanics' was confirmed to excel further.

Technological and Methodological Prospectives
During the 20th century, advances in the science of human motion seen to have been not only due to improvements in instrumentation, but also to the development of better and more creative methods of using these instruments. Methods of three dimensional cinematographic analysis were developed and refined during the 1970's, to improve the accuracy of studying complex human motions (Bergemann, 1974; Miller and Petak, 1973; Shaprio, 1978; Van Gheluwe, 1974, 1978; Walton, 1979).

Among the new techniques that have emerged in recent years are as follows:

1. Polarized Light Goniometry (Grieve, 1969; Mitchelson, 1973; Read and Reynolds, 1967).

2. Automatic image analysis in which a television image or a cine film is scanned by computer to determine the X and Y coordinated or anatomical landmarks (Winter, Greenlaw and Hobson, 1972; Kasvand, Miller and Rapley, 1974).
3. Light spot position measurement, which uses optoelectronic devices such as the SELSPOT (Lindholm, 1974) to obtain information about the three dimensional coordinates of small, active light sources attached to the human body (Woltring, 1974).

Biomechanics of Bones and Joints
The biomechanics of bones and joints mainly deals with bone mechanics, human joints under dynamic conditions, mechanical aspects of osteoarthritis, joint loads, relation to tendons and ligaments, biodynamics and bone density.

Biomechanics of Neuromuscular Aspects of Movement
After the 1980s, it is usually seen that in neuromuscular aspects of movements, biomechanics deals with the composition of skeletal muscle, utilization of elastic energy, biomechanical analysis of movements/skills, neuromuscular reflex, force and fatigue, work (mechanical) efficiency, neuromuscular adaptation, mechanical properties of human muscles, muscle architecture, force velocity relationship of different muscles, morphometrics and force length relations, prediction of muscle fiber composition by kinematics and kinetic data, biomechanical modeling, sports injuries etc.

Biomechanics of the Upper Extremity
The critical studies were the relationship of biomechanics to upper extremity, work efficiency, EMG analysis of muscle of different body parts, sports injuries and levels of skills.

Biomechanics of Lower Extremity
The critical studies reflect on biomechanical relationship to lower extremity, showing the importance of running shoes, shin splints and biomechanical fitness. Sports injuries were emphasized with greatest significance and the biomechanics of basic movements related to knees and ankle joints was also emphasized.

Biomechanics of Fitness
Critical studies reflect on biomechanics to fitness, emphasis on isokinetic strength and endurance training, flexibility by PNF, assistant and non-assistant stretching, cryo-stretching, fiber architecture, analysis on natulas for different type of exercise and polyometric (drop jumping) exercises, designing exercise devices, mode and posture of exercise etc.

Biomechanics of Balance and Stability Control
It is usually seen that biomechanical research deals with the balance in lifting works, effect of quick directional/postural changes, studies on centre of gravity etc.

Biomechanical Instrumentation
There was tremendous increase in biomechanical instrumentation, such as the evolution and use of triaxial goniometer, piezoelectric method (stress beneath the human foot), high technology, computer application, quantitative EMG, standardization of biomechanical testing for different games/sports and

also building of a better computerized video system, application of biomechanics to analyze swimming techniques etc.

Growth of Technology and Methodology in Biomechanics and Kinesiology for Physical Education and Sports (Instrumentation)

Fundamental to the study of human motion is the measurement of displacement of the body and its segments, i.e. kinematic analysis as most popular and adopted methodology. Traditionally, cinematographic analysis of relatively high-speed films has been the technique used to obtain kinematic data (Smith, 1975). However, the raw displacement data contains inherent error. For this reason, various methods of smoothing the displacement data have been employed, with the two most successful being digital filtering (Cavanagh, 1978a; Pezzack, Norman and Winter, 1977; Winter, Sidwall and Hobson, 1974) and the use of spline functions (McLaughlin, Dilman and Lardner, 1977; Wood and Jennigs, 1979; Zerbucke, Caldwell and Robers, 1976).

Methods of three-dimensional cinematographic analysis were used during the 1970s to improve the accuracy of studying complex human motions (Bergemann, 1974; Miller and Petak, 1973; Shapiro, 1978; Van Gheluwe, 1974, 1978; Walton, 1979). The use of optoelectronic devices to acquire displacement data is a promising development to replace cinematography. These new techniques include: (1) Polarized light goniometry (Grieve, 1969; Mitchelson, 1973, 1975; Reed and Reynolds, 1969); (2) Automatic image analysis in which a television image (Winter, Greenlaw and Hobson, 1972) or a cine film (Kasvand, Milner and Rapley, 1974) is scanned by computer to determine 'X', 'Y' and/or 'Z' coordinates and finally different kinematic variables may be computed to provide the necessary analysis numerically and/or graphically for clinic research and teaching/coaching purposes; (3) Light spot position measurement, which uses optoelectronic devices such as the SELSPOT (Lindholm, 1974). The area of kinematics has received relatively limited attention (Miller, 1978). Accuracy in the measurement of external forces between the ground and the foot improved during the 1970s due to advances in force platform equipment and methodology (Cavagna, 1975; Matake, 1976; Paul, 1975; Payne, Slater and Telford, 1968; Ramey, 1975).

Recent developments include a method of analyzing the centre of pressure location in conjunction with horizontal and vertical components of ground reaction (Cavangh, 1978b). While kinematic and kinetic analyses permit the explanation of the dynamics of human motion, researchers are interested in solving specific problems in human mechanics, such as determining how a given sport skill or movement can be improved by modeling, computer simulation, optimization and other statistical approaches to motion analysis. Use of a computer is essential to the efficient implementation of these methods. Also, specific data on body segment parameter (Hay, 1978b, 1974b; Jensen, 1978; Miller and Nelson, 1973) are critical in this area of research. Among the optimizing criteria that have been applied are: minimum energy, minimum time, minimal tensile stress in ligaments (Cavangh, 1978a). To identify the kinematic and anthropometric variables that were important to the successful performance of a skill, a statistical approach was employed by Hay, Wilson and Depena (1976) and Zatziorsky (1974).

Application of Biomechanics in Physical Education and Sports

The following are the importance of biomechanics bypassed by the Indians resulting to worst:

1. Knowledge of biomechanics helps to realize or understand the underline principle of efficient structure of a competitive sports performance.

Performance in any game is dependent on some basic movements and accurate application of these movements results in better performance. So, biomechanics provides the basis for effective and efficient movement as well as correct action by applications of appropriate principles and laws for the same.

2. Knowledge of biomechanics helps to improve motor qualities.

    Evaluation of exercise and activity, method of exercise, equipments or machines for exercises, posture for exercises, modalities for exercises etc. from the point of view of their effect on the human structure whether it is general/special/specific.

3. Knowledge of biomechanics helps the athletes themselves for self-evaluation and perfection.

    Knowledge of biomechanics helps in self-realization and improvement of performance. Biomechanics helps to develop the basis of some established logistics of biomechanical understanding of a technique, which help the players in self-evaluation and realization of their movements as well as effect of training with the improved self-realization of one player (due to internal motivation) who realizes and corrects his/her own faults, and thus, improves the performance to a lever that is higher, stronger and with persisted consistency.

4. Knowledge of biomechanics helps to evolve and understand new rules and regulation of games/sports, facilities etc.

    Rules and regulations of any game/sports are changed due to three reasons: (1) for security and safety; (2) to gain more challenges, excitement and entertainment in the game by introducing more and more dynamics; (3) to increase the popularity of the game, e.g. one-day cricket matches were introduced to gain more excitement. Thereafter, day-night matches and coloured kits were introduced for more entertainment. Such changes markedly depend upon biomechanical analysis and applications.

5. Knowledge of biomechanics helps in the development/acceptance of new technique/skill.

6. Knowledge of biomechanics helps in selection of the player for particular games and sports.

    Selecting players for any team is the foremost requirement in any game/sports and it includes the following:

    a. Talent Search: In developed countries, talent is spotted at very early age/stage through various biomechanical measures so that the talent can be used at the right place and at the right time.
    b. Placement or Grouping: Placement of the right people or intelligent grouping of players, considering their biomechanical characteristics is an importance consideration for success.
    c. Selection of Game: Selection of particular game, team or a particular position, e.g. forward or backward etc. very much depends on certain biomechanical/structural requirements.
    d. Planning and Strategy: Strategic planning is very useful. Thus, according to certain biomechanical performance qualities, the players are selected in respect to the environment, time and conditions of the play etc. for ultimate success.

7. Biomechanics helps in the selection/development of equipment/facilities.

    Biomechanics has helped in developing or designing various sports equipments such as, fibre hockey sticks, fibre badminton racket that have more mechanical characteristics, and thus, influence the performance, i.e. enhancement of performance etc. It has also helped in selecting equipments for particular needs, such as golfers use different sticks for different performance factors like distance or accuracy/perfection etc.

8. It helps in prevention, protection and rehabilitation of sports injuries by means of: (a) protective equipment; (b) specialized equipment for preventive purposes; (c) preventive fitness or strengthening exercises; (d) preventive skills and tactical applications; (e) appropriate motor

development; (f) fitness management; (g) injury management by improving the training and training procedure/facilities; (h) treatment; (i) rehabilitation measures; and (j) sports physiotherapy etc. Let us take an example from the most popular Indian sport, cricket. Its star, Sachin Tendulkar, was out of the team due to lateral epicondylitis (tennis elbow). He was to be examined by the BCCI physician, Anant Joshi.

Sachin Tendulkar was out till the 3rd test against Australia up to October end. Lateral epicondylitis is an 'overuse injury' caused to the extensor muscles because of repetitive movement of the forearm and wrist. The cause of the injury might have been prevented, if biomechanics had been one of the basis of Sachin Tendulkar's training and development from the very beginning.

9. It helps to determine mechanical advantages and/or disadvantages in human locomotion/ movement.

Biomechanics helps us in knowing the merits and demerits of a technique, i.e. mechanical advantage or disadvantage and moreover, its suitability for a particular person. Thus, it helps in accepting or rejecting a technique.

This can be easily explained with the help of very simple examples, such as when a person walk without swinging arms, he finds difficulty in walking in comparison to the one who walk comfortably with swinging arms, which are investigated in terms of energy cost and effectiveness of a movement.

10. The knowledge of biomechanics helps in diagnostic teaching.

Biomechanics helps in teaching any skill or movement with the understanding of cause and effect. In other words, it helps us to answer the 'Why' and 'How' of any movement, to know the causes of errors and correcting the same at the primary stage that is in the process of motor development.

Let us take the example of Zaheer Khan, who is suffering from hamstring pull and is out of the team following hamstring tweak during the recent Asia Cup. According to the team captain Saurav Ganguli, he was "Rested till the Australia and South Africa Series". On how the injury was caused, Saurav Ganguly said that it was "probably by returning to team too soon without adequate recovery, not warming-up properly prior to bowling, poor lower back flexibility, improper biomechanics". Surprisingly, the team captain Sourav Ganguly is aware of the term biomechanics!

11. The knowledge of biomechanics helps for diagnostic coaching. Specialized teaching is coaching and it is challenge oriented. As biomechanics is used in diagnostic teaching, it is used in diagnostic coaching too, to produce the best result.

Here, let us also discuss about the groin strain of Lakshmipati Balaji, who is out of Champions Trophy because of ostitis pubis (groin strain). Sourav Ganguly had stated that Balaji was "Given an injection on September 7, 2004, but no appreciable improvement". Andrew Leipus, the team's physiotherapist, said that "He will take time to recover as the injury involves bone inflammation". How it was caused via a tear/rapture to either part of or all the five adductor muscles located on the inside of the thigh? Probably the injury was caused by not warming-up adequately or sudden jerk in movement.

12. For physiotherapy, physical medicine purposes.
13. For postural analysis and correctives physical education.

Chronic movements or actions/weaknesses sometimes lead to deformity (kyphosis, lordosis, scoliosis etc.) in posture or movements, which must be corrected. Otherwise they may also lead to some major deformities or handicap, e.g. the injury caused to Lakshmipati Balaji.

References

1. Bergemann, B.W. 1947. Three Dimensional Cinematography: A flexible approach. *Research Quarterly*. **45**: 302-309.
2. Contini, R. and Drills, R. 1954. Biomechanics. *Applied Mechanics Reviews*. **7(2)**: 49-52.
3. Contini, R. and Drillis, R. 1966a. Biomechanics. In *Applied Mechanics Reviews*. H.N. Abramson, H. Libowitz, J.M. Crowley and S. Juhasz (eds.). Spartan Books, Washington, D.C.
4. Cooper, J. M. and Glassow, R.B. 1976. Kinesiology, 4th edn. C.V. Mosby Co., S. Louis.
5. Dillman, C.J. and Nelson, R.C. 1968. The Mechanical Energy Transformations of Pole Vaulting with a Fiberglass Pole. *Journal of Biomechanics*. **1**: 175-183.
6. Frankel, V.H. and Burstein, A.H. 1975. Recent Advances in the Biomechanics of Sports Injuries. *Acta Orthopaedia Scandinavica*. **46 (3)**: 484-497.
7. Fung, Y.C. 1968. Biomechanics: Its scope, history and some problems of continuum mechanics in physiology. *Applied Mechanics Reviews*. **21 (1)**: 1-20.
8. Grieve, D.W. 1969. A Device Called the Pologon for the Measurement of the Orientation of Parts of the Body Relative to a Fixed External Axis. *Journal of Physiology*. pp. 201-270.
9. Hatze, H. 1974. The Meaning of the Term: Biomechanics (A Letter). *Journal of Biomechanics*. **7**: 189-190.
10. Kasvand, T., Milner, M. and Rapley, L.F. 1974. A Computer-Based System for the Analysis of Some Aspects of Human Locomotion. *Human Locomotion Engineering*. Institution of Mechanical Engineers, London.
11. Lindholm, L.E. 1974. An Opto-Electronic Instrument for Remote On-Line Movement Monitoring. In *Biomechanics IV*. R.C. Nelson and C.A. Morehouse (eds.). University Park Press, Baltimore.
12. Miller, D.I. and Petak, K.L. 1973. Three-Dimensional Cinematography. In *Kinesiology III*. C.J. Widule (ed.). American Association for Health, Physical Education and Recreation, Washington, D.C.
13. Mitchelson, D.L. 1973. An Opto-Electronic Technique for Analysis of Angular Movement. In *Biomechanics III*. S. Cerquiglini, A. Venerando and J. Wartenweiler (eds.). University Park Press, Baltimore.
14. Nelson, R.C. and Morehouse, C.A. (eds.) 1974. *Biomechanics IV*. University Park Press, Baltimore.
15. Plagenhoef, S.C. 1968. Computer Programmes for Obtaining Kinetic Data on Human Movement. *Journal of Biomechanics*. **1**: 221-234.
16. Read, D.J. and Reynolds, P.J. 1969. A Joint Analog Detector. *Journal of Applied Physiology*. **27**: 745-748.
17. Shapiro, R. 1956. Direct Linear Transformation Method for Three-Dimensional Cinematography. *Research Quarterly*. **49**: 197-205.
18. Skarstrom, R. 1909. *Gymnastics Kinesiology*. F.A. Bassette Co., Springfield Mass.
19. Van Gheluwe, B. 1974. A New Three-Dimensional Filming Technique Involving Simplified Alignment and Measurement Procedures. In *Biomechanics IV*. R.C. Nelson and C.A. Morehouse (eds.). University Park Press, Baltimore.
20. Walton, J.S. 1979. Close-Range Cine-Photogrammetry: Another approach to motion analysis. In *Science in Biomechanics Cinematography*. J. Terauds (ed.). Academic Publishers, Del Mar, Calif.
21. Wartenweiler, J. 1973. Status Report on Biomechanics. In *Biomechanics III*. S. Cerquiglini, A. Venerando, J. Wartenweiler, J. Joki and M. Hebbelinck (eds.). University Park Press, Baltimore.
22. Wartenweiler, J., Joki, E. and Hebbelinck, M. (eds.) 1968. *Biomechanics*. S. Karger, New York.

Biomechanics
R.K. Saxena and P. Mishra (Editors)
Copyright © 2005, Anamaya Publishers, New Delhi, India

27. Role of Obesity in Genesis of Osteoarthritis

Amrita Parle
Delhi Institute of Pharmaceutical Sciences and Research, New Delhi - 110017, India

Abstract: Obesity/adiposity or corpulence is an abnormal increase in fat in connective tissues. The increasing industrialization, urbanization and mechanization caused voluntary or forced reduction in average activity without decrease in caloric intake. Sedentary lifestyles, along with high fat, high energy foods and changing mindset of people are leading to obesity, and in turn health hazards. The most common and convenient method to diagnose obesity is Body Mass Index (kg/m^2). Preventive measures especially for children and adolescents should be promoted for development of future healthy and non-obese population. Obesity is associated with health risks like cardiovascular diseases, hypertension, diabetes mellitus, colon cancer, osteoarthritis etc. The proportion of osteoarthritis attributable to obesity is estimated to be 63%. Obesity hormone, leptin, is also involved in the development of osteoarthritis. As arthritis is an incurable disease, the possible intervention for relief, is by losing body weight along with the use of NSAIDS. Weight loss considerably improves the quality of life by reduction in the pain associated with movement. Weight loss can be achieved by combination of dietary therapy, exercise, behaviour therapy, pharmacotherapy and surgery.

Introduction

Obesity/adiposity or corpulence is an abnormal increase in fat in the connective tissue. The increasing industrialization, urbanization and mechanization caused voluntary or forced reduction in average activity without decrease in caloric intake. Sedentary life styles along with high fat, high-energy foods and changing mindset of the people is leading to obesity and in turn health hazards. WHO defines obesity as globesity because of worldwide occurrence of obesity. Obesity is a metabolic disorder caused by an energy imbalance resulting from excessive intake and inadequate caloric loss. World Health Organization has estimated that there are more than 250 million prevalent cases of obesity worldwide, equivalent to 7% of the adult population. The size of the problem is expected to rise further in future. In different countries, the obesity related costs have been estimated to account for 2-8% of the total health care expenditures.

The most common and convenient method to diagnose obesity is BMI or QUETELET INDEX. BMI can be calculated by dividing patient's weight by patient's height (kg/m^2). Table 1 shows that WHO uses a BMI of 30 to define obesity, while 25 as overweight. In Asia, a BMI of 23 is recommended as a cut off point for overweight people.

Etiology

The imbalance between caloric intake and utilization can occur in the following situations:

- Inadequate pushing off oneself away from dining table causing overeating.

- Insufficient pushing off oneself out of chair leading to inactivity and sedentary life style.
- Genetic predisposition to develop obesity.
- Diet largely derived from carbohydrates and fats than protein-rich diet.
- High intake of sugar sweetened beverages and sweets.
- High intake of heavily marketed fast foods and energy dense micro-nutrient (vitamins) poor food.
- Secondary obesity may result due to underlying diseases like hypothyroidism, Cushing's disease, insulinemia and hypothalamic disorders.
- As one grows older, metabolic rate slows down and same caloric intake leads to gain of weight.
- Men have more lean body mass and thus, have higher metabolic rate than females. So with same caloric intake, women are more obese than men. The females start gaining weight after menopause because their metabolic rate decreases significantly.

Table 1. Classification of overweight in adults according to body mass index (BMI) as per WHO technical report 2003

Classification	BMI (kg/m^2)	Risk of Comorbidities
Underweight	< 18.5	Low
Normal	18.5-24.9	Average
Overweight	> 25.0	Increased
Pre obese	25.0-29.9	Increased
Obese Class I	30.0-34.9	Moderate
Obese Class II	35.0-39.9	Severe
Obese Class III	> 40	Very Severe

Factors that might promote or protect against weight gain and obesity

Decreased Risk	Increased risk
• Regular physical activity	• Sedentary life styles
• High dietary intake of non-starch polysaccharide and dietary fibre.	• High intake of energy dense micronutrient poor food.
• Home and school environment that support healthy food choices for children and increased sports activity.	• Heavy marketing of energy dense foods and fast food outlets, decreased sports activity
• Low glycaemic index foods.	• High glycaemic index foods
• Increased eating frequency	• Rigid, restrained eating pattern with less frequency
• Less alcohol	• More alcohol

For prevention of obesity in children and adolescents we must do the following:

- Promote an active life style.
- Promote intake of fruits and vegetables.
- Restrict the intake of energy dense micronutrient poor food and soft drinks.

Obesity and Related Risks

In humans when BMI reaches 30, it results in a rapid increase in mortality and morbidity. Increased morbidity associated with obesity results primarily from increased risks for cardiovascular disease, high blood pressure, diabetes mellitus, possibly some types of cancer, gallbladder disease/cholelithiasis, pulmonary dysfunction/hypoventilation syndrome and joint problems/osteoarthritis/gout.

Obesity and Osteoarthritis

Obese individuals are more prone to develop degenerative joint diseases due to wear and tear following trauma to joints because of increased weight bearing. Overweight people are 6 times as likely to end up with arthritis in both of their knees, while obese people are 8 times as likely as compared to normal weight individuals. Obesity seems to be a mechanical risk factor for osteoarthritis. Osteoarthritis occurs when the cartilage on the ends of your bones wear away with use. As this cushion degenerates the bones rub against each other causing pain.

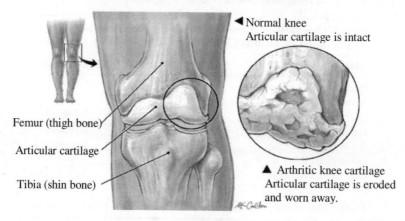

Fig. 1

Excess weight puts more force on these joints. It has been observed that weight bearing joints have to bear 1.5 times body weight while walking on level ground, 2-3 times body weight while getting out of chair, 4-5 times body weight while jumping. For patients who are overweight this means dramatically increased force on the joints. This leads to swelling and rigidity of limbs. For every 2 lb increase in weight the risk of arthritis increases by 9-13%. Females are more prone to arthritis than males after the age of 45.

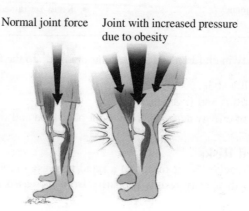

Fig. 2

Figure 3 clearly indicates that as BMI increases the chances of developing arthritis also increase.

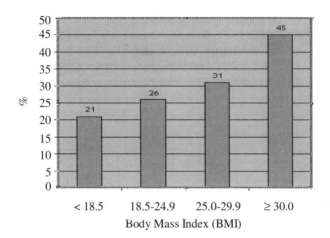

Fig. 3. Adult New Yorkers with arthritis.

Osteoarthritis in non-weight bearing joints also occurs in obese patients. This can possibly be explained as there is a correlation between high leptin concentrations with high BMI. Obesity hormone, leptin, plays an important role in development of osteoarthritis. Leptin is found in the synovial fluid of arthritic patients and leptin level is higher in heavier patients. Leptin levels were also abnormally high in osteoarthritic cartilage producing cells and number of cells producing leptin was parallel to severity of arthritis in the patients. So, leptin contributes to pathogenesis of osteoarthritis.

Weight Loss

Losing weight is the best way to help the aching and degenerating joints. Reduced weight minimizes the stress on weight bearing joints like knee. Weight loss requires a combination of therapeutic modalities including dietary therapy, exercise, behavior therapy, and if necessary, pharmacotherapy and surgery. Weight reduction programs must be designed to fit the personality, lifestyle and health status of each patient. One should target weight reduction in a series of manageable steps rather than aiming for an ideal weight from the outset, because this is easier to achieve and maintain. Weight loss should not exceed 1 kg/week. The different aspects of treatment are as follows:

Diet

Dietary treatment is the cornerstone for obesity management. To win the battle against obesity there is no substitute for good life style. The importance of learning good eating habits and familiarity with caloric content of food should be greatly emphasized. The diet programme should be directed towards slow and steady weight loss. The primary goal of any diet is to reduce the caloric intake below expenditure so excess energy stored in the form of fat can be used. Ingestion of frequent small meals with high fibre content is a way of decreasing fat intake and providing continued gastrointestinal fill.

Behaviour Modification

Behaviour therapy motivates the obese person for weight reduction and maintenance of reduced weight. The therapy can be divided into three components:

1. Self monitoring involves daily recording of type of food items eaten, and when and where they are eaten. This helps in identifying eating patterns.
2. Focusing on breaking the relationship between actual eating and external cues. Often this is done by assigning specific time and places in which meals can be eaten, chewing food a specific number of times or taking sips of water in between each bite.
3. Incorporating positive self-feedback to reinforce and maintain optimal attitude towards weight reduction. Thinking patterns should change from negative, i.e. blaming ourselves for eating, to positive, i.e. we need to exercise.

Exercise

The inclusion of physical exercise in a weight reduction program can be a valuable supplement to dieting. Regular exercise increases the degree of energy expenditure, favoring adipose tissue reduction, development of lean body mass and maintenance of weight loss. Muscles use lots of energy, so by increasing the amount of muscles in your body and by using muscles more, more calories can be burnt off. Exercise must be regular, of high quality and consistent with the patient's life style. A quality exercise regimen of 30-45 min at least 5 days/week is ideal. The selection of an activity of moderate intensity and longer duration is preferred because the longer time frame favors usage of fat stores. Conscious efforts should be made to increase energy expenditure by preferring climbing the stairs instead of taking an elevator/escalator. The mindset of people should be changed to make exercise as a natural habit. This can be done by regularly exercising for 6 months and enjoying the sense of feeling high after a bout of exercise.

Pharmacotherapy

Appetite Suppressants

Appetite suppressing drugs may be useful but should be kept reserved for individuals with a BMI above 30 kg/m^2, or above 27 kg/m^2, if associated with weight related comorbid conditions. Appetite suppressants are thought to exert their effect directly on the hypothalamic satiety center, which is under adrenergic control. The appetite suppressants can be divided into the following:

1. Adrenergic agents: Benzphetamine, Mazindol, Phenyl propanolamine etc.
2. Serotonergic agent: Fenfluramine, Fluoxetine etc.
3. Adrenergic and serotonergic agent: Sibutramine
4. Lipase inhibitors: Orlistat.
5. Bulk forming agents: Ispagula husk.
6. Thermogenic agents: Thyroxine, Thyronine.

Surgery

For individuals with BMI above 40 and 35 kg/m^2 with comorbid conditions like diabetes and heart problems, surgical manipulation of the GIT, mandibular fixation, vagotomy, liposuction can be considered.

Summary

- Prevalence of obesity greatly increases the risk of developing osteoarthritis due to increased load on weight bearing joints.

- As prevention is better than cure, life style involving less caloric intake and adequate caloric loss should be adapted.
- This can be achieved by eating fibre, micronutrient and protein rich non-fatty diet along with regular physical exercise rather than dependence on the drugs/surgery.
- Weight reduction minimizes the pain associated with daily activities of osteoarthritic patients.

References

1. Barkeling, B., Elfhag, K. et al. 2003. Short Term Effects of Sibutaramine on Appetite and Eating Behavior and Long Term Therapeutic Outcome. *Int. J. of Obesity.* **27 [6, 1]**: 693-700.
2. Cada, D.J., Covington, T.R. et al. 2003. *Drugs Facts and Comparisons, 57th edn.*, Missouri. Facts and Comparisons.
3. Cole, S.A., Martin, L.J. et al. 2003. Genetics of Leptin Expression in Baboons. *Int. J. of Obesity.* **27 [6, 2]**: 778.
4. Colin, D. 1993. *Therapeutic Drugs, Vols. 1 and 2, 2nd edn.* Chruchill Livingstone, Edinburgh.
5. Erikson, J., Forsen, T. et al. 2003. Obesity from Cradle to Grave. *Int. J. of Obesity.* **27 [6, 1]**. 722-727.
6. Hardman, J.G., Limbird, L.E. 1996. *Goodman and Gilman's: The pharmacological basis of therapeutics, 9th edn.* McGraw-Hill, U.S.A.
7. Herfindal, E.T., Gourley, D.R. 2000. *Textbook of Therapeutics, 7th edn.* Lippincot, Philadelphia.
8. Koutasari, J., Karpe, F. et al. 2003. Plasma Leptin is Influenced by Diet Composition and Exercise. *Int. J. of Obesity.* **27 [8, 1]**: 901-906.
9. Longo, F. et al. 1998. *Harrison's Principle of Internal Medicine, 14th edn.* McGraw-Hill Company, U.S.A.
10. Mohan, H. 2000. *Text Book of Pathology, 4th edn.* Jaypee Brothers, New Delhi.
11. Proctor, M.H., Moore, L. et al. 2003. Television Viewing and Change in Body Fat from Pre-school to Early Adolescence: The Framingham children's study. *Int. J. of Obesity.* **27 [6, 1]**: 827-833.
12. Roca, P., Proenza, A.M. et al. 1999. Sex Differences in the Effect of Obesity on Human Plasma Tryptoharvilage Neutral Amino Ratio. *Ann. Nutr. Metab.* **43 [3]**: 145-151.
13. Huether, S.E., McCance, K.L. 1998. *Pathophysiology, 3rd edn.* Mosby, Missouri.
14. Vas, A. 2002. Obesity Causes 30,000 Deaths. *BMJ.* **324**: 192.
15. von Eyben, F.E., Mouritsen, E. et al. 2003. Intra-abdominal Obesity and Metabolic Risk Factors: A study of young adults. *Int. J. of Obesity.* **27 [8, 1]**: 941-949.
16. Walker, R., Edward, C. 2003. *Clinical Pharmacy and Therapeutics, 3rd edn.* Elsevier Science Ltd., Edinburgh.
17. WHO Technical Reports Series 916. 2003. *Diet, Nutrition and Prevention of Chronic Diseases.* Report of a Joint FAO Expert Consultation. WHO, Geneva. pp. 54-71.
18. *Arthritis and Rheumatism.* 2003. **48**: 3118-3129.
19. *File://A:\The weight of obesity.htm*
20. *File://A:\Indiana State Department of Health.htm*
21. *File://C:\frequently asked questions about arthritis.htm*
22. www.Thismorning.co.uk/archives/series13/october/obesity.htm

Biomechanics
R.K. Saxena and P. Mishra (Editors)
Copyright © 2005, Anamaya Publishers, New Delhi, India

28. Simulation of the Coronary Blood Flow in the Domain of a Severe Stenosis Employing a Dispersed Two-phase Modeling Approach

Gladwin Philip and Nimesh Prakash
All Saints' College of Technology, Bhopal - 462001, India

Abstract: The coronary blood flow can be reduced by the formation and development of atherosclerotic plaques, by stenoses and by thrombi, which usually develops in the region of stenoses in the epicardial arteries. These pathological changes are caused by physiological processes that are strongly influenced by hemodynamic factors and are not yet very well understood. A fair knowledge of the three-dimensional blood flow in the segments of the epicardial arteries involved in these processes is a prerequisite for further research in this medical problem area. A distributed parameter model was developed of the blood flow in the epicardial arteries, which allows simulation studies of the flow field in the regions where the above-mentioned pathophysiological changes occur. In this paper, the simulation approach has been described and important simulation results presented.

Introduction

The cardiovascular system performs several essential physiological functions. One important function is to transport the substances involved in metabolic processes taking place in the cells (tissues and organs). Substances that play a major role in these processes are oxygen, various nutrients, and carbon dioxide. The cardiovascular system consists of the heart, the pulmonary circulation, and the systemic circulation. A relatively small but very important part of the systemic circulation is the coronary circulation, which is responsible for the supply of blood to the myocardium. The coronary vessels comprise the epicardial arteries, the intramyocardial arteries and arterioles, the capillary bed, the intramyocardial venules and veins, and the epicardial veins. The entire system of the coronary vessels is subdivided into the left and the right coronary system, also called the left and right coronary network, since both have a tree-like structure. The coronary blood flow must be sufficiently high to cover the metabolic requirements of the heart; otherwise, the myocardium becomes vulnerable to ischemia (Beyar et al., 1993; Spaan, 1991). The coronary blood flow can be reduced by diffuse atherosclerotic plaques and especially by stenoses in the epicardial arteries as well as by thrombi. Thrombus formation and development usually takes place in the area of a stenosis in the epicardial arterial network. It is, thus, very important to provide physicians with simulation models that:

(a) permit a quantitative assessment of the extent of such impairments and

(b) also make it possible to predict the progression of the pathophysiological changes caused by such pathophysiological processes.

The adverse effects of such stenoses and thrombi are of particular importance to cardiology and coronary surgery (Lei, Kleinstreuer and Archie, 1997). Due to the complexity of the coronary vessels, the blood flow in the entire coronary network can only be simulated on the basis of lumped parameter model. Such a lumped parameter model, which allows a quantitative assessment of the reduction in supply of blood to the myocardium caused by stenoses in the epicardial arteries was developed. However, a lumped parameter modeling approach cannot provide sufficiently detailed hemodynamic knowledge that would be required for investigations of:

(a) the development of atherosclerotic plaques,
(b) the progression of the narrowing of the lumen in the region of a stenosis, and
(c) the formation as well as the development of thrombi, which normally takes place in the immediate vicinity of a stenosis.

These medical problem areas require a fair knowledge of the three-dimensional flow in the involved epicardial artery. In this paper computer simulations of the three-dimensional flow field around a severe, eccentric stenosis has been described.

Medical Problem Area and Motivations for Our Simulation Efforts

The pathophysiological process involved (a) in the initiation and progression of atherosclerosis (atherosclerotic plaques) in the coronary arteries as well as (b) in the formation and development of thrombi, especially in the region of stenoses in the epicardial arteries, are determined by hemodynamic factors. These processes and particularly the development and progression of the resulting pathophysiological changes are not yet very well understood, and extensive research is going on in this medical problem area. A fair knowledge of the three-dimensional flow field in the regions where these processes take place is an important precondition for all further research activities in this discipline of considerable clinical relevance.

Clinical Relevance of Specific Pathophysiological Changes in the Epicardial Arteries

The formation and growth of atherosclerotic plaque may cause diffuse narrowing of the epicardial artery involved in this pathophysiological process (Gertz, Kurgan and Banai, 1995; Hunt, Hopkins and Williams, 1996). Advanced stages of atherosclerotic plaque growth may give rise to stenoses, which are localized plaques with an irregular and highly complex geometry. In general, atherosclerotic plaques are not randomly distributed along the inner wall of a particular (epicardial) artery. Instead, the sites of atherosclerosis predilection are frequently arterial segments where the blood flow is disturbed. Atherosclerotic plaques and especially stenoses cause a narrowing of the lumen and thus an increase in the resistance to flow, which results in an impairment of the coronary circulation. A mature atherosclerotic plaque or a stenosis caused by continuing lipid accumulation in the diseased epicardial artery may take many years to develop. Thereafter, however, further progression of the pathophysiological changes and the impairment of the coronary circulation may occur rapidly due to thrombus formation and development (Badimon and Badimon, 1996; Fuster and Verstraete, 1992). The high values of shear stress at the inner arterial wall and in the flow domain of the central region of a severe stenosis (apex of the stenosis) (a) promote contact between the cellular components of blood (especially platelets) and the vessel wall and (b) favor platelet activation.

This facilitates the formation and the growth of a thrombus, which is predominantly formed by platelets ("whitish" thrombus) at the apex of the stenosis (Barstad, Kierulf and Sakariassen, 1996; Mailhac et al., 1994). In immediate vicinity of the central region of the stenosis, downstream from the apex of the stenosis where shear stress and velocity are low and the flow field is highly irregular, the thrombus is composed mainly of fibrin, erythrocytes, and leukocytes ("reddish" thrombus). Once formed, the thrombus progressively enlarges, causing more and more of the lumen to become obliterated. A growing thrombus may embolize and be washed away by the blood to occlude a vessel downstream from its original location.

Motivations for our Simulation Efforts
As pointed out above, hemodynamic factors strongly influence the development of atherosclerotic plaques and stenoses in the epicardial arteries as well as the formation and development of thrombi at the sites of stenoses. The mechanisms of these pathophysiological processes and the adverse effects caused by the resulting pathophysiological changes have been thoroughly investigated. However, many questions are still unanswered, and further investigations must be carried out. A prerequisite for all future research will be a fair knowledge of the three-dimensional blood flow in the epicardial arteries, especially in the region of stenoses and other sites with disturbed blood flow. Computer simulation based on distributed parameter models is especially an advantageous way to better understand the three-dimensional blood flow in stenosed epicardial arteries. Our simulation efforts aim at the formulation of such distributed parameter models, the performance of simulation runs and the analysis of the simulation results.

Simulation of Three-Dimensional Blood Flow around a Stenosis
The simulation approach is based on a model with distributed parameters. It is thus necessary to solve partial differential equations. In doing so, the finite element method is employed. Our simulation efforts refer to the flow around a severe, eccentric stenosis in an epicardial artery (circumflex artery). Several simplifications have been made. Nevertheless, our simulation results provide insights into the specific characteristics of the flow field around such a stenosis. Other simulation approaches to the three-dimensional blood flow in stenosed vessels (Ang and Mazumdar, 1995; Ang and Mazumdar, 1997; Nakamura and Sawada, 1988) have significant imperfections; several contemporary simulation models only allow axisymmetric analyses (Ang and Mazumdar, 1995; Nakamura and Sawada, 1988).

Simplifications
At the present stage of development, the following simplifications have been made:

(a) it was assumed that the blood flowing through the stenosed epicardial artery behaves like an incompressible Newtonian fluid,
(b) it was assumed rigid walls of the stenosed artery; hence, the fluid/structure (blood/arterial wall) inter-actions was not taken into account,
(c) the authors restricted themselves to the questionable assumption of steady flow; however, this simplification can only be regarded as a rough approximation and refers solely to the blood flow at the end of the diastole, since at this point within the cardiac cycle, the variations with time of the coronary flow are moderate.

Governing Equations

The three-dimensional blood flow in a stenosed epicardial artery can be formalized in terms of partial differential equations, which can be solved numerically using the finite element method. In this case, the partial differential equations are the three-dimensional continuity equations and the three-dimensional Navier-Stokes equations. These partial differential equations are the governing equations of the problem area.

Solution of the Governing Equations Using the Finite Element Method and the Software Package "FIDAP"

The finite element method allows the computation of an approximate solution, provided that (a) well-defined geometrical boundaries of the domain and (b) essential boundary conditions are specified.

The finite element method is characterized by a spatial discretizing of the three-dimensional flow domain under consideration. This is achieved by the generation of a mesh, usually by exploiting the facilities of contemporary mesh generators. In this way, the authors arrived at a distributed parameter model of the three-dimensional blood flow in the stenosed artery. In this paper, the three-dimensional flow in a stenosed epicardial artery is investigated. The circumflex artery with a severe stenosis (90% stenosis) has been chosen. The assumed geometry and dimensions are completely realistic; the stenosis is markedly eccentric.

The governing partial differential equations has been solved with the software package "FIDAP" (Fluid Dynamics Analysis Package) of Fluent, Inc., which is a complete, integrated computational fluid dynamics (CFD) package for simulating fluid flow. For the semi-automatic generation of the mesh, the "FIMESH" mesh generator was used incorporated in this package. Finite element analysis of a fluid flow problem produces a wealth of numerical data. However, long lists of numerical data are difficult to analyze. Therefore, our simulation results are represented as "fishnet" surface plots, contour plots, diagrams, and other graphics, which are adequate to analyses. To create them, the graphics postprocessor "FIPOST" was used, which is also included in the "FIDAP" software package.

Generation of the Mesh

In the investigations, the shape (geometry) of the flow domain is irregular and highly complex. The geometry and the dimensions of the stenosed artery as a boundary representation defined by boundary elements, namely by faces, edges and vertices were specified. To obtain an accurate solution, a mesh with tens of thousands of elements is required. Hence, the generation of the mesh is a demanding task. A mesh composed of hexahedra was generated by using a semi-automatic method, which required some interaction with the user. The method used is based on the so-called multiblock approach (George 1996). In this approach, the user subdivides the flow domain into several suitable blocks. The shape of the blocks must be such that an automatic (local) mesh generation method can be applied to each block. This automatic method is based on a discretizing of the boundary of each block, which must be specified by the user.

Figure 1 depicts the central region of the severe, eccentric stenosis. In this figure, the subdivision into blocks can easily be recognized. The edges are subdivided by points that specify the discretizing of the boundaries of the block as required by the already-mentioned automatic mesh generation method.

Figures 2 and 3 show a detail of Fig. 1 (magnified), Fig. 2 before and Fig. 3 after the automatic generation of the mesh of one block.

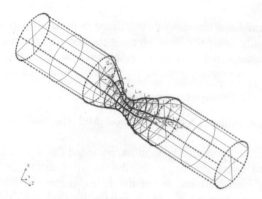

Fig. 1. Geometry and dimensions of the flow domain around the chosen eccentric stenosis in the circumflex artery, subdivided into blocks.

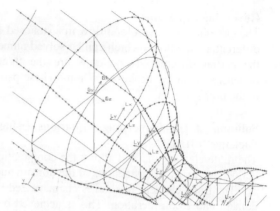

Fig. 2. Detail of geometry and dimensions of the flow domain around the chosen eccentric stenosis in the circumflex artery, subdivided into blocks, before the generation of the mesh (magnified).

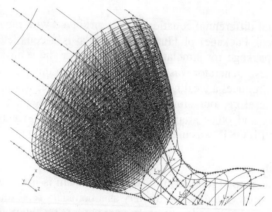

Fig. 3. Detail of geometry and dimensions of the flow domain around the chosen eccentric stenosis in the circumflex artery, subdivided into blocks, after the generation of the mesh in one block (magnified).

Boundary Conditions

In order to perform computer simulations of a (steady) flow problem, it is necessary to specify boundary conditions. It is logical to assume (a) "no slip" conditions at the wall of the stenosed artery and (b) "natural" boundary conditions at the outlet.

This term means that the normal stress is equal to zero. As the contribution of viscosity to the normal stress is relatively small, "natural" boundary conditions force the pressure to be close to zero at the outlet. It is also logical to assume a paraboloid velocity profile at the inlet, bearing in mind that this is only a qualitative characteristic. However, an appropriate quantitative specification of the boundary conditions at the inlet is also necessary. In general, it is difficult to arrive at such quantitative specification, since the model describes the flow in a particular stenosed artery that is only one of many sections of the coronary system.

The authors of previous models of three-dimensional coronary flow in the domain of a stenosis made plausible assumptions but could not prove whether or not these assumptions were really meaningful for an individual patient. In contrast to all other contemporary modeling concepts, this problem can be solved rather easily by using lumped parameter model (Quatember and Veit 1995), to calculate the required boundary conditions.

Simulation Results

In all the following graphical representations of the simulation results (Figs. 4 to 9), a three-dimensional Cartesian coordinate system (XYZ coordinates) was used. The z-axis is the longitudinal axis of the stenosed artery; the x-axis lies in the longitudinal (sectional) plane that contains the longitudinal axis z and in which the stenosis has the greatest eccentricity.

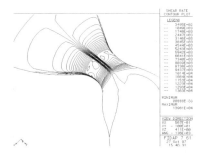

Fig. 4. Contour plot of the variation of the shear stress along the inner arterial wall; range: 0.0 to 13081.0 dyn/cm^2.

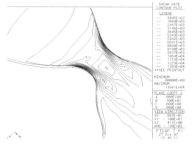

Fig. 5. Contour plot of the variation of the shear stress on longitudinal sectional plane (X-Z plane); range: 0.0 to 13981.0 dyn/cm^2.

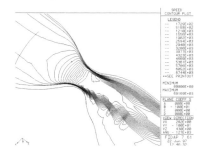

Fig. 6. Contour plot of the variation of absolute value of velocity (speed) on longitudinal sectional plane (X-Z plane); range: 0.00 to 69.17 cm/s.

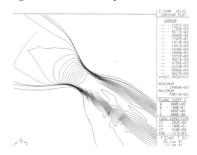

Fig. 7. Contour plot of the variation of the Z component of velocity on longitudinal sectional plane (X-Z plane); range: −24.95 to 68.61 cm/s.

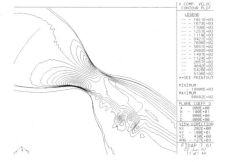

Fig. 8. Contour plot of variation of the X component of velocity on longitudinal sectional plane (X-Z plane); range: −18.80 to 8.88 cm/s.

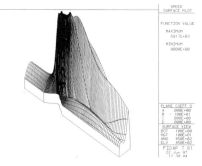

Fig. 9. Surface plot of variation of the absolute value of velocity (speed); range: 0.00 to 69.17 cm/s.

The spatial variation of shear stress is shown in Figs. 4 and 5. As described, high values of shear stress promote the activation and aggregation of platelets and the platelet/vessel wall interactions; these processes are important in the formation and growth of a thrombus.

Figure 4 depicts the spatial variation of shear stress over the inner wall of epicardial artery (only stenosed region) as a contour plot with contour lines in the range of 0.0 to 13081.0 dyn/cm^2. Figure 5 refers to the spatial variations of the shear stress in the flow domain around the stenosis; more specifically, the contour lines of the shear stress (0.0 to 13981.0 dyn/cm^2) show the spatial variation of shear stress on the X-Z plane (longitudinal sectional plane). The contour plots in Figs. 6 to 8 show the spatial variation of the velocity around the stenosis on the abovementioned longitudinal sectional plane (X-Z-plane).

Figure 6 depicts the spatial variation of the absolute value of velocity (speed), whereas Fig. 7 represents the Z component and Fig. 8 the X component of the velocity.

Another way of representing of the spatial variations of the absolute value of the velocity (speed) around the stenosis can be seen in Fig. 9. The spatial variations of the values of the speed on the above-mentioned longitudinal sectional plane are represented as a so-called surface ("fishnet") plot.

Conclusions

This paper deals with important clinical problems caused by stenoses in the epicardial arteries. The pathophysiological processes related to these problems comprise the development of atherosclerotic plaque and formation of thrombi in the stenosed epicardial arteries. These processes are strongly influenced by hemodynamic factors. At the present stage of development, several restrictive simplifications have been made. Nevertheless, simulation results have already been achieved that provide useful insights into the three-dimensional flow field around a severe, eccentric stenosis in an epicardial artery. In future, the authors would continue to refine the simulation method. In doing so, they will fully consider the fluid/structure interaction (Dankelman 1989; Spaan 1985) and thus, deal with moving boundaries. They would end the restriction of simulations to the end of the diastole and to the questionable assumption of steady flow conditions.

References

1. Ang, K.C. and Mazumdar, J. 1995. Mathematical Modelling of Triple Arterial Stenoses. *Australasian Physical & Engineering Sciences in Medicine.* **18(2)**: 89-94.
2. Ang, K.C. and Mazumdar, J.N. 1997. Mathematical Modelling of Three-Dimensional Flow through an Asymmetric Arterial Stenosis. *Mathematical and Computer Modelling.* **25(1)(January)**: 19-29.
3. Badimon, L. and Badimon, J.J. 1996. Interaction of Platelet Activation and Coagulation. In: *Atherosclerosis and Coronary Artery Disease (Vol. 1).* Fuster, V., Ross, R. and Topol, E.J. (Eds.). Lippincott-Raven, Philadelphia, New York. pp. 639-656.
4. Barstad, R.M., Kierulf, P. and Sakariassen, K.S. 1996. Collagen Induced Thrombus Formation at the Apex of Eccentric Stenoses-A time course study with non-anticoagulated human blood". *Thrombosis and Haemostasis.* **75(4)**: 685-692.
5. Beyar, R., Caminker, R., Manor, D. and Sideman, S. 1993. Coronary Flow Patterns in Normal and Ischemic Hearts: Transmyocardial and artery to vein distribution. *Annals of Biomedical Engineering.* **21(4)**: 435-458.
6. Dankelman, J. 1989. On the Dynamics of the Coronary Circulation. *Ph.D. Thesis.* University of Technology Delft, The Netherlands.
7. Fuster, V. and Verstraete, M. (Eds.). 1992. *Thrombosis in Cardiovascular Disorders.* W.B. Saunders, Philadelphia.

8. George, P.L. 1996. Automatic Mesh Generation and Finite Element Computation. In: *Handbook of Numerical Analysis (Vol. IV)*. Ciarlet, P.G. and Lions, J.L. (Eds.). Elsevier, Amsterdam. pp. 69-190.
9. Gertz, S.D., Kurgan, A. and Banai, S. 1995. Pathogenesis of Coronary Atherosclerosis. In: *Physiology and Pathophysiology of the Heart. Third Edition. Developments in Cardiovascular Medicine (Vol. 151)*. Sperelakis, N. (Ed.). Kluwer Academic Publishers, Boston, Dordrecht
10. Hunt, S.C., Hopkins, P.N. and Williams, R.R. 1996. Genetics and Mechanisms. In: *Atherosclerosis and Coronary Artery Disease (Vol. 1)*. Fuster, V. Ross, R. and Topol, E.J. (Eds.). Lippincott-Raven, Philadelphia, New York. pp. 209-235.
11. Spaan, J.A.E. 1985. Coronary Diastolic Pressure-Flow Relation and Zero Flow Pressure Explained on the Basis of Intramyocardial Compliance. *Circulation Research*. **56(3)(March)**: 293-309.
12. Spaan, J.A.E. (Ed.). 1991. Coronary Blood Flow: Mechanics, distribution, and control. In: *Developments in Cardiovascular Medicine (Vol. 124)*. Kluwer Academic Publishers, Dordrecht, Boston.

Biomechanics
R.K. Saxena and P. Mishra (Editors)
Copyright © 2005, Anamaya Publishers, New Delhi, India

29. Beta-Alanine Protects Mice from Memory-Impairment

Milind Parle and Dinesh Dhingra
Pharmacology Division, Department of Pharmaceutical Sciences,
Guru Jambheshwar University, Hisar - 125001, India

Abstract: The present study was undertaken to investigate the effects of beta-alanine (a glycine agonist) on learning and memory in mice. Beta-alanine (5, 10, 20 and 40 mg/kg i.p.) was administered for 6 successive days to separate groups of young and aged mice. The learning and memory parameters were assessed using elevated plus-maze and passive-avoidance apparatus. Beta-alanine *per se* at the dose of 10 and 20 mg/kg i.p. significantly improved learning and memory of both young and aged mice. Beta-alanine reversed scopolamine-, ethanol- and diazepam-induced impairments in learning and memory of young mice. The probable underlying mechanism of memory enhancing effect of beta-alanine appears to be related to its antioxidant, anti-amyloid or NMDA receptor modulatory activity. This study highlights the potential of beta-alanine as a memory-enhancer in treating the dementia seen in elderly patients.

Introduction

Glycine is an inhibitory transmitter between spinal inter-neurons and motor-neurons. It is the most abundant amino acid with inhibitory activity in the ventral-quadrant gray matter of the spinal cord and in the reticular formation. The NMDA receptor seems to play an important role in glutamate excitotoxicity, a process thought to be involved in a number of neurodegenerative disorders such as focal cerebral ischaemia (stroke), Parkinson's, Huntington's, Alzheimer's disease, schizophrenia and epilepsy. The unique glycine site on the NMDA receptor represents an interesting target for the development of neuroprotective compounds (Dannhardt and Kohl, 1998). Loss of glutamatergic function appears to be responsible for the deposition of amyloid beta-peptide plaques found in the brains of individuals suffering from Alzheimer's disease (Butterfield and Pocernich, 2003). The NMDA receptor complex may be looked upon as the biological target for modulating learning and memory processes. This complex can be manipulated in a number of ways, one of which is through the strychnine-insensitive glycine receptor coagonist site. Glycine (bioglycin) significantly improved retrieval of information in both young and middle-aged adults, without however, affecting attention (File et al., 1999). Beta-alanine is an inhibitory amino acid as well as a glycine agonist (Mori et al., 2002). In the light of above, the present study was undertaken to assess the potential of beta-alanine as a memory-enhancing agent employing exteroceptive and interoceptive behavioral models in mice.

Materials and Methods

Animals
Swiss male albino mice, younger ones (3 months old and weighing around 25-30 g) and aged ones (around 16 months old and weighing around 30-40 g) were used in the present study. Animals were procured from disease-free animal house of CCS Haryana Agriculture University, Hisar (Haryana, India). They were acclimatized to the laboratory conditions for 5 days before behavioral studies. The animals had free access to food and water and were stored under natural light-dark cycle. All experiments were carried out during daytime from 0900 to 1500 hrs. Institutional Animals Ethics Committee (IAEC) approved the experimental protocol.

Drugs
Beta-alanine (Hi-Media, Mumbai, India), Diazepam injection (CALMPOSE® Ranbaxy Labs., Gurgaon, India) and Scopolamine Hydrobromide (Sigma-Aldrich, USA) were used in the present study. Beta-alanine and scopolamine hydrobromide were dissolved in normal saline. Diazepam injection and ethanol were diluted with normal saline. All drugs were injected intraperitoneally (i.p.). Volume of injection was 1 ml/100g body weight of the mouse.

Laboratory Models
The experimental models used for studying learning and memory are broadly classified into:

1. Exteroceptive behavioral-models (wherein the aversive stimuli lie outside the body)

    (a) Elevated plus-maze: The procedure and end-point applied in this paradigm for testing learning and memory was as per the criteria described by the investigators working in the area of psychopharmacology and behavioral pharmacology (Dhingra et al., 2004; Itoh et al., 1990; Dhingra et al., 2003; Parle et al., 2004).
    (b) Passive-avoidance apparatus: It was another sensitive-model employed in the present study for testing learning and memory (Dhingra et al., 2004; Parle et al., 2004).

2. Interoceptive behavioral-models (wherein aversive stimuli lie within the body)

    (i) Ageing-induced amnesia
    (ii) Scopolamine-induced amnesia
    (iii) Diazepam-induced amnesia and
    (iv) Ethanol-induced amnesia.

Drug Protocol: The animals were divided into 34 groups. Each group comprised of a minimum of 5 animals. In the present investigation, dose selection was based on the pilot study and earlier reports.

Using Elevated Plus-Maze

Young Mice
Group I served as the control group for young mice.
Group II, III, IV and V: Beta-alanine (5, 10, 20, 40 mg/kg i.p., respectively) was injected for 6 days to young mice. TL was noted after 60 min of injection on 6th day and again after 24 h, i.e. on 7th day.
Group VIII: Scopolamine hydro bromide (0.4 mg/kg i.p.) before training.
Group IX: Beta-alanine (10 mg/kg i.p.) and scopolamine hydrobromide (0.4 mg/kg) before training.

Group X: Scopolamine hydrobromide (0.4 mg/kg i.p.) was injected 45 min prior to retention test.
Group XI: Beta-alanine (10 mg/kg) and scopolamine hydrobromide (0.4 mg/kg) prior to retention test.
Group XII: Diazepam (1 mg/kg).
Group XIII: Beta-alanine (10 mg/kg) and diazepam (1 mg/kg).
Group XIV: Ethanol (1 g/kg) before training.
Group XV: Ethanol (1 g/kg i.p.) prior to retention test.
Group XVI: Beta-alanine (10 mg/kg i.p.) and ethanol (1 g/kg) before training.
Group XVII: Beta-alanine (10 mg/kg i.p.) and ethanol (1 g/kg) prior to retention test.

Aged Mice
Group XVIII, XIX and XX were control, beta-alanine (10 and 20 mg/kg) respectively.

Using Passive-Avoidance Paradigm
Group XXI to XXXII: SDL was measured.
Group XXXIII and XXXIV: Effect of beta-alanine (20 mg/kg i.p. administered for 6 successive days) on locomotor function of young and aged mice respectively was studied using photoactometer (INCO, Ambala, India).

Statistical Analysis
The results were expressed as mean ± standard error of mean (SEM). The data were analyzed using the Kruskal-Wallis ANOVA and Mann-Whitney U-test. The data for locomotor activity scores and tail flick latencies were subjected to paired student's t-test. In all tests, the criterion for statistical significance was $P < 0.05$.

Results

Using Photoactometer
There was no significant effect on the locomotor activity of young (211 ± 25.3) or aged mice (201.5 ± 14.5) when treated with 20 mg/kg of beta-alanine for 6 successive days as compared to control values (164.6 ± 18.4 and 253.8 ± 19.4, respectively).

Using Elevated Plus-Maze
Beta-alanine (10 and 20 mg/kg i.p.) injected for 6 consecutive days to young mice significantly decreased transfer latencies as compared to their control group, indicating significant improvement of learning and memory. On the other hand, lower dose (5 mg/kg) and highest dose (40 mg/kg) of beta-alanine for 6 consecutive days did not have any significant effect on transfer latencies of mice as compared to control group. Scopolamine hydrobromide (0.4 mg/kg) significantly increased transfer latencies on 1st as well as 2nd day as compared to their respective control groups, when injected either before training or after training, indicating significant impairment of learning and memory. Beta-alanine (10 mg/kg) for 6 consecutive days significantly reduced the amnesia induced by scopolamine injected before or after training as indicated by decreased TL when compared to scopolamine treated group. Diazepam (1.0 mg/kg) significantly increased transfer latencies on 1st as well as 2nd day as compared to respective control groups indicating significant impairment of learning and memory. Beta-alanine (10 mg/kg) for 6 consecutive days significantly reversed the amnesia induced by diazepam, as indicated by decrease in TL as compared to diazepam treated group. Ethanol (1 g/kg)

significantly increased transfer latency as compared to control group, when injected either before training or after training indicating significant impairment of learning and memory. Beta-alanine (10 mg/kg) for 6 consecutive days significantly reversed the amnesia induced by ethanol injected before or after training (Table 1). Higher dose of beta-alanine (20 mg/kg i.p.) injected for 6 consecutive days to aged mice significantly decreased transfer latencies as compared to their control group, indicating significant improvement of both learning and memory (Table 2).

Table 1. Transfer-latency (TL) of young mice using elevated plus-maze

Group No.	Treatment	Dose (kg^{-1})	TL on Last Day of Treatment (s)	TL after 24 h (s)
I	Normal saline (Control)	10 ml	22.3 ± 1.36	15.4 ± 1.11
IV	Beta-alanine for 6 days	5 mg	24.9 ± 3.66	18.98 ± 3.12
V	Beta-alanine for 6 days	10 mg	9.36 ± 1.86*	7.98 ± 1.67*
VI	Beta-alanine for 6 days	20 mg	12.3 ± 0.97*	7.28 ± 1.57*
VII	Beta-alanine for 6 days	40 mg	26.7 ± 7.5	20.3 ± 3.96
VIII	Scopolamine (before training)	0.4 mg	52.2 ± 6.3*	30.4 ± 6.6
IX	Beta-alanine 6 days + Scopolamine (before training)	10 mg 0.4 mg	17.3 ± 1.77[a]	9.66 ± 1.58
X	Scopolamine (before retention)	0.4 mg	20.7 ± 1.95	40.6 ± 5*
XI	Beta-alanine 6 days + Scopolamine (before retention)	10 mg 0.4 mg	15.1 ± 1.79	16.04 ± 4[a]
XII	Diazepam (before training)	1 mg	49.7 ± 7.03*	30.4 ± 6.79*
XIII	Beta-alanine for 6 days + Diazepam (before training)	10 mg 1 mg	18.1 ± 2.26[a]	12.8 ± 1.5
XIV	Ethanol (before training)	1 g	36.8 ± 2.9*	14.8 ± 2.7
XV	Ethanol (before retention)	1 g	25.6 ± 0.5	37.5 ± 1.1*
XVI	Beta-alanine for 6 days + ethanol (before training)	10 mg 1 g	9.9 ± 1.45[a]	8.74 ± 1.18
XVII	Beta-alanine for 6 days + ethanol (before retention)	10 mg 1 g	11.6 ± 0.93*	11.4 ± 1.2[a]

N = 5 in each group, values are in mean ± S.E.M.
* $P < 0.05$ as compared to control group for young mice (Mann-Whitney U-test).
[a] $P < 0.05$ as compared to scopolamine, diazepam or ethanol alone (Mann-Whitney U-test).

Table 2. Effect on transfer-latency (TL) of aged mice using elevated plus-maze

Group No.	Treatment	Dose (kg^{-1})	TL on Last Day of Treatment (s)	TL after 24h (s)
I	Control (young)	10 ml	22.3 ± 1.4	15.4 ± 1.1
XVIII	Control (aged mice)	10 ml	36.4 ± 0.88*	16.2 ± 1.47
XIX	Beta-alanine (6 days)	10 mg	15 ± 1.45[b]	15.8 ± 0.77
XX	Beta-alanine (6 days)	20 mg	12.9 ± 0.74[b]	7.66 ± 0.45[b]

Using Passive-Avoidance Paradigm

Beta-alanine (10 and 20 mg/kg i.p.) administered for 6 successive days to young and aged mice significantly increased step-down latency (SDL) when exposed to passive avoidance apparatus on the 7th day as compared to their respective control groups, indicating significant improvement of memory. Ageing, scopolamine, ethanol and diazepam *per se* significantly decreased SDL as compared to the control group for young mice, indicating significant impairment of memory. Furthermore, beta-alanine (10 mg/kg) injected for 6 successive days reversed scopolamine-, diazepam- and ethanol-induced amnesia, indicated by increased SDL as compared to their respective control groups (Table 3).

Table 3. Effect of beta-alanine on step-down-latency (SDL) using passive-avoidance apparatus

Group	Mice	Treatment	Dose (kg^{-1})	SDL after 24 h (s)
XXI	Young	Control (Saline)	10 ml	135.6 ± 13.4
XXII	Young	Beta-alanine for 6 days	10 mg	209.6 ± 20.8*
XXIII	Young	Beta-alanine (6 days)	20 mg	197.4 ± 14.3*
XXIV	Young	Scopolamine	0.4 mg	15.6 ± 2.21 *
XXV	Young	Beta-alanine + Scopolamine	10/0.4 mg	29.8 ± 3.46^a
XXVI	Young	Ethanol	1.0 g	21.2 ± 5.56*
XXVII	Young	Beta-alanine + Ethanol	10 mg/1 g	219.2 ± 34.7^a
XXVIII	Young	Diazepam	1.0 mg	35.4 ± 4.15*
XXIX	Young	Beta-alanine + Diazepam	10/1 mg	229.9 ± 33.8^a
XXX	Aged	Control (Saline)	10 ml	81.6 ± 7.4*
XXXI	Aged	Beta-alanine (6 days)	10 mg	202.8 ± 40.6^b
XXXII	Aged	Beta-alanine (6 days)	20 mg	273 ± 13.2^b

Values are in mean ± S.E.M.
* $P < 0.05$ as compared to control group for young mice (Mann-Whitney U-test).
^{a} $P < 0.05$ as compared to scopolamine, diazepam or ethanol alone (Mann-Whitney U-test).
^{b} $P < 0.05$ as compared to control group for aged mice (Mann-Whitney U-test).

Discussion

The findings of the present study revealed that beta-alanine improved learning and memory of both young and aged mice, when tested on elevated plus-maze and passive-avoidance paradigms. Beta-alanine crossed the blood brain barrier via secondary active transport mechanism that is common to beta-amino acids (Komura et al., 1996). High concentrations of beta-alanine, together with the inhibitory amino acids taurine and GABA released simultaneously with excitatory amino acids in the hippocampus constituted an important protective mechanism against neuronal death due to excitotoxicity (Saransaari and Oja, 1999). NMDA type glutamate receptors are ligand-gated ion channels activated by co-agonists glutamate and glycine (Chen et al., 2004). Beta-alanine reversed amnesia produced by scopolamine (anti-cholinergic agent), ethanol (GABA agonist) and diazepam in this study. Since beta-alanine reversed scopolamine-induced amnesia, it is likely that beta-alanine improved memory, perhaps by increasing acetylcholine levels in the brain. Benzodiazepine- induced amnesia appears to be mediated through benzodiazepine receptors, since flumazenil (benzodiazepine receptor antagonist) and beta-carbolines (benzodiazepine inverse agonist) reversed benzodiazepine-induced amnesia (Jensen et al., 1987). A role for GABA (gamma amino butyric acid) in the memory enhancing activity of beta-alanine might explain the reversal of ethanol-induced amnesia by beta-alanine observed in this study. This is substantiated by the findings of Zarrindast et al. (2002) wherein, activation of GABA receptors have been shown to impair memory (Zarrindast et al., 2002).

Normal ageing is known to deteriorate memory in human beings. In the present study, aged animals showed impaired learning and memory due to ageing process. Oxygen free-radicals are implicated in the process of ageing and may be responsible for the development of Alzheimer's disease in elderly persons (Sinclair et al., 1998). Beta-alanine improved memory of both young and aged animals, in the exteroceptive models employed in this study. This effect was more pronounced in aged animals, probably because aged animals were already suffering from memory impairment due to ageing. Antioxidant-rich diets improved cerebellar physiology and motor learning in aged rats (Bickford et al., 2000). Furthermore, beta-alanine has also been reported to possess antioxidant property (Klebnov et al., 1998), thereby enhancing memory. Beta-alanine also provided protection against toxic effects of neurotoxin beta-amyloid (the main culprit for impaired neurotransmission) on rat brain vascular endothelial cells (Preston et al., 1998). Thus, memory-enhancing activity exhibited by beta-alanine in the present study appears to be related to its antioxidant or anti-amyloid effects. However, there is a possibility that positive modulation of the NMDA receptor through occupation of glycine-binding site by beta-alanine may be responsible for the observed memory enhancing activity.

References

1. Bickford, P.C., Gould, T., Briederick, L., Chadman, K., Polloch, A., Young, D., Shukitt Hale, B. and Joseph, J. 2000. Antioxidants rich Diets Improve Cerebellar Physiology and Motor Learning in Aged Rats. *Brain Res.* **866**: 211-217.
2. Butterfield, D.A. and Pocernich, C.B. 2003. The Glutamatergic System and Alzheimer's Disease: Therapeutic implications. *CNS Drugs.* **17**: 641-652.
3. Chen, N., Li, B., Murphy, T.H. and Raymond, L.A. 2004. Site within N-Methyl-D-aspartate Receptor Pore Modulates Channel Gating. *Mol. Pharmacol.* **65**: 157-164.
4. Dannhardt, G. and Kohl, B.K. 1998. The Glycine Site on the NMDA Receptor: Structure-activity relationships and possible therapeutic applications. *Curr. Med. Chem.* **5(4)**: 253-263.
5. Dhingra, D., Parle, M. and Kulkarni, S.K. 2004. Memory Enhancing Activity of *Glycyrrhiza glabra* in Mice. *J. Ethnopharmacol.* **91(2-3)**: 361-365.
6. Dhingra, D., Parle, M. and Kulkarni, S.K. 2003. Effect of Combination of Insulin with Dextrose, D (-) - Fructose and diet on learning and memory in mice. *Indian J. Pharmacol.* **35**: 151-156.
7. File, S.E., Fluck, E., Fernandes, C. 1999. Beneficial Effects of Glycine (bioglycin) on Memory and Attention in Young and Middle-aged Adults. *J. Clin. Psychopharmacol.* **19**: 506-512.
8. Itoh, J., Nabeshima, T. and Kameyama, T. 1990. Utility of an Elevated Plus Maze for the Evaluation of Nootropics, Scopolamine and Electro Convulsive Shock. *Psychopharmacol.* **101**: 27-33.
9. Jensen, L.H., Stephens, D.N., Sarter, M. and Petersen, E.N. 1987. Bidirectional Effects of Beta-carbolines and Benzodiazepines on Memory Processes. *Brain Res. Bull.* **19**: 359-364.
10. Klebnov, G.I., Teselkin, Yu O., Babenkova, I.V., Lyubitsky, O.B., Rebrova, Oyu, Blodyrev, A.A. and Vladimirov, Yua. 1998. Effect of Carnosine and its Components on Free-radical Reactions. *Membr. Cell Biol.* **12**: 89-99.
11. Komura, J., Tamai, I., Senmaru, M., Terasaki, T., Sai, Y. and Tsuji, A. 1996. Sodium and Chloride Ion-dependent Transport of Beta-alanine across the Blood-brain Barrier. *J. Neurochem.* **67**: 330-335.
12. Mori, M., Gahwiler, B.H. and Gerber, U. 2002. Beta-alanine and Taurine as Endogenous Agonists at Glycine Receptors in Rat Hippocampus *in vitro. J. Physiol.* **539**: 191-200.
13. Parle, M., Dhingra, D. and Kulkarni, S.K. 2004. Improvement of Mouse Memory by *Myristica fragrans* Seeds. *J. Medicinal Food.* **7(2)**: 157-161.
14. Preston, J.E., Hipkiss, A.R., Himsworth, D.T., Romero, I.A. and Abbott, J.N. 1998. Toxic Effects of Beta-amyloid (25-35) on Immortalized Rat Brain Endothelial Cell: Protection by carnosine, homocarnosine and beta-alanine. *Neurosci. Lett.* **242**: 105-108.
15. Saransaari, P. and Oja, S.S. 1999. Beta-alanine Release from the Adult and Developing Hippocampus is Enhanced by Ionotropic Glutamate Receptor Agonists and Cell-damaging Conditions. *Neurochem. Res.* **24**: 407-414.
16. Sinclair, A.J., Bayer, A.J., Johnston, J., Warner, C. and Maxwell, S.R. 1998. Altered Plasma Antioxidant Status in Subjects with Alzheimer's Disease and Vascular Dementia. *Int. J. Geriatr. Psychiatry.* **13(12)**: 840-855.
17. Zarrindast, M.R., Bakhsha, A., Rostami, P. and Shafaghi, B. 2002. Effects of Intrahippocampal Injection of GABAergic Drugs on Memory Retention of Passive Avoidance Learning in Rats. *J. Psychopharmacol.* **16(4)**: 313-319.

Biomechanics
R.K. Saxena and P. Mishra (Editors)
Copyright © 2005, Anamaya Publishers, New Delhi, India

30. Finite Element Analysis of Composite Hip Prosthesis

M. Sivasankar, D. Chakraborty and S.K. Dwivedy
Mechanical Engineering Department, Indian Institute of Technology Guwahati, Guwahati - 781039, India

Abstract: In this work the stress analysis of the hip prosthesis is carried out using 3D finite element method. The model consists of a conical stem with a spherical head, cement layer and cortical bone. Initially stress analysis of a prosthesis model consisting of a conical stem fitted inside the cortical bone with cement layer has been studied when a static load is applied at the head. The stem and head consists of Ti6Al4V material, the cement layer is made of UHMWPE-Al_2O_3 and cortical bone is of AS4/PEEK material. Literature review on artificial hip joints reveals that in most of the cases, interfacial shear stresses (at the interface of implant and bone) are the main reasons for the failure of such joints. In this work, the complete hip joint has been modeled using 3D finite element in ANSYS 7.0. From the finite element analysis, interfacial shear stresses have been evaluated and failure of the hip joint has been assessed. Design optimization has been carried out for minimizing the interfacial shear stresses and effects of various geometric parameters such as sphere radius, stem radius, stem length, neck radius, neck length, inclination of neck, cortical bone length, cortical bone inner and outer radii, and cement layer length, inner and outer radii on the performance of the joint has been studied.

Introduction

Total hip replacement (THR) is a successful operation for the restoration of normal daily activity of patients affected by the hip diseases. The THR was performed almost 300,000 times in United States of America in 1997 and it is expected in UK will rise from 46000 in 1996 to 65000 in 2026 (Phillips, 2001). Research in this area is going on since a long time to develop new material and techniques, which will replace affected hip without incurring much pain to the patient and last for a longer period. Recently many researchers working in this area applied finite element method to determine the failure of the artificial hip prosthesis. Prendergast (1997) has reviewed the finite element (FE) modeling in three major areas such as analysis of a skeleton, analysis and design of orthopaedic devices and analysis of tissue growth. Scifert et al. (1999) applied nonlinear FE modeling technique to analyze the stability a hip prosthesis consisting of a convex-curved acetabular lip extending from the hemispherical articulating surface to the outer edge of the cup and a femoral component with a reverse curve. They have studied the recurrent dislocation, which is the second leading cause of total hip failure next to late loosening. A theoretical protocol to predict the maximum stress induced in the stem by the ISO experimental test set-up has been investigated by Baleani et al. (2000). They used both beam theory and FE analysis for the custom made hip prosthesis and experimentally measured strain and fatigue behaviour. It is observed that FE results are more accurate than those obtained by the beam theory but both the theories overestimate the maximum equivalent stress acting on the stem. An efficient statistical predictor to optimize a flexible hip implant defined by a midstem reduction, subjected to multiple environmental conditions has been studied by Chang et al. (2001) using FE analysis. An integrated approach to the material optimization of the femoral components of hip replacements was suggested by Katoozian et al. (2001).

Recently, composite materials are mostly used for hip prosthesis applications because of their tailorability in stiffness and strength property, which will help in reduction of stress shielding (Phillips, 2001). Senapati and Pal (2002) used FE analysis for a three-dimensional cemented hip prosthesis model to study the influence of stem materials on stem-cement interface. Stolk et al. (2002) compared the FE analysis results for strains in artificial composite hip with the experimental bone and cement strains and observed that most FE bone strains corresponded to the mean experimental strains within two standard deviations and most FE cement strains were within one standard deviation. Li et al. (2003) studied the progressive failure of thick laminated composite femoral components and demonstrated how analytical and numerical models may be used before conducting extensive experimental tests as initial tools to evaluate components for the design of composite hip implants.

In most literatures cylindrical stem is used for simplification purpose. Here a conical stem is used in the FE analysis, which is similar to the actual prosthesis. The effect of different system parameters on maximum shear stress has been studied and a design optimization is carried out to minimize the maximum shear stress that developed at the proximal end of the stem cortical bone interface.

Modeling

Figure 1 shows the FE mesh for the model of a hip prosthesis consisting of a spherical head, conical stem and cortical bone. The cortical bone and the stem are separated by a thin cement layer.

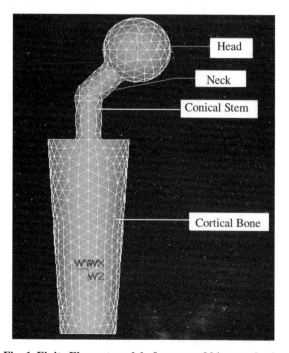

Fig. 1. Finite Element model of cemented hip prosthesis.

FE modeling was done using ANSYS Software with solid 92 element, which is a higher order 3-D 10 noded element with three degrees of freedom at each node (i.e., translations in the nodal x, y and z directions) (Fig. 2). It has quadratic displacement behaviour and is well suited to modelling irregular meshes. The material properties used for different parts of the model are described in Table 1.

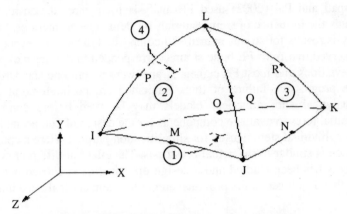

Fig. 2. Solid 92 element used in finite element analysis.

Table 1. Material properties used for analysis of total hip prosthesis

Parts	Material	Young's Modulus (MPa)	Poisson's Ratio	Geometrical Parameter (All dimensions are in mm)
Head and Stem	Ti_6Al_4V	110×10^3	0.33	Sphere radius 25 Stem radius 10 Stem outer radius 10 Stem inner radius 7.5
Cement Layer	$UHMWPE-Al_2O_3$	1×10^3	0.39	Inner radius 10.5 Outer radius 12.2182 Length 100
Cortical Bone	AS4/PEEK	3×10^3	0.30	Inner radius 20.5 Outer radius 30

In this case, considering 100 kg as the maximum weight of a patient, the maximum load acting on the two femoral head will be twice the weight, which may occur due to sudden impact. Hence a concentrated load of 1000 N is applied at the contact point between acetabulum and femoral head.

The FE model of prosthesis consists of 11127 nodes and 7162 elements and the acetabulum consists of 1593 nodes and 850 elements. The boundary condition for the prosthesis is the displacements in all directions at the cortical bone are arrested and the point load is applied over the top of the femoral head. The top surface of the acetabulum is fully constrained and the displacements in all directions were arrested over the acetabulum and the load is given in the inner middle part of the acetabulum and the stresses were calculated. The behaviors of acetabulum particularly the impacted morcelised bone graft following revision of hip arthroplasties has been studied with the help of a finite element model of acetabulum by Phillips (2001).

Table 2. Dimensions of the hip prosthesis before optimization

Parts	State Variables		Design Variables	
Femur	Sphere radius	25 mm	Stem outer radius	10 mm
			Stem inner radius	7.5 mm
	Neck inclination	45°	Stem length	145.5 mm
			Neck length	50 mm

Results and Discussions

Shear Stresses at the Interface

Since shear stresses at the interface between the stem and the cortical bone are responsible for bone resorption, they are the critical parameters to be observed during analysis.

Figure 3 shows the deformed stem and cortical bone portion of the prosthesis under static loading condition. It has been observed that the region where the stem enters inside the cortical bone (marked AA) is subjected to maximum shear stress. Hence this interfacial region is the critical zone, explaining the cause of many cases of stem failure in the artificial hip prosthesis. The shear (SXY, SYZ and SZX) stresses around the circumference of the proximal end of the stem-cortical bone interface are shown in Fig. 4. Point S of Fig. 3 corresponds to the starting point (i.e. 0) in Fig. 4.

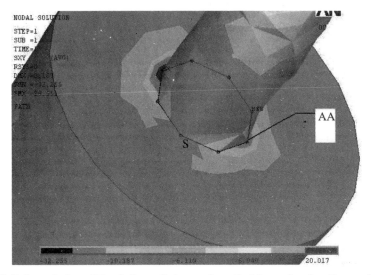

Fig. 3. Enlarged view of the deformed stem and cortical bone showing the maximum shear stress region (Path AA).

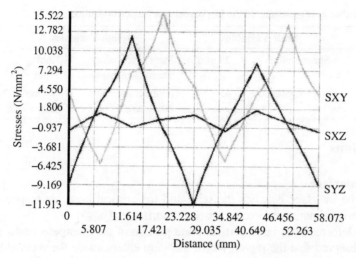

Fig. 4. Shear stresses in the interface of stem and cortical bone at the proximal end.

Table 3 shows the variation of maximum shear stress with different system parameters viz., stem length, neck length, stem inner radius and neck inclination. It may be observed that variation of individual parameter is not showing a particular trend. Hence one should go for a design optimization to find the optimum system parameters, which will minimize the maximum shear stress in the hip prosthesis. In the next section the design optimization of the prosthesis is carried out.

Table 3. Variation of maximum shear stress with system parameters

Systems Parameters	Maximum Shear Stress (MPa)
Stem length (mm)	
145	17.314
145.5	15.522
147.5	21.033
150	20.919
152.5	17.144
155	20.262
Neck length (mm)	
45	13.337
47.5	17.376
50	17.314
52.5	25.363
Stem inner radius (mm)	
7.5	17.314
8.0	20.655
8.5	19.443
Neck inclination (°)	
45	17.314
47.5	20.383
50	22.964

Design Optimization of the Prosthesis

In order to minimize the maximum shear stress at the critical zone, random design optimization procedure has been adopted. Here stem length, neck length, stem outer radius, stem inner radius are taken as the design variables and neck inclination and sphere radius are taken as the state variables and the minimization of maximum shear stress is taken as the objective function.

These variables are decided based on physical constraints, which may vary from patient to patient. For each set of random variables ANSYS routine will form the mesh and calculate the nodal displacements and stresses. The maximum shear stress among the three components is calculated for all the random sets of design variables. After performing required number of random iterations routine will give out the design set, which is the optimum solution for which objective function is minimum. In the present analysis 100 random iterations were performed for giving 20 feasible solutions. Among these feasible solutions, the minimum value of objective function, is the optimum solution.

The design variables after optimization are shown in Table 4. The variation of shear stress components in the three mutual perpendicular planes in the critical zone is shown in Fig. 5. It could be observed that the maximum shear stress is reduced from 15.522 MPa (Fig. 4) to 14.979 MPa (Fig. 5). So optimum design helps in reducing the interfacial shear stresses and thus reducing the chance of failure in the stem of the prosthesis.

Table 4. Optimum design variables of femoral components

Design Variables	Dimension (mm)
Stem outer radius	9.9301
Stem inner radius	8.0405
Stem length	153.22
Neck length	50.975

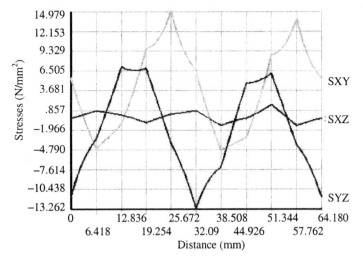

Fig. 5. Shear stresses in the interface of stem and cortical bone at the proximal end after optimization.

Conclusion

A 3D finite element analysis has been done for analysis of composite hip prosthesis. Location and magnitude of shear stresses show the region of failure, which is in agreement with the earlier published results. Here, a conical shaped stem made of Ti6Al4V implanted in the cortical bone with cement layer has been considered and the effect of stem length, neck length, stem inner radius and neck inclination on maximum shear stress in the prosthesis is investigated. As the variation of these parameters do not show a particular trend, design optimization has been carried out to minimize the magnitude of maximum shear stress. The optimum dimensions obtained from the present analysis show considerable reduction in shear stress thus justifying the efficacy of the design optimization for fixing the dimensions of stem and head.

References

1. Baleani, M., Viceconti, M., Muccini, R. and Ansaloni, M. 2000. Endurance Verification of Custom-made Hip Prostheses. *International Journal of Fatigue.* **22**: 865-871.
2. Chang, P.B., Williams, B.J., Bhalla, K.S.B., Belknap, T.W., Santner, T.J., Notz, W.I. and Bartel, D.L. 2001. Design and Analysis of Robust Total Joints Replacements: Finite element model experiments with environmental variables. *ASME, Journal of Biomechanical Engineering.* **123**: 239-246.
3. Katoozian, H., Davy, D.T., Arshi, A. and Saadati, U. 2001. Material Optimization of Femoral Component of Total Hip Prosthesis using Fiber Reinforced Polymeric Composites. *Medical Engineering and Physics.* 23: 503-509.
4. Li, C., Granger, C., Schutte, H.D. Jr., Biggers, S.B. Jr., Kennedy, J.M. and, Latour., R.A. Jr. 2003. Failure Analysis of Composite Femoral Components for Hip Arthroplasty. *Journal of Rehabilitation Research and Development.* **40(2)**: 131-146.
5. Phillips, A. 2001. *Finite Element Analysis of the Acetabulum after Impaction Grafting.* The University of Edinburgh.
6. Prendergast, P.J. 1997. Review Paper-Finite element models in tissue mechanics and orthopaedic implant design. *Clinical Biomechanics.* **12(6)**: 343-366.
7. Scifert, F., Brown, T.D. and Lipman, J.D. 1999. Finite Element Analysis of a Novel Design Approach to Resisting Total Hip Dislocation. *Clinical Biomechanics.* **14**: 697-703.
8. Senapati, S.K. and Pal, S. 2002. UHMWPE-ALUMINA Ceramic Composite: An improved prosthesis material for an artificial cemented hip joint. *Trends in Biomaterials Artificial Organs.* **16(1)**: 5-7.
9. Stolk, J., Verdonschot, N., Cristofolini, L., Toni, A. and Huiskes, R. 2002, Finite Element and Experimental Models of Cemented Hip Joint Reconstructions can Produce Similar Bone and Cement Strains in Pre-clinical Tests. *ASME, Journal of Biomechanics.* **35**: 499-510.

31. Some Biomechanical Aspects of Fitness Training for Olympic Weightlifters

Sudhir Kumar
Department of G.T.M.T., Netaji Subhash National Institute of Sports,
Patiala - 147001, India

Abstract: Weightlifting is a sport where competitors attempt to lift heavy weights on steel bars (barbell) overhead. To achieve winning performance in the weightlifting contest, the competitors should have high level of fitness. The objective of this study is to address the role of the biomechanics and exercise physiology in design and development of fitness training for the competitor. It also tries to explore some of the biomechanical aspects related to body muscles, leverages and weightlifting process. It also discusses some of the concepts of biomechanics, which are helpful in designing exercises for the weight lifter. It also suggests the role of some of the safety devices, such as lifting belt etc. in Olympic weight lifting. The study is based on the practical experience of the women weight lifters national coaching camp (Olympic probable) held during 2004 at Netaji Subhash National Institute of Sports Patiala, India.

Introduction

Weightlifting is a competitive strength sport in which men and women compete at the highest level by lifting a barbell overhead. The sport has existed on an international – the Commonwealth and Olympic level, in its current form, for more than 100 years. Weightlifting is the only Olympic sport involving weights, which is why it is sometimes referred to as Olympic lifting or as Olympic-style lifting or Olympic-style weightlifting. A weight lifter is a person whose sport is the competitive lifting of weights.

Olympic weightlifting is a highly technical event, which requires scientifically designed weight training and facilities. The knowledge of the biomechanics may be very useful in designing such weight training. The concepts of biomechanics may help in understanding the factors related to the body structure of the weight lifters specially its mechanical structure, coordination, movement, mechanics involved in weightlifting, properties and structure of the body muscles etc.

Biomechanical Aspects of Fitness Training

The ultimate goal of a comprehensive fitness training program/coaching camp for the Olympic probables is to aid the competitors in achieving their fullest potential through the acquisition of knowledge and skills of the weight lifting event. Such comprehensive scientific fitness training is designed and developed on the basis of the knowledge of biomechanics, exercise physiology, sports psychology, sports nutrition sports medicine, sports vision, and facilities. Some of the components of the weightlifting fitness training performance are given in Fig. 1.

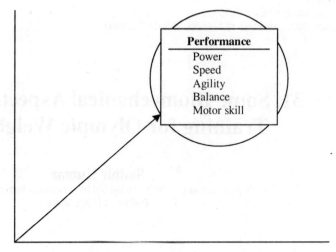

Health Related
Body composition
Cardiovascular endurance
Muscular strength
Muscular strength
Flexibility

Psychological
Motivation
Habit

Heredity, Age, Gender

Somatic state
Dimensions, composition

Social
Family, work, religion

Training, Nutrition, Technical conditions, Environment

Fig. 1. Components of weightlifting fitness.

Biomechanical Aspects of Weightlifting Process

Biomechanics is the study of the mechanics of living things (McNeil, 1987). Biomechanics deals with the principles and methods of mechanics applied to the structure and functions of living systems including any artificial interventions thereof. The working definition is based on the work of Shaw (1998), Fung (1981), Halze (1974), Wartenweiter (1973) etc. Galileo as a substitute to his book "Two New Science" used the word mechanics, 1932 to describe the force, motion and strength of materials (Fung, 1981). Mechanics is the study of the behavior of the system under the action of forces. Statics deals with cases where the forces either produce no motion or the motion are of no interest. Dynamics deals with the motion under forces (Goodman, 1987). Kinetics is the study of forces that produce or change the state of motion. Kinematics is the study of geometry of motion. It describes the motion of bodies in terms of time, displacement, velocity and acceleration. It discusses the cause(s) of force. Mechanical means for the conversion of motion. Mechanical are the core of the working of any machine (Freudestein, 1987).

The concept of biomechanics and exercise physiology is also very helpful in understanding the lifting dynamics, muscle activity and spinal load in the Olympic weightlifting procedures – the Snatch and the Clean and Jerk (C & J). The first event, the snatch, involves lifting a barbell from the platform to arm's length overhead in one continuous motion. It is one of the most difficult, explosive and elegant events in sport. The second event, called clean and jerk (C & J) involves lifting a barbell from the floor to the shoulders in one continuous motion and then, in a second motion, bringing the weight to arms's length overhead. A brief description of process of weightlifting is given as follows:

Snatch and Clean

We can divide snatch and clean in six phases of movements:

1. First phase of the pull: Pro-left off. In this phase the weight lifter stands with his/her toes directly under the bar.

2. Second phase of the pull: Preliminary acceleration. The lifter bends to hold to bar with arms out wide (about the distance from elbow to elbow when the arms are held perpendicular to the body).
3. Third phase of the pull: Adjustment. The back and arms remain straight and the shoulders should be forward of the bar.
4. Fourth phase of the pull: Final acceleration. Once in this correct starting position, the lifter pulls the bar off the floor while pushing backwards with the knees so that the bar can rise vertically. Once the lifter has pulled the bar to just above the knees, the knees move forward once more. Keeping the back straight and shoulders over the bar, the lifter explosively drives upwards, pulling the bar to hip height.
5. Fifth phase of the pull: The unsupported squat under. Having finished the 'pull', the lifter then jumps under the bar to catch it overhead in a deep squat position. In order to successfully get under large weights, it is imperative that the lifter is able to comfortably and quickly execute a deep squat. At this stage, it is also important that a lifter is able to hold the weight firmly overhead with the arms straight. If the lifter does not manage to 'lock out' his/her elbows as soon as the bar has reached arms length, the lift will be disqualified.
6. Sixth phase of the pull: The supported squat under. In the final phase, having successfully caught the bar overhead in a squat position, the lifter pushes up out of the squat until the knees are locked, the feet are in line and the body is standing motionless.

The Jerk
The Jerk can also be divided in six stages as follows:

1. First phase of the Jerk: The start. The lifter, with his/her toes directly under the bar (as in the snatch) bends down to grip the bar. Hands are slightly wider than shoulder width. With the arms kept straight, the lifter pulls the bar off the floor, pushing the knees backward at the same time.
2. Second phase of the Jerk: The dip. Once the bar is lifted past the knees, the knees come slightly forward once more. The lifter then drives upward explosively keeping the arms straight.
3. Third phase of the Jerk: The breaking phase. When the pull is finished (when the bar has reached the top of the thighs), the lifter pulls himself or herself under the weight to catch the bar in a low squat position. The weight is fixed firmly on the shoulders and the elbows are held high.
4. Fourth phase of the Jerk: The thrust or explosion. Finally, the lifter drives up out of the clean, ready for the jerk.
5. Fifth phase of the Jerk: The unsupported jerk under. Once standing, the lifter begins the jerk element of the clean and jerk by dipping down slowly, keeping the torso vertical. The body is then violently driven upwards, so that the lifter rises up onto his/her toes and the bar begins to rise in the air.
6. Sixth phase of the Jerk: The supported squat under. As soon as the bar leaves the shoulders, the lifter drives himself/herself under the bar by splitting his/her legs apart. The weight is held overhead with straight arms locked out at the elbows. The knee of the front leg must not be positioned over the ankle and the back leg should be bent at the knee with the heel off the floor. Finally, the lifter stands up out of the lift by first bringing his/her front leg back and then stepping forward with the back leg. When the lifter stands with feet inline and the weight steady, the lift is deemed complete.

Weightlifting is also a work. The following aspects of biomechanics may be useful in designing and development of the fitness-training program for the Olympic weightlifters probable:

Effort
It is influenced by the duration of contraction, i.e. the continuous period of muscle contraction or total amount of time per minute muscles are active, and frequency, i.e. the number of work cycles per minute, hour, or shift which defines the pattern of work; and intensity, i.e. heavy, moderate, or light work according to oxygen demands or strength requirements.

Energy Requirements
Energy consumption is greater with more intense contractions, longer strenuous effort and more frequent repetitions. Carbohydrates (sugars and starches) and fats (free fatty acids) are the two nutrients most involved in providing energy for muscular work. Nutrients are burned to provide energy for muscular work via two processes:

Anaerobic Processes
They are the muscular processes, which don't need oxygen. There are two ways muscles can contract without oxygen:

Anaerobic Glycolysis: Breakdown of glucose leads to lactic acid. If glycolysis is the only energy source then work cannot be sustained for very long because of lactic acid accumulation leads to cramps. Energy produced is stored in adenosine triphosphate (ATP).

High Energy Phosphates: Creatinine phosphate (CP) and ATP stored in muscles can release energy for work. As CP and ATP levels decline it leads to rapid fatigue of muscles. CP is a storage system for ATP and it replenishes ATP supply as it is used.

Aerobic Processes
They are muscular processes, which use oxygen. Small amounts of oxygen are stored bound in muscle protein, myoglobin, which can be utilized in short, intense periods of work. All other oxygen is supplied via the circulatory system. It can take up to 1 min for blood flow to respond to sudden increases in workload so there may not be sufficient oxygen at the start of work.

Aerobic Glycolysis: After CP and ATP levels decline, aerobic glycolysis takes over and by measuring the amount of oxygen consumed by a person during work we can determine the amount of aerobic metabolism taking place for that work. Aerobic glycolysis generates almost 20 times as much ATP as anaerobic glycolysis.

Oxygen Debt: This is the amount of oxygen required by muscles after the beginning of work over and above that which is supplied to them by the circulatory system during their activity. Following work, there is a recovery period during which the oxygen debt must be repaid.

Muscle Fatigue: It occurs because of decreased energy stores in muscle along with the accumulation of lactate (waste metabolic product).

Endurance Time: It is the time taken for a muscle to fatigue and this is a function of contraction force.

Stretch Reflex: As strain on muscle increases so does the reflex contraction. This stretch reflex is maintained until the load is removed. The reflex controls posture as muscles respond to being strained by gravity.

Muscle Tone: This is the result of the stretch reflex and the elastic properties of the muscle as well as the ability to resist passive movement.

Force and Torque
In the weightlifting process, muscles contract, causing bones such that bones act as a series of levers. The weight acts as a load (or resistance) on the forearm (lever).

Force
Force is a function of the strength of muscle contraction, which is translated into torque. Forces developed by muscle are results of muscles' contractile and connective tissue, and arrangements of muscles on the skeleton. Contractile elements are the individual myofibrils, which change length to produce tension. Connective tissue determines the stiffness of the muscle by acting as a parallel elastic component (like an elastic band). This component is slack at the resting length of a muscle so the length-tension curve is like that for the contractile elements alone. As the muscle is lengthened the non-linearity of the connective tissue is increased and the tissues increase in tension. When a muscle is stimulated by a single nerve action potential, a contraction called a twitch occurs. The duration of the periods during muscle contraction varies greatly between muscles. When several neural stimuli arrive at the muscle this continues summation of contraction until at a certain rate. The number of stimuli leads to tetanus, the maximal sustained contraction. Force from a contracting muscle is transmitted by the tendon. When this occurs at a joint there is movement.

The mechanical properties of muscle are:

1. Isometric: Contraction doesn't change muscle length. It produces muscle tension but no useful mechanical work.
2. Isotonic: Contraction changes muscle length and produces useful mechanical work.

Torque (Moment)
This is the tendency for forces developed by muscle-tendon complexes to cause rotation at joints. When a perpendicular force is applied on a lever arm at some distance from its axis of rotation (fulcrum) this causes a rotational tendency (torque or moment). Torques generated by the body translate muscle contraction into mechanical work.

Weightlifting and Newton's Law of Motion
The Newton's law of motion is also equally applicable to weightlifting. For example:

Newton's 1st Law of Motion states that "a body that is originally at rest will remain at rest, or a body moving with constant velocity in a straight line will maintain its motion until an external resultant force is applied".

Newton's 2nd Law of Motion states that "acceleration of a body is directly proportional to the net force acting on the body and inversely proportional to its mass, i.e. $F = ma$ where F = force, m = mass and a = acceleration".

Newton's 3rd Law of Motion states that "for every action there is an equal and opposite reaction", i.e. Velocity = Position/Time; Acceleration = Velocity/Time. Unit of force: System in the US = Pound (lb); SI system = Newton (N); c.g.s. system = dyne (dyn). $1 N = 10^5$ dyn = 0.225 lb.

Lever Systems in Weightlifting Process
In the weightlifting process, all the three classes of lever systems are used.

First Class Lever
Fulcrum (center of rotation) is the joint, and it is located between the load (resistance) and force (muscle). When Force Arm length is greater than Resistance Arm length there is a mechanical advantage, e.g. when a person is looking down into a microscope the fulcrum is at the lanto-occipatal joint connecting the head and spinal column. The mass of head is the resistance and the force is the contractions of muscles in the back of the neck. Fatigue of neck muscles from static contraction can lead to neck pain.

Second Class Lever
The Force Arm is always larger than the Resistance Arm. There are only a few examples in the body, e.g. opening mouth against a resistance.

Third Class Lever
Here, lever systems are always at a mechanical disadvantage because Resistance Arm always exceeds the Force Arm, so greater force is needed to move the load. There are many examples in the body such as raising a coffee cup.

Strength and Endurance
The performance in weightlifting is directly co-related to the level of strength and endurance. Endurance = low weight, high reps. Strength = high weight, low reps. Endurance is the ability to maintain or repeat a contraction without getting tired. Strength training makes the muscles grow thicker. In such muscles the contractions are stronger. Endurance training makes muscle better at using oxygen and burning fat for energy. It makes more capillaries grow around the muscles. All muscle training involves lifting, pushing or pulling a load or resistance. One can use the same exercises to improve strength or endurance: just change weight and repetitions (reps). Strength may be of three kinds, viz. (i) static: the strength required to hold an object above the head, (ii) dynamic: the strength required to keep a load moving, and (iii) explosive: the strength needed for a single explosive act, e.g. high jump.

Role of Biomechanics in Designing and Development of Exercise and Training Programs for Weight Lifters
To develop the skill, strength and endurance related to weightlifting, knowledge of the biomechanics and exercise physiology play an important. It helps in understanding the mechanics of muscles, types of muscular activity associated with the particular event, the movement pattern involved, the type of strength required etc. Such knowledge will be helpful in identifying the exercises that will produce the desired development. Although specificity is important, it is necessary in every schedule to include exercises of a general nature, e.g. power clean, power snatch, bench press, back squats, sit ups, shoulder press, chest press, lat pull downs, lower back extensions, triceps press, calf raise, bicep curls, leg curls, leg extension and leg press.

These general exercises give a balanced development and provide a strong base upon which highly specific exercise can be built.

Biomechanical Aspects of Weightlifting Belt

There are studies, which support that use of weightlifting belt may help in reducing injuries and at the same time also improve performance. It also helps in protecting the back from stress and strain damage, and also reduces the likelihood of pain or injury for a variety of activities. There are numerous studies indicating the use of back belts, weight belts and lumbar corsets improves performance, endurance and reduces chances of injury.

Here, it may also be mentioned that available research also demonstrates that belts are unable to stabilize the spine at a segmental level, therefore, only stabilizing the torso. Gross stabilization, as provided by belts, may allow a lifter to lift more weight than he/she could without the belt, indicating a stabilizer dysfunction within your body. The increased weight being lifted as afforded to the lifter by the belt will likely serve to traumatize the spine due to increased levels of impression, torsion and sheer, increasing the potential for a serious injury.

Caution should be exercised by those using belts to increase "proprioception", as a belt is clearly a form of "exteroceptive stimuli". When the belt is removed, it is likely to have accomplished little in improving proprioception, leaving the lifter with an increased risk of injury secondary to belt usage.

Conclusion

The objective of the discussion was to present the role of biomechanics in understanding the process of weightlifting. It also advocated that knowledge of biomechanics can play an important role in designing and development of the fitness training programs for the weight lifters. In the end, it also tried to discuss in brief the biomechanical aspects of safety devices used in weightlifting.

Acknowledgements

The study is an outcome of the project work assigned by the Sports Authority of India (SAI), under its fellowship grant to the author at Department of GTMT, Netaji Subhash National Institute of Sports (NSNIS), Patiala, India. The author expresses his gratitude to the Principal Supervisor of the project Dr. Hardyal Singh, Head, GTMT, NSNIS, as well as Dr. Samarjeet Singh, co-Supervisor, GTMT, NSNIS under whose supervision and kind help the efforts of writing this article were made.

References

1. Drechsler, A. 2002. *The Weight-Lifting Encyclopedia, 2002.* A & A Communication, New York.
2. Freudestein, F. 1987. *Mechanism. McGraw-Hill Encyclopedia of Science and Technolgy, Vol. 10.* p. 551.
3. Fung, Y.C. 1981. *Biomechanics. Mechanical Properties of Living Tissues.*
4. Goodman, B. 1987. *Mechanics. McGraw-Hill Encyclopedia of Science and Technology, Vol. 10.* p. 551.
5. Halze, H. 1974. The Meaning of Term Biomechanics. *Journal of Biomechanics.* **7**: 189-190.
6. McNeil, A.R. 1987. *Biomechanics. McGraw-Hill Encyclopedia of Science and Technolgy, Vol. 2.* pp 553- 555.
7. Shaw, D. 1998. *Biomechanics and Kinesiology of Human Motion.*
8. Singh, H. 1993. *Science of Sports Training.* D.V.S. Publication, New Delhi.
9. Singh, H. 1993. *Sports Training. General Theory and Methods.* 1984 NIS Publication.
10. Wartenweiter, J. 1973. *Biomechanics III.* Status Report in Biomechanics, 1979.

Biomechanics
R.K. Saxena and P. Mishra (Editors)
Copyright © 2005, Anamaya Publishers, New Delhi, India

32. Effect of Alcohol on Mechanical Properties of Bone

Reeva Gupta, A. Koul and D.V. Rai
Department of Biophysics, Panjab University, Chandigarh - 160014, India

Abstract: In the present study, rats were fed ethyl alcohol (EA) orally for 8 weeks. Since the effect of EA on bone mineral metabolism is poorly understood, the changes in mineral content and mechanical strength of fibula were studied after sacrificing the animals. There was no significant change in body weight of control and experimental groups. A decrease in bone strength in three-point bending was observed in the two subgroups of experimental animals as compared to the control animals. However, no significant change was seen in one of the subgroups.

Introduction

Bones consist of two-phase porous natural composite material (Park, 1992) comprised of collagen and mineral, which contribute to the support of human body. The organic component, mainly collagen, gives bone its form and contributes to its ability to resist tension, while the inorganic, or mineral component primarily resists compression. The orientation of these fibres seems to be load related (Singh et al., 1981, Rai et al., 1986). The tensile strength of bone is nearly equal to that of cast iron, bone is 3 times lighter and 10 times more flexible. Previous studies of Gaarcia et al. (2002) have shown that bone is 4 times as strong as concrete.

Certain biological parameters at macrostructural and microstructural levels which have been investigated as determinants of cortical bone are apparent density (proportional to porosity), histology (number of osteons, primary versus secondary bone), collagen composition and content and accumulation of microcracks in and around osteons (Evans, 1976; Currey, 1988, 1990). In addition, they contain haversian and volkman canals that are functionally needed but are structurally weak (Ravaglioli and Krajewski, 1992; Sadananda, 1991; Burr, 2002). However, it is believed that they play an important role on the fracture toughness due to absorption capability of the crack propagation energy. Although test on normal bones have correlated the mechanical strength of bone with its microstructure and mineral content (Evans, 1973; Saha and Hayes, 1977; Saha and Gorman, 1981), such information is lacking for pathological samples as induced by dietary disturbances. Recent evidences (Marcus et al., 2002) also suggest that bone mineral density is correlated with bone strength and fracture risk (Veenland et al., 1997).

The fibula is a long, slender bone that articulates with the lateral condyle of the tibia and runs parallel to the shaft of the tibia. Technically, it is not part of the knee joint, but it does contribute to the functioning of the joint in some important ways. The fibula provides an attachment site for some of the muscles that support the knee joint, and it is indirectly involved in stabilizing the joint. Fibula has been used as the site of experimentation for analysis of the mechanical strengths of various vascularised and non-vascularised grafts (Dell et al., 1985).

Several environmental factors are involved in the etiology and progression of osteoporosis. Smoking, alcoholism, glucocorticoids and various other medications have been noted as causative factors in the pathophysiology of osteoporosis. The cause of osteoporosis is a multifactorial entity in which alcohol consumption is known to play a part (Rico, 1990; Moniz, 1994). Clinical studies have demonstrated that chronic alcohol consumption is associated with osteopenia and increased risk of fractures (Bikle, 1993; Klein, 1997; Nyquist et al., 1997; Chakkalalkal et al., 2002). However, there are few studies that support the effect of alcohol on osteogenesis but the studies on the mechanical properties of bone are very scanty. Therefore, there is a strong need to reveal the various biological factors involving the strength of bones.

The present study is designed to investigate the effect of differential doses of ethanol on body weight, mineral content and mechanical strength of bone. An attempt has also been made to establish the relationship between mineral content and the mechanical strength.

Materials and Methods

Adult male rats of Wistar strain with body weight ranging from 170 to 210 g were procured from the central animal house of Panjab University, Chandigarh. The rats were monitored for health and body weight every week till the end of the experiment. Ethanol (Dehydrated alcohol B.P.), given during the treatment period was provided by Bengal Chemicals and Pharmaceuticals Ltd. Bombay, India.

Forty eight animals were divided into two major groups:

Group 1 (Control): Control rats were fed on standard animal chow and water ad-libitum.
Group 2 (Ethanol Treated): Animals were further divided into 3 subgroups.

Group 2 (I) (10% v/v): Animals were given 10% ethanol orally for the period of 8 weeks.
Group 2 (II) (20% v/v): Animals were given 20% ethanol orally for the period of 8 weeks.
Group 2 (III) (30% v/v): Animals were given 30% ethanol orally for the period of 8 weeks.

The rats of each control and experimental group were sacrificed to take out the fibula. Soft tissues were removed from the intact surfaces of the bone (fibulae). The bone marrow was flushed out with normal saline after giving an incision on diaphyseal ends of the bone.

Mineral Content

The powdered samples prepared from control and experimental groups were placed in crucible and subjected to 800°C for 24 h in a muffle furnace (Laboratory Model, manufactured by Narang Scientific Works Pvt. Ltd., New Delhi). The ash content for each sample was monitored. The ash content left, gave the mineral content while the weight lost gave the organic content of the bone samples (Rai and Singh, 2000).

Mechanical Study

The length and area of cross section of all the fibulae bones were measured. Three-point bending tests for both fibulae of control and treated groups were performed. The bones were held from two opposite ends by hard resinous material to prevent axial rotation and to give an accurately measurable span. The load was applied transverse to the long axis of the bone. The actual dimensions of the apparatus depended on the length of the bones. This set-up was laid horizontally on two solid steel supports. A strong wire was hooked over the bone at mid-shaft region, which carried a pan for weights. The weights were increased gradually until the bone broke. The load, defined as force in kilograms per square millimeter of cross-sectional area, represented the bone strength (Nyquist et al. 2002).

Results and Discussion

The results of body weight are shown in Fig. 1. There was slight but non-significant increase in the body weights in treated groups as compared to control. This increase in the body weight with ethanol doses could be related to caloric content. In terms of caloric content ethanol is a unique drug whose oxidation produces 7.1 kcal/g of ethanol oxidized. However, it is not known to what extent these calories contribute to body weight gain. There seems to be no consensus about the data from animals and human studies. Some studies show that end weights of male and female alcohol-fed rats were significantly lower that both control groups (Wezeman and Gonz, 2001; Tavares et al., 2000). However, one study (Brzoska et al., 2002) has demonstrated that there was no net gain in final weight after the intake of 10% ethanol in drinking water There is controversy on whether regular alcohol intake contributes to weight gain and risk of obesity.

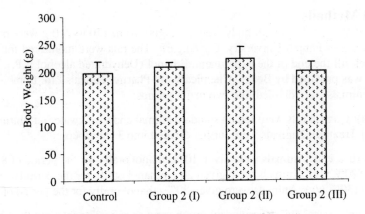

Fig. 1. Body weight of control and treated groups.

Aguiar et al. (2004) observed that moderate ethanol consumption favours an increase in fat storage and could result in weight gain.

The ash content was measured and the control bone contained 30% organic matrix and 70% inorganic matrix. A relative increase in the values (73.2%) of mineral content of Group 2 (I) was observed in comparison to control. This may be due to the recently believed fact that bone mineral density is augmented under the effects of moderate doses of ethanol (Williams et al., 2004).

However, in case of Group 2 (II) and Group 2 (III) the loss of inorganic matrix increased with the increasing doses of ethanol. The mineral content was found to be 65.6% and 62.5% respectively. The bone tissue became more flexible and soft with increasing demineralization making the bone mechanically incompetent. These observations support earlier studies (Rai and Singh 2000; Shah et al., 1995; Boskey and Posner, 1985) that mineral content is an important parameter, which determines the bone strength.

In present investigation, no significant increase in bone strength of Group 2 (I) was observed when compared to control (Fig. 2). However, a significant decrease in bone strength was found at higher doses of ethanol (Group 2 (II) and Group 2 (III) (Fig. 2)). This significant decrease can be attributed to the chronic consumption of ethanol as demonstrated by earlier studies of Nqyuist et al. (1997) and Chakkalalkal et al. (2002).

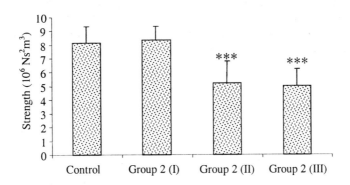

Fig. 2. Bone strength of control and treated groups.

Our findings in this study are consistent with the consensus in the literature on ethanol-induced osteopenia (Bikle et al., 1985; Klein et al., 1997; Nyquist et al., 1997).

The results of Reed et al. (2002) suggest that alcohol consumption weakens the skeleton and increases the incidence of endurance-exercise-related bone injuries. This might be due to the decrease in mineral density caused by higher doses of ethanol consumption. Baran et al. (1980), Sampson et al. (1997, 1998) have demonstrated that chronic ethanol feeding of rats causes deficiency in bone matrix synthesis and mineralisation, resulting in inferior micro-architectural and mechanical properties of bone. However, lower doses do not show any significant change, which might be due to increase in mineral content under the effect of moderate doses (Fig. 3).

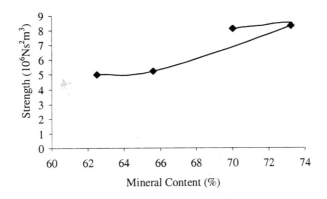

Fig. 3. Relationship of mineral content and bone strength.

The relationship between mineral content and bone strength was calculated. It was found that mineral content and bone strength were statistically co-related for which regression value ($r = 1.557$) was positive i.e. as mineral content decreases, a concomitant reduction in bone strength was observed. These results were supported by earlier studies of Shah et al. (1995) and Marcus et al. (2002).

Conclusion

The results show that the changes in mineral content and bone strength are alcohol dose dependent. Ethanol consumption at higher doses exerts toxic effects on the bone, thus, hampering biomineralization processes resulting in decrease in bone strength. However, lower doses may not inflict much damage. Thus, we conclude that individuals who participate in endurance exercise and also consume alcohol may be at greater risk for biomechanical failures of bone.

References

1. Aguiar, A.S., Da-Silva, V.A. and Buawentura, G.J. 2004. *Braz. J. Med. Biol. Res.*, **Vol. 37(6)**: 841-846 (Short Communication).
2. Boskey, A.L. and Posner, A.S. 1985. Bone Structure, Composition and Mineralization. *Orthop. Clinc North Am.* **15(4)**: 597-612.
3. Baran, D.T., Teitelbaum, S.L., Bergfeld, M.A., Parker, G., Cruvant, E.M. and Avioli, L.V. 1980. Effect of Alcohol Ingestion on Bone and Mineral Metabolism in Rats. *American Journal of Physiology.* **238**: E507-E510.
4. Bikle, D.D., Genant, H.K., Cann, C., Recker, R.R., Halloran, B.P. and Strewler, G.J. 1985. Bone Disease in Alcohol Abuse. *Annals of Internal Medicine.* **103**: 42-48.
5. Bikle, D.D. 1993. Alcohol-induced Bone Disease. *World Review of Nutrition and Dieteticts.* **73**: 53-79.
6. Brzoska, M.M., Moniuszko-Jakoniuk, J., Jurczuk, M. and Sidorezuk Galazyn, M. 2002. Cadmium Turnover and Changes of Zinc and Copper Body Status of Rats Continuously Exposed to Cadmium and Ethanol. *Alcohol and Alcoholism.* **37(3)**: 213-221.
7. Burr, D.B. 2002. The Contribution of the Organic Matrix to Bone's Material Properties. *Bone.* **31(1)**: 8-11.
8. Chakkalakal, D.A., Novak, J.R., Fritz, E.D., Mollner, T.J., McVicker, D.L., Lybarger, D.L., McGuire and Donohue, T.M. 2002. Chronic Ethanol Consumption Results in Deficient Bone Repair in Rats. *Alcohol and Alcoholism.* **37(1)**: 13-20.
9. Currey, J.D. 1988. The Effect of Porosity and Mineral Content on the Young's Modulus of Elasticity of Compact Bone. *J. Biomech.* **21**: 131-139.
10. Currey, J.D. 1990. Physical Characteristics Affecting the Tensile Failure Properties of Compact Bone. *J. Bone Joint Surg. Am.* **67(1)**: 105-12.
11. Dell, P.C., Burchardt, H., Glowczewskie, F.P. Jr. 1985. A Roentgenographic, Biomechanical, and Histological Evaluation of Vascularized and Non-vascularized Segmental Fibular Canine Autografts. *J. Bone Joint Surg. Am.* **67(1)**: 105-12.
12. Evans, F.G. 1973. Mechanical Properties of Bone. Charles C. Thomas, Springfield, Illinois.
13. Evans, F.G. 1976. Mechanical Properties and Histology of Cortical Bone from Younger and Older Men. *Anat. Rec.* **185**: 1-12.
14. Gaarcia, J.M., Doblare, M., Cegonine, J. 2002. Bone Remodelling Simulation: A Tool for Implant Design. *Computational Mater. Sci.* **25**: 100-114.
15. Moniz, C. 1994. Alcohol and Bone. *British Medical Journal.* **50**: 67-75.
16. Nyquist, F., Karlsson, M.K., Obrant, K.J. and Nilsson, J.A. 1997. Osteopenia in Alcoholics after Tibia Shaft Fractures. *Alcohol and Alcoholism.* **32**: 599-604.
17. Nyquist, F., Duppe, H., Obrant, K.J., Bondeson, L. and Nordsletten, L. 2002. Effects of Alcohol on Bone Mineral and Mechanical Properties of Bone in Male Rats. *Alcohol and Alcoholism.* **37(1)**: 21-24.
18. Park, J.B. and Lakes, R.S. 1992. In: *Biomaterials: An Introduction, 2nd Edn.* Plenum Press, NY.
19. Rai, D.V., Behari, J. and Saha, S. 1986. The Effect of Mineral Deficient Diet on the Structural and Mechanical Properties of Long Bones. *Biomedical Engineering V Recent Development,* Saha, S. (Ed.). Pergamon Press NY. pp 456-460.
20. Rai, D.V. and Singh, K.V. 2000. Effect of Mineral Loss on Elemental Composition and Thermostability of Bone Collagen. *JPAS.* **2(1)**: 15-18
21. Rapuri, P.B., Gallaher, J.C., Balhorn, K.E. and Ryschon, K.L. 2000. Alcohol Intake and Bone Metabolism in Elderly Women. *American Journal of Clinical Nutrition.* **72**: 1206-1213.
22. Ravaglioli, A. and Krajewski, A. 1992. *Bioceramics.* Chapman & Hall, NY. p. 16.

23. Reed, A.H., McCarty, H.L., Evans, G.L., Turner, R.T., Westerlind, K.C. 2002. The Effects of Chronic Alcohol Consumption and Exercise on the Skeleton of Adult Male Rats. *Alcohol Clin. Exp. Res.* **26(8)**: 1269-74.
24. Rico, H. 1990. Alcohol and Bone Disease. *Alcohol and Alcoholism.* **25**: 345-352.
25. Sadananada, R.A. 1991. Probabilistic Approach to Bone Fracture. *J. Mater. Res.* **6(1)**: 202.
26. Saha, S. 1987. Skeletal Biomechanics and Ageing. *Orthop. Res. Clin. North. Am.* **13(2)**: 1-22.
27. Saha, S. and Gorman, P.H. 1981. *Trans. 27th Ann. Mtg. Of Orthopaedic Res. Soc.* **6**: 217 (also in Orthopaedic Transactiions, 5, #2, 323-324).
28. Saha, S. and Hayes, W.C. 1977. *Calcif. Tissu. Res.* **24**: 65-72.
29. Shah, K.M., Goh, J.C.H., Karunanithy, R., Low, S.L., Das, S. and Bose, K. 1995. Effect of Decalcification on Bone Mineral Content and Bending Strength of Feline Femur. *Calcif. Tiss. Int.* **56**: 78-82.
30. Sampson, H.W., Chaffin, C., Lange, J. and Deffee, B. 1997. Alcohol Consumption by Young Actively Growing Rats: A histomorphometric study of cancellous bone. *Alcoholism: Clinical and experimental research.* **21(2)**: 352-359.
31. Sampson, H.W., Hebert, V.A., Boone, H.L. and Champney, T.H. 1998. Effect of Alcohol Consumption on Adult and Aged Bone: Composition, morphology, and hormone levels of a rat animal model. *Alcoholism: Clinical and Experimental Research.* **22**: 1746-1753.
32. Singh, S., Behari, J. and Rai, D.V. 1981. Elastic Constants Relating to Two Phase Bone System. *Ultrason. Int.* June 30- June 2, Brighton, U.K.
33. Tavares, D.C., Cecchi, A.O., Jordao, Jr. A., Vannuchi, H. and Takahashi, C.S. 2000. Cytogenetic Study of Chronic Ethanol Consumption in Rats. *Cancer Detection and Prevention.* **24 (Supplement 1)**.
34. Üçisik, A.H., Göksan, M.A., Üçok, I. and Bindal, C. 1985. Comparative Mechanical Properties of Human Bones. *Proc. of 7th Conf. on Strength of Metals and Alloys*, Quebec, Canada, **3**: 2143.
35. Veenland, J.F., Link, T.M., Konermann, W., Meier, N., Grashuis, J.L. and Gelsema, E.S. 1997. Unraveling the Role of Structure and Density in Determining Vertebral Bone Strength. *Calcif. Tissue Int.* **61**: 474-479.
36. Wezeman, F.H. and Gonz, Z. 2001. Bone Marrow Triglyceride Accumulation and Hormonal Changes during Long-term Intake in Male and Female Rats. *Alcoholism: Clinical and Experimental Research.* **25(10)**: 1515-22.
37. Williams, F., Cherkas, L.F., Spector, T.D. and MacGregor, A.J. 2004. The Effect of Moderate Alcohol Consumption on Bone Mineral Density: A study of female twins. *Ann Rheum Dis*. [E-pub ahead of print].

Biomechanics
R.K. Saxena and P. Mishra (Editors)
Copyright © 2005, Anamaya Publishers, New Delhi, India

33. Effect of Moderate Cyclic Load Stress on Biochemical Properties of Articular Cartilage of Bovine Synovial Joints

Ritu Rathi and R.K. Saxena
Centre for Biomedical Engineering, Indian Institute of Technology Delhi, New Delhi - 110016, India

Abstract: Earlier synovial joints articular cartilage studies reveals that biochemical properties of synovial joints are responsible for load bearing properties and good health of weight synovial joints. Articular cartilages moderately loaded bovine knee joints showed increase in proteoglycan content, decrease in water and appeared "healthier" with better structural integrity. Loading and articulation significantly alters the tissue biomaterial property, which in turn affects the biomechanical load bearing properties. Mechanical loading of the order of physiological levels in a way improves the tissue health and weight bearing property. At moderate (100 kg) cyclic loading of articular cartilage an increase of 24% in proteoglycan content was shown, whereas the percentage of water content decreased by 15% in the load tested articular cartilage. The collagen contents did not show any variation in percentage with loading. The results of the study suggest that there is an improvement of the health of articular cartilage with given 100 kg of loading and produced beneficial effect on the joint physiological properties.

Introduction

Locomotion is a dynamic functional property of living beings and performs several means and activities. In human system, weight related locomotion functions are usually performed by movable joints called as synovial joints. The remarkable performance of load bearing human joints is well known. Several abnormal mechanical stresses i.e., overloading, postures, obesity etc. have been found to cause early degeneration of articular cartilage and may lead to a degenerative disease commonly known as osteoarthritis. Epidemiological and clinical studies showed that the mechanical forces are the main factors in the genesis of osteoarthritis and failure of load bearing joints (Peyron, 1987). Articular cartilage tissue of synovial joints is influenced by ranges of motion and loading and causes changes in physio-pathological properties of joints (Kiviranta et al., 1986; Jurveline et al., 1987). Supraphysiological joint loading can lead to degeneration or atrophy of articular cartilage, however moderate loading has been found to improve cartilage thickness and proteoglycan concentration of the joint articular cartilage. Certain range of loading to cartilage may be harmless however the tissue above and below can suffer (Sokoloff, 1966). The objective of the present article is to understand the phenomenon of cyclic loading in freshly slaughtered amputated young bovine metatarsal joint *in vitro*. In this acute animal model study, moderate cyclic loading of 100 kg will be put during articulation by a specially designed knee joint articulating machine. The changes in biochemical properties of articular cartilage samples will be measured and variations in proteoglycan, collagen content and percentage water content will be monitored. Experimental load tested biochemical values will be correlated with control values in order to understand the cause of alteration. Matrix of articular cartilage consists of water content, collagen fibers and hygroscopic proteoglycans aggregates. The strength of articular cartilage comes from tough protein fibers called collagen. Collagen fibres meshwork provides the articular cartilage tissue and tensile strength and forms a major ingredient in the cartilage solid component. Several studies have found that proteoglycan aggregates attach to collagen fibres by covalent and noncovalent interactions (Mow et al., 1984, Akizuki et al., 1986). Articular cartilage is a highly hydrated tissue with its water content ranging from about 65 to 80%. The water is important in maintaining the tissues resiliency load bearing and shock absorbing properties as well as contributing to the almost frictionless movement of the articulating cartilage (Linn, 1967; Mccutchen, 1978).

In order to understand the changes in biochemical properties of load bearing synovial joint articular cartilage. The present study has been done in a simulated acute (60 min.) *in vitro* environment on young freshly amputated bovine knee joints for the damage in biochemical contents i.e. proteoglycan, collagen and water contents.

Material and Methods

Bovine knee joint from freshly slaughtered healthy young animals were procured. The left knee joint served as the control joint while the right was used for load testing and was called experimental joint.

Mechanical Load Testing on Bovine Knee Joint

Bovine knee joints were subjected to a load of 100 kg for 1 hr at 'knee joint articulating machine' developed by Saxena et al. (1990). The experimental joints were articulated for 1 h at 40 cycles/min immediately after completing 1 h of loaded articulation, the experimental knee joint was removed from the knee joint articulating machine. The control joint duly rapped in cotton soaked in physiological saline was kept in vertical position duly clamped on a stand and was not subjected to any load or articulation. The joints were frequently bathed in isotonic saline to prevent drying of joint tissues. The complete loading experiment was carried out at room temperature of 25°C.

Preparation of Cartilage Samples

After 1 h of articulated loading of the bovine knee joint, the test joint and the control joint were brought to the dissection table. First the control joint was dissected and the joint capsule was carefully opened. The tibial bone with its articular cartilage was carefully separated. After opening the joint, the bony configuration and cartilage surfaces were examined for any visible gross defects like malformations, discolorations, fibrillation, pitting or other irregularities. If any of these abnormal irregularities were observed in the control joint or in the corresponding experimental joint, the experiment was discarded. Next, the load tested knee joint was dissected to separate the tibial and femoral ends. After gross examination for any of the abnormalities as stated above, the tibial head cartilage was subjected to microscopic examination with a naked eye. The articular cartilage surface of the load-tested joint was compared to that of control for visible change in color, nature and texture.

Biochemical Analysis of Articular Cartilage Specimen

Articular cartilage samples from control and load tested bovine knee joints were tested and analyzed to determine the material properties and tissues and to measure the load related changes in the properties of the tissues. The biochemical properties of the articular cartilage obviously depend upon the biochemical composing of the cartilage matrix. Hence any change in the biochemical properties should be due to corresponding alteration in the biochemical features of the tissue.

Estimation of Proteoglycan Content
Articular cartilage obtained samples from the control and load tested bovine knee joints were cut into very thin pieces by a sharp scalpel blade. Small pieces of cartilage samples were weighed on an electric balance and put in a solution of extraction mixture in the ratio 1: 5 (w/v) in a conical flask to extract the proteoglycan content into solution. The extraction mixture was made of 4.0 M guanidine hydrochloride, 0.05 M sodium acetate, 0.10 M 6-amino-corpic acid (sigma), 0.01 M disodium EDTA, 0.005 M benzamidine hydrochloride (sigma), 0.004 M N-ethylmalemide The cartilage sample was placed in a constant temperature-refrigerating shaker at 4°C for 48 h under constant stirring. The extraction solution was then centrifuged at 4000 rpm for 20 min and then separated. This aliquot was used for the quantitative content in the tissue. The procedure to estimate the proteoglycan content is based on precipitation of alcian blue-glycosaminoglycan complexes described by Whiteman, 1973. The optical density was measured at 620 nm of resultant dye solutions.

Estimation of Water Content
The water content of the cartilage tissue was measured by the method described by Radin et al. (1982).

Estimation of Collagen
1 g of articular cartilage tissue was suspended in the collagen in 5 ml of extraction buffer in the conical flask. 0.2 µg/ml of collagenase enzyme put into the extraction solution. Extraction solution containing conical flask was placed in the water bath shaker for 18 h at room temperature. After extraction it was centrifuged at 6000 rpm for 20 min. Aliquot was used for the quantitative estimation of collagen content in tissue. Collagen content was estimated by Ehrlich's and series of supernatants tubes were prepared ranging from 5 to 10 and 2 N NaOH was put for hydrolyzing the samples by boiling in water-bath for 90 min. Samples are made up to 2 ml by adding assay buffer. Then 500 µl of chloramine T reagent was put and kept for 20 min at room temperature. Then again 500 µl freshly prepared Ehrlich reagent was put and mix thoroughly and incubated the tubes at 60ºC for 25 min. it was cooled in tap water for 5 min. The absorbance spectra of the resulting materials were measured at 550 nm by a spectrophotometer.

Results
Moderate mechanical loading of 100 kg and frequencies of 40 cycle/min by synovial joint articulating machine was given to 5 freshly amputated bovine metatarsal stumps obtained from the local slaughter house. Immediately after loading, experimental joints were aseptically obtained and articular cartilage was removed gently by sharp scalpel blade and kept on saline soaked filter paper for biochemical (proteoglycan content, collagen content) and percentage water content. In the study, left synovial joint was used as experimental and right synovial joint was used as control. Articular cartilage are made in small pieces and the following parameters were studied:

1. Proteoglycan content was measured by the method of Whiteman (1973) and modified by Carney et al. (1986).
2. Collagen content measured by digesting collagen by collagenase enzyme and measuring the hydroxy-proline content by the method of Radin et al. (1984).
3. Percentage water content done by the method described by Radin et al. (1982).

The study found an increase in proteoglycan concentration compared to the control i.e. 16.84 µg/ml to control, i.e. 13.48 µg/ml and showed 24% improvement in proteoglycan concentration. There was no change in the collagen content and had been seen in the experimental articular cartilage compared to the control. The control synovial joint cartilage showed 17.1 µg/ml whereas experimental load tested cartilage showed a value of 16.28 µg/ml. Results of experimental and control were almost similar. In order to measure water content, articular cartilage was put at a temperature of 60°C and 2 to 3 consistent readings were taken as % water content. Results showed a decrease in water content, i.e. 73.74. The study observed a 15% decrease in water content to control values (Table 1) (Fig. 1).

Table 1. Mean/ SD± changes in proteoglycan, collagen and water content

	Control (Mean/SD±)	Experimental (Mean/SD±)
Proteoglycan	13.48 ± 2.19	16.84 ± 1.93
Collagen	17.1 ± 1.66	16.28 ± 1.56
Water content	73.74 + 3.78	64.03 ± 3.21

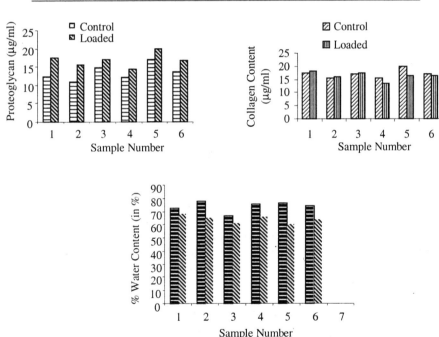

Fig. 1. Comparative value of proteoglycan (µg/ml), collagen (µg/ml) and % water content (in %) of control and loaded (100 kg), articulate cartilage knee joint sample.

Discussion

Effect of moderate cyclic loading of synovial joint articular cartilage has been the subject of this investigation. In this investigation, freshly amputated bovine metatarsal joints have been employed to study and measure the biochemical properties of articular cartilage when the metatarsal joints are subjected to moderate loading condition. Moderate loadings have been applied on the knee joints to induce the stress related changes in the articular cartilage that is equal to the body weight of the animal (moderate load). The right metatarsal joint served as the control while the left metatarsal joint was subjected to load testing. Articular cartilage specimens from load tested and control metatarsal joints have been analyzed for changes in the biochemical properties under loading and compared with their corresponding control values. The metatarsal joints of 5 animals each have been employed to study the load related changes occurring in the articular cartilage properties under moderate loadings. A total of 5 left metatarsal joints were used as experimental and were analyzed under identical situations. The proteoglycan content, collagen content and percentage water property of articular cartilage from control and experimental metatarsal joints have been analyzed. The mean proteoglycan content found in this study was 13.48 and 16.84 µg/ml for controls and moderately loaded groups, respectively (Table 1). The mean percentage water content for the same groups has been 73.74 and 64.03, respectively (Table 1). The mean collagen content has been found to be 16.69 for which the controls were the same (Table 1). About 20-40% of the tissue weight to be comprised of solid matrix of which 70-72% by dry weight is collagen and 15-20% by dry weight is proteoglycan content (Lipshitz et al., 1975). The proteoglycan content for normal cartilage to be between 3-10% and water content to be 65-80% of the total tissue weight (Maroudas, 1979). The composition of these components vary over a wide range depending on the age, species and joint from which extracted and the depth of tissue from surface. The variation of water content with age and degeneration with respect to normal cartilage report similar values for the normal human cartilage (Armstrong and Mow, 1982). The cartilage tissue is made up of 65-85% water (Torzilli et al., 1983). The findings obtained in this study for control samples of the two groups for proteoglycan content is 24% and water content 15% which compares well with those of other investigators. Gross examination of the cartilage tissue from moderately loaded metatarsal joints showed no damage on the cartilage surface and was normal in appearance. The articular cartilage surfaces of the condyles appeared to be stiffer and more firm than its control.

The femoral head cartilage of rabbits under SEM (scanning electron microscope), after subjecting the rabbits to prolonged sub-maximal running. The prolonged sub-maximal running produced the most regular surfaces in the articular cartilage (Condolin and Videman, 1986). It is also observed in our study. Saxena et al., (1994), have also not observed any difference in the morphologies of the cartilage samples obtained from moderately loaded (150 kg) bovine knee joints and from their controls when examined microscopically. Biochemical parameters during moderate loading showed changes in proteoglycan content, water content and collagen property. Proteoglycan content in the matrix of the cartilage increased from 13.48 to 16.84 µg/ml corresponding to a rise of 19% with respect to the control value (Table 1). The percentage water content of the tissue decreased to 15% and the collagen content property showed almost no change in its material property (Table 1).

The proteoglycan content had increased in case of non-strenuous exercise whereas the proteoglycan content decreased in strenuous exercise (Saamanen et al., 1986; Kiviranta et al., 1987). *In vitro* the proteoglycan synthesis responses to the loading on bovine knee joint cartilage observed that cyclic loads of moderate levels (0.5 MPa) resulted in enhanced incorporation of radiosulphide tracer implying an increase in the proteoglycan synthesis (Parkkinen et al., 1989). The bovine knee joint articular cartilage for changes in the biochemical and morphological properties after subjecting the joint to moderate loading (150 kg) and articulation found that the cartilage proteoglycan content had increased by 24% and water content had reduced by 15% (Saxena et al., 1994) which is quite similar to the findings of the present study. Results of this showed that moderate loading of 100 kg has caused an increase in proteoglycan content of the articular cartilage of metatarsal synovial joint and proteoglycan has been found to enhance the load bearing property of weight bearing joints and life of locomotory joints and it is a significant finding of this study.

References

1. Akizuki, S., Mow, V.C., Muller, F., Pita, J.C., Howell, D.S. and Manicourt, D.S. 1986. Tensile Properties of Knee Joint Cartilage: Influence of ionic conditions, weight bearing, and fibrillation on the tensile modulus. *J. Orthop. Res.* **4**: 379-392.
2. Armstrong, C.G. and Mow, V.C. 1982. Variations in the Intrinsic Mechanical Properties of Human Articular Cartilage with Age, Degeneration and Water Content. *J. Bone Jt. Surg.* **64**: 88-94.
3. Condolin, T. and Videman, T. 1986. Effects of Running on the Scanning Electron Microscopic Appearance of Surface of Rabbit. *XV Symposium of European Soc. of Osteoarthrology Kuopio,* Finland. p. 27.
4. Jurvelin. J., Kiviranta, I., Arakoshi, J., Tammi, M. and Helminen, H.J. 1987. Indentation Study of the Biomechanical Properties of Articular Cartilage in the Canine Knee. *Engineering Med.* **16**: 15-22.
5. Kiviranta, I., Jurevelin, J., Sammanen, A.M., Arokoski, J., Tammi, M. and Hellminen, H.J. 1987. Strenuous Running Exercise Attenuates the Increase of Articuler Cartilage Proteoglycans Observed after Moderate Running Programme of Young Beagle Dogs. *Proc. XVth Symposium of the European Society of Osthrology Kuopio,* Finland. p. 20.
6. Linn, F.C. 1967. Lubrication of animal joints. I: The arthrotripsometer. *J. Bone Jt. Surg.* **49**: 1079-1098.
7. Lipshitz, H.; Etherredge, R. and Glimcher, M.J. 1975. *In vitro* Wear of Aarticular Cartilage. *J. Bone Surg.* **57**: 527-534.
8. Maroudas, A. 1979. Physical Chemistry of Articular Cartilage and the Intervertebral Disc. In: *The Joints and Synovial Fluid.* Sokoloff (Ed.). Acadamic Press, New York. pp. 239-291.
9. McCutchen, C.W. 1978. Lubrication of Joints. In: *The Joints and Synovial Fluid.* Sokoloff (Ed.). Acadamic Press, New York. **1**: 438-483.
10. Mow, V.C., Holmes, M.H. and Lai, W.M. 1984. Fluid Transport and Mechanical Properties of Articular Cartilage: A review. *J. Biomech.* **17**: 377-394.
11. Parkinnen, J.J., Lammi, M.J., Helminen, H.J. and Tammi, M. 1989. A Mechanical Apparatus with Microprocessor Controlled Stress Profile for Cyclic Compression of Cultured Articular Cartilage Explants by Dynamic Compression *in vitro. J. Orthop. Res.* **10**: 610-620.
12. Peyron, J.G. 1987. Risk Factors in Osteoarthritis: How do they work. *J. Rheumatol.* **14**: 1-2.
13. Radin, E.L., Swan, D.A., Paul, I.L. and Megrath, P.J. 1982. Factors Influencing Articular Cartilage Wear *in vitro. Arthritis Rehum.* **25**: 974-989.
14. Radin, E.L., Martin, R.B. and Burr, D.B. 1984. Effects of Mechanical Loading on the Tissues of the Rabbit Knee. *J. Orthop. Res.* **2**: 221-234.
15. Sammanen, A.M., Tammi, M., Kiviranta, I., Jurvelin, J. and Helminen, H.J. 1986. Moderate Running Increased but Strenuous Running Prevented Elevation of Proteoglycan Content in the Canine Articular Cartilage. In: *Proc. XVth Symposium of the European Society of Osteoarthrology,* 25-27 June 1986, Kuopio, Finland. p. 23.
16. Saxena, R.K., Sahay, K.B. and Guha, S.K. 1990. Design and Development of Knee Joint Articulating Machine. *J.I.D.* **71**: 52-54.
17. Saxena, R.K., Sahay, K.B. and Guha, S.K. 1994. Biomechanical, Biochemical and Morphological Correlates of Bovine Knee Joint Articular Cartilage. *J.I.D.* **75**: 8-16.
18. Sokoloff, L. 1966. Elasticity of Aging Cartilage. *Fed. Proc., Fed. Am. Soc. Exp. Biol.* **25**: 1089-1095.
19. Torzilli, P.A., Dethmers, D.A. and Rose D.A. 1983. Movement of Interstitial Water through Loaded Articular Cartilage. *J. Biochem.* **16**: 169-179.
20. Whiteman, P. 1973. The Quantitative Measurement of Alcian Blue-glycosaminoglycan Complexes. *Biochem. J.* **131**: 343-350.

Biomechanics
R.K. Saxena and P. Mishra (Editors)
Copyright © 2005, Anamaya Publishers, New Delhi, India

34. Effect of Physical Exercise on the Relationship between Selected Kinanthropometric Variables and Percentage Height of Centre of Gravity of Female

Seema Kaushik[1] and Dhananjoy Shaw[2]

[1]Department of Physical Education, Lakshmibai College, University of Delhi, Delhi - 110007, India

[2]Biomechanics Laboratory, Department of Natural/Medical Sciences,
I.G.I.P.E.S.S., University of Delhi, New Delhi - 110018, India

Abstract: The purpose of the present investigation was to study the effect of exercise on the relationship between selected kinanthropometric variables and percentage height of centre of gravity on female students of University of Delhi. 90 female students of University of Delhi, age ranging between 17 and 25 yrs were selected randomly for the purpose of the study. The sample was classified into three groups viz. conditioning female (n_1 = 33), non-conditioning female (n_2 = 35) and sedentary female (n_3 = 22). The conditioning program consisted of three meso cycles viz. M-1, M-2 and M-3 of six weeks duration each with the target training intensity (HR) of 130 ± 10, 150 ± 10 and 170 ± 10 beats/min, respectively for a session of 1 h duration (45 min for general conditioning programme and 15 min for warming up and cooling down) regularly for five sessions per week. The subjects were tested four times viz. T-1 (at zero weeks of training), T-2 (after 6 weeks of training), T-3 (after 12 weeks of training) and T-4 (after 18 weeks of training). Body weight, height, sitting height, leg length, length of arm, length of foot, biacromian breadth, bicristal breadth, bitrochanterion breadth, wrist diameter, knee diameter, ankle diameter, elbow diameter, chest circumference, upper-arm circumference, fore-arm circumference, thigh circumference, calf circumference, biceps skinfold, triceps skinfold, forearm skinfold, subscapular skinfold, suprailiac skinfold, thigh skinfold, calf skinfold and percentage height of centre of gravity were the selected variables for the purpose of the study. Equipments included lever based weighing scale, a wooden board of 183×100×2 cm (fitted within a steel frame and fixed fulcrum), anthropometer, steel tape and skinfold caliper. The data was statistically analyzed while computing mean, standard deviation and percentage etc. The study concluded that physical exercise affects the relationship between selected kinanthropometric variables and percentage height of centre of gravity of female.

Introduction

Humans are meticulously designed for physical activity, yet our modern mechanical age has eliminated many of the opportunities we once had to incorporate moderate physical activities as a natural part of our lives. Today, getting moving is a challenge because today physical activity is less a part of our daily lives. We have mechanically mobile society, relying on machines rather than muscles to get around.

Machines have shifted many of the burdens of man from his muscles to his intellect. Automation results in fewer hours of work and more hours of leisure. During extended leisure time, people choose sedentary forms of recreation and entertainment such as playing cards, reading, watching movies, enjoying television or listening to the radio. In the large cities, lack of space and facilities inside and outside the home, inhibit to a large extent, the play life of children and adults. One of the challenges of physical education activities in modern life is to alleviate some of the evils of unnatural living.

One of the most significant detrimental effects of modern day technology has been an increase in chronic conditions, which are related to a lack of physical activity (e.g. hypertension, heart disease, chronic low back pain and obesity etc.). Among the number of sufferings/problems, persons by and large suffers from posture, stability and balance related problems. The balance, posture and stability in the whole spectrum of life are important for work, working environment and sports performances. For such circumstances, the location of CG always found to be the most attributing factor. The height of centre of gravity and location of centre of gravity (CG) plays a vital role in controlling the posture through maintaining the natural alignments, which may positively be effected by physical training/ conditioning. Hence, the role of location of centre of gravity in solving the problems of the society, job conditions, working environment as well as sports performance can't be overlooked.

The resultant of centre of gravities of various bodily segments, depends upon its own structural dimension and mass which is likely to change due to certain changing factors, such as height of the individual, mass of various body segments, posture, width of bone, varying length dimension of each segment of the body, body composition (lean body mass, percentage of fat, percentage of water in the body), and type of physique etc., which are under direct influence of nature of activities and nutrition etc. (Charles, 1985; Burleigh et al., 1994; Kawamura et al., 1987; Kaushik and Shaw, 1995; Oyster, 1979).

For achieving excellence, one must be habitual of vigorous and strenuous training to meet the demands of hardships of life. For instance, in sports, this aim is fulfilled by the means of physical training/conditioning that leads to various physical, physiological and psychological changes (Blomquist, 1983; Shephord, 1983; Monahan, 1987; Harris et. al., 1989; McHenery et. al., 1990). Though the rate of change may vary from one body part to another or from one variable to another, for instance, the lengths and skeletal diameters are likely to be affected to a less extent especially at college level of students, where growth doesn't have much influence. Whereas, skinfold variables, circumferences, fat percentage, water percentage and lean body mass in the body etc. are likely to change in different ratios, may be at a faster or a slower pace due to their nature of activity involvement or other related factors, thereby affecting the somatotypes, body composition and different indices. Such changes may lead to certain biomechanical changes also (Jensen, Schultz and Bamgartner, 1984). Any such momentarily change is likely to deviate the location of any of the three planes and axes. Due to the little deviation of centre of gravity, the unstable equilibrium of the body is disturbed, leading to the postural disturbances and further may cause postural deformities or functional impairment.

Hence, the purpose of the present investigation was to study the effects of physical exercise on the relationship between selected kinanthropometric variables and percentage height of centre of gravity of female students of University of Delhi.

Material and Method

Sample
Ninety (90) female students of University of Delhi, age ranging between 17 to 25 years were selected randomly for the purpose of the study. The sample was classified into three groups namely conditioning female ($n_1 = 33$), non-conditioning female ($n_2 = 35$) and sedentary female ($n_3 = 22$).

Selection of Variables

Body weight, height, sitting height, leg length, length of arm, length of foot, biacromian breadth, bicristal breadth, bitrochanterion breadth, wrist diameter, knee diameter, ankle diameter, elbow diameter, chest circumference, upper-arm circumference, fore-arm circumference, thigh circumference, calf circumference, biceps skinfold, triceps skinfold, forearm skinfold, subscapular skinfold, suprailiac skinfold, thigh skinfold, calf skinfold and percentage height of CG were the selected variables for the purpose of the study.

Experimental Protocol

The conditioning programme consisted of three meso cycles, M-1, M-2 and M-3 of 6 weeks duration each with the target training intensity (HR) of 130 ± 10, 150 ± 10 and 170 ± 10 beats/min, respectively, for a session of 1 h duration (45 min for general conditioning programme and 15 min for warming up and cooling down) regularly for 5 sessions/week where, conditioning group participated in the described conditioning programme of physical exercises along with the regular physical education programme of Indira Gandhi Institute of Physical Education and Sports Sciences (IGIPESS) curriculum, however, non-conditioning group participated in the regular physical education programme of IGIPESS curriculum only, whereas, sedentary group did not participate in any kind of physical exercise for the period of experimentation. Table 1 gives the detailed experimental protocol.

Table 1. Experimental protocol

S. No.	Meso Cycles	Target HR/Intensity (beats/min)	Target Components
1	Meso cycle one (M-1)	130 ± 10	Flexibility, cardio-respiratory endurance
2	Meso cycle two (M-2)	150 ± 10	Muscular endurance, strength
3	Meso cycle three (M-3)	170 ± 10	Speed, power (explosive strength)

Testing Protocol

Each subject was tested four times during 4½ month of physical exercise programme viz. T-1 (at zero weeks of training), T-2 (after 6 weeks of training), T-3 (after 12 weeks of training) and T-4 (after 18 weeks of training). The detailed testing protocol is exhibited in Table 2.

Table 2. Testing protocol

S. No.	Tests	Test Code	Time of Testing
1	Testing one (pre-test)	T-1	Before the start of physical training/ conditioning at zero weeks of training i.e. 3rd and 4th week of July.
2	Testing two (first post test)	T-2	At the end of first meso-cycle (M-1) after six weeks of training i.e. 1st and 2nd week of September.
3	Testing three (second post test)	T-3	At the end of second meso-cycle (M-2) after twelve weeks of training i.e. 4th week of October.
4	Testing four (third post test)	T-4	At the end of third meso cycle (M-3) after eighteen weeks of training i.e. 2nd and 3rd week of December.

Collection of Data

For the purpose of collection of data, the method explained by Sen and Ray (1983), Das and Ganguli (1982), scientifically authenticated by Shaw, Kaushik and Kaushik (1998) and validated by Shaw et al. (1998) was strictly administered. A lever based weighing machine (Avery) with a range from 0 to 100 kg with the balancing accuracy of ±10 g, a wooden board of 183 cm in length, 100 cm in width and 2 cm in thickness were among the equipment used to collect the data for determination of centre of gravity.

The partial body weight of the wooden board was determined (S_0). The two knife edged wooden blocks, one on the platform of the balance and other on the wooden box (placed opposite end to the balancing scale) below which the footrest was fabricated. Thereafter, the scale reading was recorded (S_0). The subjects were instructed to be flat in supine position on the board in such a way that the imaginary vertical plane passed through the longitudinal axis of symmetry of the board and the sagittal axis of the body are coincident. The feet were kept flush at the end of the board (foot rest), opposite to the direction of the weighing scale was placed. The scale reading in this position of the body was recorded (S_1).

The following equation was used to calculate the height of centre of gravity

$$\text{Height of CG} = \frac{(S_1 - S_0)}{M} \times L$$

where S_0 = scale reading of the system; S_1 = scale reading when subject is lying flat on the system; L = length of the reaction board and M = mass of the body.

The height of centre of gravity was converted to percentage height of centre of gravity by

$$\text{Percentage Height of CG} = \frac{\text{Height of CG (m)}}{\text{Stature/Height (m)}} \times 100$$

All the kinanthropometric measurements were taken from the left side of the individual. Standard landmarks and measurement protocols were followed to measure the selected fundamental variables (Martin and Carter, S. P. Singh, H. S. Sodhi and D. K. Kansal).

Analysis of Data

The collected data was analyzed while computing mean, standard deviation and percentage. The analysis of data pertaining to the effect of physical exercise on the relationship between selected kinanthropometric variables and percentage height of centre of gravity of male have been presented in Table 3 and illustrated vide Figs. 1 to 4.

Table 3. Effect of physical exercise on selected variables of female students of University of Delhi

S. No.	Variable	Percentage Change		
		Conditioning Group	Non-Conditioning Group	Sedentary Group
1	Body weight	1.18% (Decrease)	0.49% (Decrease)	3.04% (Increase)
2	Height	0.18% (Increase)	0.13% (Increase)	0.12% (Increase)
3	Sitting height	0.16% (Increase)	0.10% (Increase)	0.07% (Increase)
4	Leg length	0.21% (Increase)	0.15% (Increase)	0.18% (Increase)
5	Length of arm	0.27% (Increase)	0.23% (Increase)	0.35% (Increase)
6	Length of foot	0.61% (Increase)	0.44% (Increase)	0.36% (Increase)
7	Biacromian breadth	0.60% (Increase)	0.74% (Increase)	0.51% (Increase)
8	Biiliocristal breadth	0.92% (Increase)	0.50% (Increase)	0.58% (Increase)
9	Bitrochanterion breadth	0.64% (Increase)	0.53% (Increase)	0.60% (Increase)

Contd.

Table 3. Contd.

S. No.	Variable	Percentage Change		
		Conditioning Group	Non-Conditioning Group	Sedentary Group
10	Wrist diameter	2.31% (Increase)	2.13% (Increase)	1.99% (Increase)
11	Knee diameter	1.82% (Increase)	1.08% (Increase)	1.70% (Increase)
12	Ankle diameter	1.80% (Increase)	1.48% (Increase)	1.35% (Increase)
13	Elbow diameter	1.76% (Increase)	1.61% (Increase)	1.50% (Increase)
14	Chest circumference	0.37% (Decrease)	0.50% (Increase)	1.44% (Increase)
15	Upper-arm circumference	1.04% (Decrease)	0.62% (Increase)	2.16% (Increase)
16	Forearm circumference	0.46% (Decrease)	0.46% (Increase)	2.28% (Increase)
17	Thigh circumference	1.21% (Decrease)	0.53% (Increase)	2.60% (Increase)
18	Calf circumference	0.92% (Decrease)	0.57% (Increase)	1.56% (Increase)
19	Ankle circumference	0.87% (Increase)	1.13% (Increase)	1.14% (Increase)
20	Biceps skinfold	6.88% (Decrease)	10.87% (Increase)	14.43% (Increase)
21	Triceps skinfold	8.73% (Decrease)	4.92% (Increase)	5.58% (Increase)
22	Forearm skinfold	9.41% (Decrease)	12.69% (Increase)	6.55% (Increase)
23	Subscapular skinfold	2.74% (Decrease)	2.45% (Increase)	7.43% (Increase)
24	Suprailiac skinfold	8.13% (Decrease)	4.35% (Increase)	8.90% (Increase)
25	Thigh skinfold	4.78% (Decrease)	4.82% (Increase)	12.30% (Increase)
26	Calf skinfold	4.47% (Decrease)	1.91% (Increase)	4.60% (Increase)
27	Percentage height of centre of gravity	3.85% (Decrease)	2.05% (Decrease)	2.72% (Increase)
	Mean (%)	2.46	2.13	3.19
	S.D. (%)	2.80	3.11	3.75

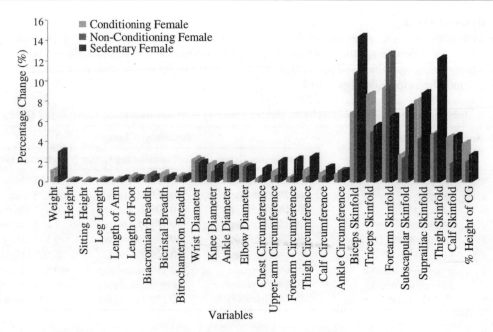

Fig. 1. Effect of physical exercises on selected kinanthropometric variables of female students of University of Delhi.

Effect of Physical Exercise on the Relationship between Selected Kinanthropometric Variables 245

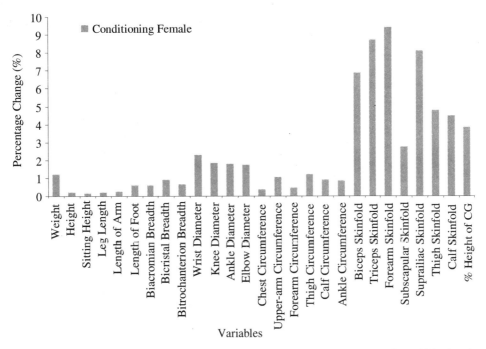

Fig. 2. Effect of physical exercises on selected kinanthropometric variables of conditioning female students of University of Delhi.

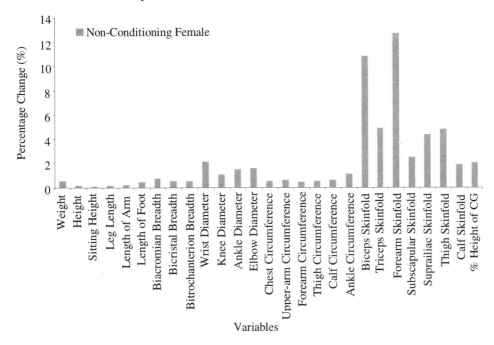

Fig. 3. Effect of physical exercises on selected kinanthropometric variables of non-conditioning female students of University of Delhi.

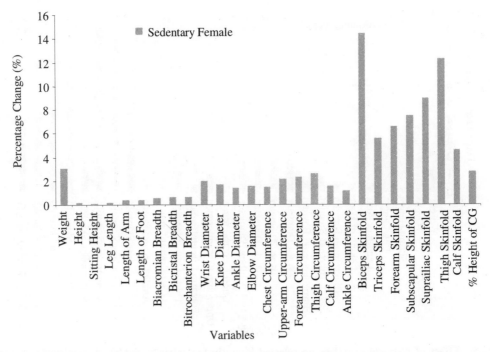

Fig. 4. Effect of physical exercises on selected kinanthropometric variables of sedentary female students of University of Delhi.

Further, a seven-point scale was prepared i.e. extremely high, very high, above average, average, below average, very poor and extremely poor to understand the relationship among the selected variables. The three seven-point scales developed for each category of the sample namely, conditioning female (CF), non-conditioning female (NCF) and sedentary female (SF) are presented in Tables 4, 5 and 6 respectively.

The analysis of data in Table 3 pertaining to the effect of physical exercise on selected variables of female students of University of Delhi reveals that the variables namely weight, chest circumference, upper-arm circumference, forearm circumference, thigh circumference, calf circumference, biceps skinfold, triceps skinfold, forearm skinfold, subscapular skinfold, supraliac skinfold, thigh skinfold, calf skinfold and percentage height of centre of gravity were observed to be decreasing in the conditioning female (CF) whereas, the variables namely height, sitting height, leg length, length of arm, length of foot, biacromial breadth, biiliocristal breadth, bitrochanterion breadth, wrist diameter, knee diameter, ankle diameter, elbow diameter and ankle circumference were observed to be increasing.

In regard to non-conditioning female (NCF), the variables namely weight and percentage height of centre of gravity were observed to be decreasing whereas, the variables namely height, sitting height, leg length, length of arm, length of foot, biacromial breadth, biiliocristal breadth, bitrochanterion breadth, wrist diameter, knee diameter, ankle diameter, elbow diameter, chest circumference, upper-arm circumference, forearm circumference, thigh circumference, calf circumference, ankle circumference, biceps skinfold, triceps skinfold, forearm skinfold, subscapular skinfold, supraliac skinfold, thigh skinfold and calf skinfold were found to be increasing.

Table 4. Seven-point scale for conditioning female

S. No.	Group for Relationship	Range (Min. to Max. Value)	Variables
1	Extremely high	9.47 onwards	
2	Very high	6.67 to 9.46	Biceps skinfold, triceps skinfold, forearm skinfold, suprailieac skinfold
3	Above average	3.87 to 6.66	Thigh skinfold, calf skinfold
4	Average	1.06 to 3.86	Weight, thigh circumference, subscapular skinfold, % height of centre of gravity
5	Below average	3.85 to −1.74	Height, sitting height, leg length, length of arm, length of foot, biacromial diameter, biiliocristal diameter, bitro-chanterion diameter, wrist diameter, knee diameter, ankle diameter, elbow diameter, chest circumference, upper-arm circumference, forearm circumference, thigh circumference, calf circumference, ankle circumference
6	Very poor	−1.75 to −4.54	Wrist diameter, knee diameter, ankle diameter, elbow diameter
7	Extremely poor	−4.55 downwards	

Table 5. Seven-point scale for non-conditioning female

S. No.	Group for Relationship	Range (Min. to Max. Value)	Variables
1	Extremely high	9.92 onwards	
2	Very high	6.81 to 9.91	
3	Above average	3.70 to 6.80	
4	Average	0.58 to 3.69	% height of centre of gravity
5	Below average	3.68 to −2.54	Weight, height, sitting height, leg length, length of arm, length of foot, biacromial diameter, biiliocristal diameter, bitrochanterion diameter, wrist diameter, knee diameter, ankle diameter, elbow diameter, chest circumference, upper-arm circumference, forearm circumference, thigh circumference, calf circumference, ankle circumference, subscapular skinfold, calf skinfold
6	Very poor	−2.55 to −5.65	Triceps skinfold, suprailiac skinfold, thigh skinfold
7	Extremely poor	−5.66 downwards	Biceps skinfold, forearm skinfold

Table 6. Seven-point scale for sedentary female

S. No.	Group for Relationship	Range (Min. to Max. Value)	Variables
1	Extremely high	12.58 onwards	
2	Very high	8.83 to 12.57	
3	Above average	5.08 to 8.82	
4	Average	1.32 to 5.07	
5	Below average	5.06 to −2.44	Weight, height, sitting height, leg length, length of arm, length of foot, biacromial diameter, biiliocristal diameter, bitrochanterion diameter, wrist diameter, knee diameter, ankle diameter, elbow diameter, chest circumference, upper-arm circumference, forearm circumference, thigh circumference, calf circumference, ankle circumference
6	Very poor	−2.45 to −6.19	Triceps skinfold, calf skinfold, % height of centre of gravity
7	Extremely poor	−6.20 downwards	Biceps skinfold, forearm skinfold, subscapular skinfold, suprailiac skinfold, thigh skinfold

In regard to sedentary female (SF), all the selected variables namely weight, height, sitting height, leg length, length of arm, length of foot, biacromial breadth, biiliocristal breadth, bitrochanterion breadth, wrist diameter, knee diameter, ankle diameter, elbow diameter, chest circumference, upper-arm circumference, forearm circumference, thigh circumference, calf circumference and ankle circumference, biceps skinfold, triceps skinfold, forearm skinfold, subscapular skinfold, supra iliac skinfold, thigh skinfold, calf skinfold and percentage height of centre of gravity were observed to be increasing.

Conclusions

On the basis of developed scales, the following conclusions have been drawn:

The percentage change in the variables namely weight, thigh circumference and subscapular skinfold are corroborating with percentage height of centre of gravity in conditioning female, falling in the average category.

The percentage change in the variables namely biceps skinfold, triceps skinfold, forearm skinfold and supra iliac skinfold (falling in very high category), thigh skinfold and calf skinfold (falling in above average category) indicating a higher change than the percentage height of centre of gravity (falling in the average category) in conditioning female.

The percentage change in the variables namely height, sitting height, leg length, length of arm, length of foot, biacromial diameter, biiliocristal diameter, bitrochanterion diameter, chest circumference, upper-arm circumference, forearm circumference, calf circumference, ankle circumference (falling in below average category), wrist diameter, knee diameter, ankle diameter and elbow diameter is observed to be lower (falling in very poor category) than the change in percentage height of centre of gravity (falling in average category) in conditioning female.

The percentage height of centre of gravity is observed to be decreasing in conditioning female. Similar trends have been observed by the variables namely weight, chest circumference, upper-arm circumference, forearm circumference, thigh circumference, calf circumference, biceps skinfold, triceps skinfold, forearm skinfold, subscapular skinfold, supra iliac skinfold, thigh skinfold and calf skinfold, whereas, reverse trends have been observed by the variables namely, height, sitting height, leg length, length of arm, length of foot, biacromial diameter, bicristal diameter, bitrochanterion diameter, wrist diameter, knee diameter, ankle diameter and ankle circumference.

None of the selected variable corroborated with percentage height of centre of gravity in non-conditioning female, falling in the average category.

None of the variable's percentage change was found to be higher than the percentage height of centre of gravity (falling in the average category) in non-conditioning female.

The percentage change in the variables namely weight, height, sitting height, leg length, length of arm, length of foot, biacromial breadth, biiliocristal breadth, bitrochanterion breadth, wrist diameter, knee diameter, ankle diameter, elbow diameter, chest circumference, upper arm circumference, forearm circumference, thigh circumference, calf circumference, ankle circumference, subscapular skinfold and calf skinfold (falling in below average category), triceps skinfold, supra iliac skinfold and thigh skinfold (falling in very poor category), biceps skinfold and fore-arm skinfold (falling in extremely poor category) was lower than the change in percentage height of centre of gravity (falling in the average category) in non-conditioning female.

The percentage height of centre of gravity is observed to be decreasing in non-conditioning female. Similar trends have been observed by the variable weight, whereas, reverse trends have been observed by the variables namely, height, sitting height, leg length, length of arm, length of foot, biacromial diameter, biiliocristal diameter, bitrochanterion diameter, wrist diameter, knee diameter, ankle diameter, elbow diameter, chest circumference, upper-arm circumference, forearm circumference, thigh circumference, calf circumference, ankle circumference, biceps skinfold, triceps skinfold, forearm skinfold, subscapular skinfold, suprailiac skinfold, thigh skinfold and calf skinfold.

The percentage change in the variables namely triceps skinfold and calf skinfold are corroborating with percentage height of centre of gravity in sedentary female, falling in the very poor category.

The percentage change in the variables namely weight, height, sitting height, leg length, length of arm, length of foot, biacromial diameter, bicristal diameter, bitrochanterion diameter, wrist diameter, knee diameter, ankle diameter, elbow diameter, chest circumference, upper-arm circumference, forearm circumference, thigh circumference, calf circumference and ankle circumference (falling in below average category) found to be higher than the percentage height of centre of gravity (falling in very poor category) in sedentary female.

The percentage change in the variables namely biceps skinfold, forearm skinfold, subscapular skinfold, suprailiac skinfold and thigh skinfold (falling in extremely poor category) observed to be less than the change in percentage height of centre of gravity (falling in extremely poor category) in sedentary female.

The percentage height of centre of gravity was observed to be increasing in sedentary female. Similar trends were observed by the all the selected variables namely weight, height, sitting height, leg length, length of arm, length of foot, biacromial diameter, bicristal diameter, bitrochanterion diameter, wrist diameter, knee diameter, ankle diameter, elbow diameter, chest circumference, upper-arm circumference, forearm circumference, thigh circumference, calf circumference, ankle circumference, biceps skinfold, triceps skinfold, forearm skinfold, subscapular skinfold, suprailiac skinfold, thigh skinfold and calf skinfold.

References

1. Blomquist, C.G. 1983. CV Adaptation to Physical Training. *Annual Review of Physiology.* **45**: 169.
2. Bunn, J.W. 1957. *Basket Methods.* The McMillan Company, New York.
3. Bunn, J.W. 1972. *Scientific Principles of Coaching 2nd Edn.* Prentice Hall, Inc., Englewood Cliffs, NJ.
4. Burleigh L., Horak, F.B. and Malquin, F. 1994. Modification of Postural Responses and Step Initiation: Evidence for goal-directed postural interactions. *Journal of Neuro Physiology.* **72(6)**: 2892-2901.
5. Charles, S. 1976. *Fundamentals of Sports Biomechanics.* Kendall/Hunt Publishing Company, Dubuque, Ia.
6. Das, R.N. and Ganguli, S. 1982. Mass and Centre of Gravity of Human Body and Body Segments. *Journal of the Institution of Engineers.* **62: IDGE-3**.
7. Das, R.N. and Ganguli, S. 1982. Moment of Inertia of Living Body and Body Segments Geometrically Shaped Timber and Sleep Specimens. *Journal of the Institution of Engineers.* **62: IDGE-3**.
8. Harris, S.S., Caspersen, C.J., Defriese, G.H. and Estes, E.H. 1989. Physical Activity Counselling for Health Adults as a Primary Preventive Intervention in the Clinical Setting. *JAMA.* **261**: 3590-98.
9. Hoeger and Hoeger. 1990. *Fitness and Wellness.* p. 3.
10. Jensen, C.R., Schultz, G.W., and Bangerter, B.L. 1984. *Applied Kinesiology 3rd Edn.* McGraw Hill, Singapore.
11. Johnson, P. and Strolberg, D. 1971. *Conditioning.* Prentice Hall, Englewood Cliffs, N.J.
12. Kaushik, R., Kaushik, S. and Shaw, D. 1995. Changes in the Location of Three-Dimensional Centre of Gravity as a Result of Six-weeks Conditioning Programme on Male Athletes. *Souvenir: 5th National Conference of NAPESS and GANSF,* October 28-29, Delhi. p. 33.
13. Kawamura, T. et al. 1984. An Analysis of Somatotypes and Postures of Judoist. *Bulletin of the Scientific Studies on Judo: Report VI,* Kodokan, Tokyo. pp. 107-116.
14. Logan, G.A. 1976. *Adaptation of Physical Activities.* Prentice Hall, Englewood Cliffs, NJ.

15. McHenery, P.L. et al. 1990. Statement of Exercise. *Special Report Circulation 51 (Jan., 1990).* p. 1.
16. Monahan, T. 1987. Is Activity As Good As Exercise. *The Physician and Sports Medicine.* **15(10)**: 181.
17. Ray, G.G. and Sen, R.N. 1983. Determination of Whole Body Centre of Gravity in Indians. *Journal of Human Ergo.* **12**: 3-4.

35. Finite Element Modeling and Analysis of $L_{2/3}$ Functional Spine Unit

J. Manjusha[1], Venkatesh Balasubramanian[2] and C. Sujatha[1]
[1]Department of Mechanical Engineering, [2]Department of Biotechnology,
Indian Institute of Technology Madras, Chennai - 600036, India

Abstract: Bilaterally symmetric 3D finite element model of the lumbar motion segment $L_{2/3}$ was developed using ANSYS 7.0. Loads were applied on the top of the L_2 vertebra, fully constraining the bottom of L_3 vertebra. This represents a simulation of laboratorial boundary conditions of $L_{2/3}$ functional spine unit (FSU). The model was analyzed for compression, flexion, torsion and lateral bending loads. The model was analyzed for a static load of 400 N (compressive). For all bending simulations, a moment of 4.5 Nm was used. Modal analysis of $L_{2/3}$ segment was carried out by fixing the bottom of L_3 and applying a load of 400 N on top of L_2 vertebra.

For all the loads maximum displacement was found in the L_2 vertebra. Maximum displacement of 2.1mm was obtained in the anterior region for a compressive load of 400 N and the intervertebral disc (IVD) is bulged. For flexion, maximum displacement of 0.09 mm was observed in the anterior region, whereas for extension it was 0.08 mm in the posterior region. For lateral bending, displacement of 0.34 mm was observed in the left sagittal plane. Similarly, for torsion the maximum displacement was observed to be 0.26 mm in right sagittal plane. The natural frequencies obtained from modal analysis are in the range of 4-8 Hz. The harmonics were 4.5, 4.7, 5.1, 7.2 and 7.4 Hz. This is a preliminary study of static response of the $L_{2/3}$ FSU. A chronic change in response, stresses and IVD bulge especially at the resonance frequency, beyond the base values may trigger the remodeling process leading to spinal degeneration/disorders associated with chronic vibration exposure.

Introduction

Epidemiological investigations have suggested that whole body vibration (WBV) significantly contributes to injuries and functional disorders of the skeleton and joints including spine (Kelsey and Hardy, 1975). Radiographic evidence of degenerative and geometric changes of the vertebral bodies in the lumbar region has been documented as a consequence of remodeling due to chronic occupational vibration exposure (Sandover, 1988). Several investigators have quantified vibration response of the human spine in terms of its resonant frequency, transmissibility, mechanical impedance and relative motion between two vertebral bodies using noninvasive and invasive measurement techniques. A comprehensive review of literature is provided by Brinckmann and Pope (1990) and Kasra et al. (1992). It has been found that the first resonant frequency of the human spine is in the range of 4-8 Hz (Panjabi et al., 1986). The transmissibility and impedance response of the spine show increases at the resonant frequency. Kasra et al. (1992) analyzed the changes in biomechanical parameters, including the resonant frequency, of a ligamentous $L_{2/3}$ motion segment.

Literature indicates that vibration exposure does change the biomechanical response of the human spine and thus may contribute to spinal disorders. Furthermore, in a vibration environment, these changes are maximum at the resonant frequency of the spine. This is based on the assumption that the abnormal relative movements across a segment are closely related to the injurious effects. A chronic alteration of stresses and strains in, and loads transmitted across, various structures of the spine under cyclic loads, especially in the vicinity of the resonant frequency of the spine, in comparison with the static load response, may lead to changes in the hard and soft tissue structure (remodeling) clinically manifested as degenerative changes, disk prolapse or increased incidences of low back pain.

Modeling

The present model of the $L_{2/3}$ functional spine unit (FSU) was developed, using *in vitro* measurements as shown in Fig. 1. To reduce the computational load resulting from the complex geometry of the realistic vertebra, the main load-carrying functional units have been approximated using simplified geometrical parts.

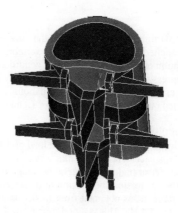

Fig. 1. Finite element model of $L_{2/3}$ FSU.

The software package ANSYS 7.0 was used to develop a geometrical model of the FSU. The details of the model in terms of type of elements used to represent the various regions of $L_{2/3}$ FSU are presented in Table 1. The material properties assigned to various structures are provided in Table 2.

Table 1. Elements used to simulate different structures of the FSU

Spinal Elements	Element Type
Vertebral body	3-d 10 node tetrahedral solid
Posterior elements	3-d 10 node tetrahedral solid
Annulus	3-d 10 node tetrahedral solid
Nucleus	3-d fluid
Facet joints	3-d contact

Table 2. Material properties assigned to various structures of the FSU

Materials	Young's Modulus (MPa)	Poisson's Ratio	Density (kg/mm^2)
Cortical bone	12,000.0	0.30	1.70×10^{-6}
Cancellous bone	100.0	0.20	1.10×10^{-6}
Bony posterior elements	3,500.0	0.25	1.40×10^{-6}
Annulus	4.2	0.45	1.05×10^{-6}
Nucleus polposus	1,666.7*	–	1.02×10^{-6}

The model is meshed using the free mesh option. A finite element (FE) model is made up of a total of 34,664 nodes and 21,528 ten-noded tetrahedral solid elements with each node having 6 degree of freedom. Ligaments were not taken into account in the model.

Boundary and Loading Condition

Loads were applied on the top of the L_2 vertebra, fully constraining the bottom of L_3 vertebra. This represents a simulation of laboratorial boundary conditions of $L_{2/3}$ FSU. Studies of lumbar vertebral specimens from adult humans have shown that specimens failed for loads in the range of 5000-8000 N. The simulation was maximum load. Since the average mass of adult humans fall within the range of 40-100 kg, corresponding to a compressive load of 400-1000 N, the model was analyzed initially for a static load of 400 N.

The model was analyzed for compression, flexion, torsion and lateral bending loads. The moments were simulated as force couples. For flexion two nodes diametrically opposite in the anterior-posterior region were chosen and opposite forces with same magnitude were applied on the model. The FSU was studied under extension in a similar fashion as in flexion by changing the direction of load to anterior-posterior direction. The model was studied for lateral bending by applying loads on two nodes diametrically opposite in the right/left lateral edge of the L_2 vertebra. To simulate the condition of torsional load in the FSU, two concentrated forces of equal magnitude and opposite in direction were applied on the top edge of the FSU.

Modal analysis of $L_{2/3}$ segment was carried out by fixing the bottom of L_3 and applying a load of 400 N on top of L_2 vertebra.

Results

A maximum displacement of 2.1 mm was obtained in the anterior region for a compressive load of 400 N and the intervertebral disc (IVD) bulges as shown in Fig. 2. Table 3 shows von Mises stress and disc bulge of the FSU for the 400 N compressive load. For flexion, the deformations are about 5° and for extension around 2.5°. For torque the deformations are well below 1° and for lateral bending the deformations averaged to 1.2°. Deformations for different types of bending were compared with the results from Panjabi et al. (1994) and are shown in Fig. 3. The von Mises stresses developed in the IVD for a load of 400 N compression, 4.5 Nm flexion and extension are shown in Fig. 4. Modal analysis of the L_2-L_3 segment was carried out and the natural frequencies obtained are as shown in Table 4.

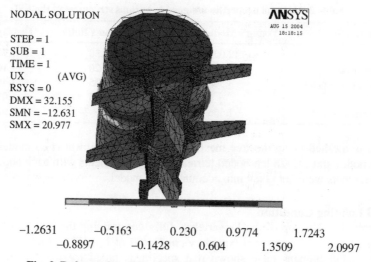

Fig. 2. Deformation of $L_{2/3}$ FSU for a compressive load of 400 N.

Table 3. Deformation and stresses for a static compressive load of 400 N

Calculated Data	Result
Deformation	2.1 mm
Maximum von Mises stress	1.24 MPa
Intradiscal pressure	214.8 kPa
Disc bulge	
$L_{2/3}$ anterior node	0.413 mm
$L_{2/3}$ posterior node	0.351 mm
$L_{2/3}$ lateral node	0.219 mm

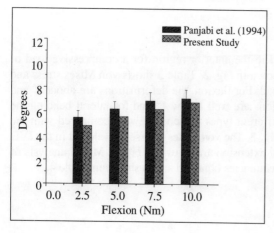

(a)

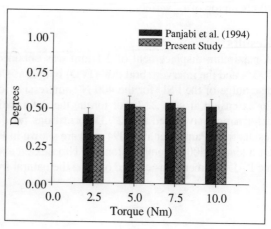

(b)

Finite Element Modeling and Analysis of $L_{2/3}$ Functional Spine Unit 255

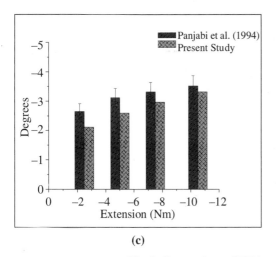

(c)

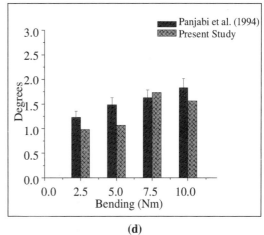

(d)

Fig. 3. Comparison of FEA results with Panjabi et al. (1994).

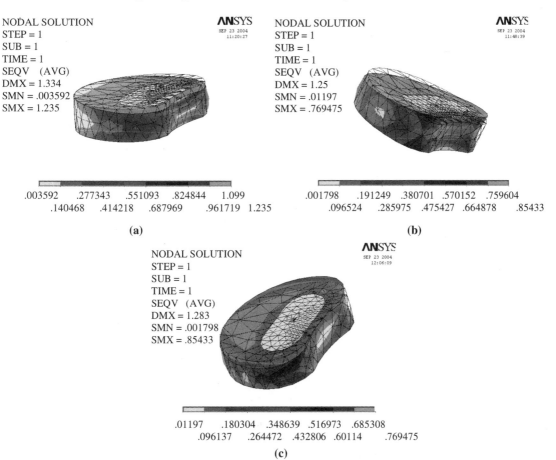

Fig. 4. von Mises stresses in IVD for (a) 400 N compression, (b) 4.5 Nm flexion and (c) 4.5 Nm extension.

Table 4. Modal analysis of $L_{2/3}$

Mode	Frequency (Hz)
1	4.5
2	4.7
3	5.1
4	7.2
5	7.4

Discussion

A three-dimensional finite element model of $L_{2/3}$ functional spinal unit was developed using ANSYS 7.0. Loads were applied on the top of the L_2 vertebra, fully constraining the bottom of L_3 vertebra. This represents a simulation of laboratorial testing boundary conditions. The model was analyzed for compression, flexion, torsion and lateral bending loads. Limitations of this study were, the mass distribution of the model in the FE model was assumed to be constant, but in reality, the mass distribution was more complex. Non-linearity of the material properties was not taken into account. Ligaments had not been included in the present model. In addition this model did not include muscles (active or passive) or overlying soft tissue, posture, fluid flow through disk and the vertebral body, viscoelastic effects including the damping characteristics, creep, and complex loads.

In situ experimental studies do have their limitations in the form of variation in, experimental set-ups concerning loading devices and fixators, geometry of the vertebral segments specimens, measuring technique of the response characteristics, anatomical level of applying the experimental test apparatus. However for the lack of other information, *in situ* experimental result presented by Panjabi et al. (1994) was used for comparing the performance of our model in structural analysis. The deformations predicted by our model are less and varied within one standard deviation from the reported values. It should also be noted that some difference would be expected between our and experimental results depending on the experimental and biological variation of study.

Modal analysis of $L_{2/3}$ segment was carried out by fixing the bottom of L_3 and applying a load of 400 N on top of L_2 vertebra. The natural frequencies obtained from modal analysis are in the range of 4-8 Hz. The resonant frequencies obtained are in good agreement with that reported in literature by Smith (2000). As structural parameters are frequency dependent, cyclic loading, especially in the vicinity of the resonant frequency of the spine could result in disk prolapse or increased incidences of low back pain.

Conclusion

For all the load cases cortical bone absorbs stresses and transmits it to the annulus. The structural parameters values are in good agreement with literature. The resonant frequencies of the model are found in the range of 4-8 Hz. The model developed, although with certain limitations, compared fairly well with experimental results published in the literature (Panjabi et al., 1994). This model can be used to perform further studies on the dynamic response of the FSU to sinusoidal input and random vibratory input. For studying the structural behavior of the FSU, this model could be improved further by incorporating the non-linear properties of structures and ligaments.

References

1. Brinckmann, P. and Pope, M.H. 1990. Effects of Repeated Loads and Vibration. *The Lumbar Spine.*
2. Cowan, S. and Lee, F. 1999. A Biomechanical Model of Lumbar Spine. *Proceedings of ASME.*
3. Frymoyer, J.W., Pope, M.H., Rosen, J.C., Goggin, J. and Constanza, M. 1991. Epidemilogical Studies of Low Back Pain. *Spine.* **Vol. 13**: 419-423.
4. Kasra, M., Shirazi-Adl and Drouin, G. 1992. Dynamics of Human Lumbar Intervertebral Joints-Experimental and Finite Element Investigations. *Spine.* **Vol. 102**: 63-73.
5. Kelsey, J.L. and Hardy, R.J. 1975. Driving Motor Vehicles as a Risk Factor for Acute Herniated Lumbar Interverbral Disc. *Am. J. Epid.* **Vol. 102**: 63-73.
6. Lavaste, F.W., Skalli, S., Robin, C. and Mazel, R.C. 1992. Three Dimensional Geometrical and Mechanical Modeling of the Lumbar Spine. *Journal of Biomechanics.* **25**: 1153-1164.
7. Panjabi, M.M., Anderson, G.B.J., Jorenus, L., Hullt, E. and Mattson, L. 1986. In Vivo Measurements of Spinal Column Vibrations. *Journal of Bone and Joint Surgery.* **Vol. 68A**: 695-702.
8. Panjabi, M.M., Oxland, T.R., Yamamoto and Crisco, T. 1994. Mechanical Behavior of the Human Lumbar and Lumbosacral Spine as Shown by Tree-Dimensional Load-Displacement Curves. *Spine.* **Vol. 76-A**: 413-424.
9. Sandover, J. 1988. Behaviour of Spine under Shock and Vibration: A Review. *Clinical Biomechanics.* **Vol. 3**: 249-256.
10. Smith, S.D. 2000. Modeling Differences in the Vibration Response Characteristics of the Human Body. *Journal of Biomechanics.* **Vol. 33**: 1513-16.
11. Ueno, K. and Liu, Y.K. 1987. A Three Dimensional Nonlinear Finite Element Model of Lumbar Intervertebral Joint in Torsion. *Transactions of ASME.* **Vol. 109**: 201-209.

Biomechanics
R.K. Saxena and P. Mishra (Editors)
Copyright © 2005, Anamaya Publishers, New Delhi, India

36. Effects of Noise on Blood Pressure and Heart Rate

V.K. Katiyar[1], A.K. Guptar[1] and Jaipal[2]

[1]Department of Mathematics, Indian Institute of Technology Roorkee, Roorkee - 247667, India
[2]Department of Mathematics, D.V.S. College, Dehradun, India

Abstract: Stress factors in working environment have recently been discussed as risk factors for cardiovascular diseases. Acute exposure to noise has been shown to cause a temporary rise in blood pressure but repeated and prolonged exposure to noise is a cause of permanent rise in blood pressure. In this paper we have studied the effect of noise on blood pressure on male industrial workers at Bharat Heavy Electricals Limited, Hardwar, India in connection with the noise-induced raised blood pressure.

We proposed a mathematical model for the above observation and solved the basic fluid dynamical equations to find the pressure, wall shear stress and velocity distribution. Results for various parameters have also been compared with experimental values.

Introduction

The most frequent occupational diseases hazard in industrial work is probably hearing loss caused by exposure to noise in the industry. Relatively little attention is paid to the non-specific and non-auditory effects of noise (a distinction can be made between the specific and non-specific effects of noise exposure according to the Fig. 1). Mental stresses have also been associated with a temporary rise in blood pressure. Moller (1978) and Peterson (1981) showed in animal studies that repeated exposure to noise exceeding 90 decibel caused a permanent rise in blood pressure. Renal blood flow decreases during the first few minutes of exposure to vibrations and noise but becomes significantly elevated for the remaining exposure period and heartbeat remains increased throughout (Stephan et al. 1983). Jones and Fronek (1988) used a physical model to test the effect of vibration of position of transition to turbulence (z_t) downstream of a constriction. They predicted that for a given constriction and flow velocity there was a band of frequencies, which caused z_t to move upstream and the value of z_t decreased with increasing vibration amplitude. A relationship between repeated stressful stimuli and a permanent rise in blood pressure is not easy to establish in human beings, because environmental stress is difficult to measure and quantify. Chen et al. (1991) suggested in their short-term experimental study of 25 men, that there was a linear relationship with a high correlation between the evoked systolic blood pressure and sound pressure level (SPL) of white noise. A peak of blood pressure rise was found at about 10 sec after the onset of noise. Several studies have shown that workers exposed to long-term industrial noise suffer from high blood pressure and hypertension (Jonsson et al. 1977 and Talbott et al. 1985). Dijk (1986, 1987) suggested that other environmental working conditions also influence blood pressure. Many non-auditory effects of noise can be described in stress model, for example somatic effects, vestibular effects, activity interference, cardiovascular damage and physiological effects. In human studies, increased diastolic blood pressure was observed and industrial epidemiological surveys have confirmed increased blood pressure after long term noise exposure (Harris, 1979). In this paper we studied the effect of noise on blood pressure on male industrial workers at Bharat Heavy Electricals Limited, Hardwar, India. A mathematical model is proposed and the pressure, wall shear stress and velocity profile in the aorta have been obtained for different values of noise induced heartbeat.

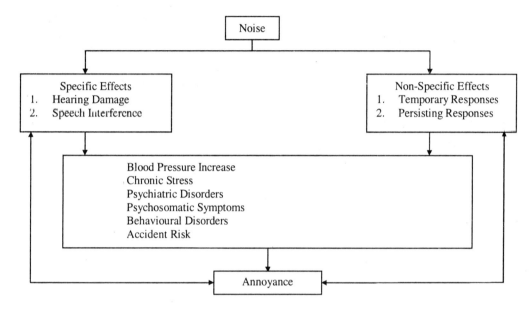

Fig. 1. Scheme for describing noise exposure effects.

Method of Data Collection

The subjects were 23 male workers taken from two different sections of Bharat Heavy Electricals Limited (BHEL), Hardwar, India. Of these, two persons were hypertensive. SPL was measured at 2 m distance from the source of noise and at different locations with the help of Sound Level Meter B and K 2231 and decibel values on A-weighted network dB (A) was measured for the continuous sound pressure level denoted by L_{EQ}. Octave band frequency analysis of the continuous noise level was conducted with the help of B & K 1625. Blood pressure and heartbeat were measured with the subject in a sitting position. Aneroid Manometer (Sphygmomanometer) with a cuff containing 13×45 cm^2 rubber band was used to measure the blood pressure. Disappearance of Korotkoff sound was taken as diastolic blood pressure. Each measurement value was taken as an average of two consecutive and each subject was measured twice a day. An average of 10 days continuous measurements is shown in Tables 1 and 2.

Table 1

S. No.	Age (yrs)	Systolic Blood Pressure		Diastolic Blood Pressure		Heart Rate (min^{-1})	
		Morning	Evening	Morning	Evening	Morning	Evening
1	32	135.0	140.0	88.0	103.30	102.0	98.0
2	34	137.0	138.6	98.0	102.00	84.0	80.6
3	36	127.0	133.3	89.0	98.60	84.0	80.0
4	37	136.0	140.0	93.0	96.60	87.0	90.0
5	38	111.0	117.3	79.0	87.60	93.0	96.6
6	38	121.0	122.0	88.0	88.60	75.0	76.0
7	39	126.0	136.0	96.0	99.00	90.0	90.0
8	39	123.0	129.0	80.0	80.50	82.5	79.3
9	39	129.0	130.0	96.0	100.50	98.5	99.0
10	39	121.5	122.5	86.0	92.50	92.0	90.5
11	41	118.5	118.5	89.0	87.50	87.0	98.5
12	42	121.0	124.5	84.0	88.50	90.0	90.0
13	42	121.0	122.5	89.5	92.00	79.5	101.0
14	43	121.5	124.0	85.5	86.25	82.5	89.5
15	45	132.0	133.5	86.0	90.00	79.0	78.0
16	48	137.0	138.0	95.0	97.00	93.0	94.5
17	48	142.5	140.0	98.0	95.00	90.5	97.0
18	48	145.0	149.5	108.5	107.50	84.0	96.0
19	48	191.0	188.0	130.0	130.00	101.0	98.6
20	50	187.5	191.0	120.8	126.30	88.5	92.5
21	51	116.5	117.5	87.5	89.00	86.0	88.5
22	53	130.5	130.0	82.5	88.00	81.0	80.0
23	56	119.5	122.6	85.5	83.50	82.0	88.5

Table 2

Frequency	Measurement Site	
	Grinding and Fettling	Foundry
31.5	U	U
63.0	U	69.7
124.0	68.0	79.0
150.0	74.2	85.9
500.0	84.1	93.6
1000.0	90.1	97.4
2000.0	102.0	95.9
4000.0	101.6	93.2
8000.0	91.3	91.6
16000.0	76.4	79.0
Linear	101.2	101.6
Max. L	110.9	102.5
Min. L	U	U
LEQ	95.6	100.6

Sound pressure level: decibel (dB); Weighting: A; Preset time: 10 min;
First scale deflection: 120.8 dB, STEP 1/1, BW 1/1.

Mathematical Modeling

A sound starts as a disturbance of the air, which produces sound waves. The visible ear (outer ear) collects sound waves and channels them down the outer ear canal, so that they hit the eardrum and make it vibrate. The vibrations pass through the hammer, anvil, stirrup and oval window in the fluid of the cochlea. Tiny hairs that line the cochlea change the vibration in the fluid into nerve impulses, which are transmitted to the brain along the auditory nerves. The acute effects of these nerve impulses that are mediated via the sympathetic nervous system are those associated with peripheral circulation and heart activation, as well as that medicated by circulating adrenaline and noradrenaline produced by the adrenal medulla in response to sympathetic nervous stimulation. Constriction of blood vessels, contraction of peripheral arteries, an increase in heart rate and an increase in electrical muscular activity are produced, caused by the activation of sympathetic nervous system to raise blood pressure (Ahrlin, 1978, Miyakita, 1991 and Yammamoto, 1980). The constriction phenomena have been considered in the aortic value (Fig. 2) and the flow has been taken pulsatile to model the problem of noise induced heart beat and aortic value vibration, pressure and wall shear stress.

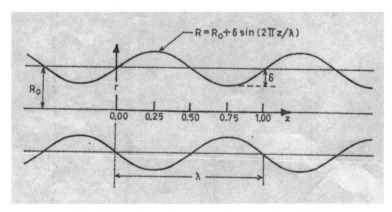

Fig. 2. Stimulation in aorta (axisymmetric model).

The blood is taken as an incompressible Newtonian fluid and the aorta as an axisymmetric tube whose boundary is given by

$$R = R_0 + a\sin\left(\frac{2\pi z}{\lambda}\right) \quad (1)$$

where z, a, λ and R_0 denote axial distance, height of the wall constriction, wavelength and radius of unconstricted aorta, respectively. The equations governing the fluid flow in three-dimensional cylindrical coordinates system (r, θ, z) for axisymmetric case can be written as

$$\frac{\partial u_r}{\partial r} + \frac{\partial u_z}{\partial z} + \frac{u_r}{r} = 0 \quad (2)$$

$$\frac{\partial u_r}{\partial t} + u_r\frac{\partial u_r}{\partial r} + u_z\frac{\partial u_r}{\partial z} = -\frac{1}{\rho}\frac{\partial p}{\partial r} + \upsilon\left(\frac{\partial^2 u_r}{\partial r^2} + \frac{\partial^2 u_r}{\partial z^2} - \frac{u_r}{r^2} + \frac{1}{r}\frac{\partial u_r}{\partial r}\right) \quad (3)$$

$$\frac{\partial u_z}{\partial t}+u_r\frac{\partial u_z}{\partial r}+u_z\frac{\partial u_z}{\partial z}=-\frac{1}{\rho}\frac{\partial p}{\partial z}+\upsilon\left(\frac{\partial^2 u_z}{\partial r^2}+\frac{\partial^2 u_z}{\partial z^2}+\frac{1}{r}\frac{\partial u_z}{\partial r}\right) \quad (4)$$

Eliminating pressure gradient term from Eqs. (3) and (4) and then substituting

$$u_z=-\frac{1}{r}\frac{\partial \psi}{\partial r},\ u_r=\frac{1}{r}\frac{\partial \psi}{\partial z} \quad (5)$$

we get

$$\left(\frac{\partial}{\partial t}+\frac{1}{r}\frac{\partial \psi}{\partial z}\frac{\partial}{\partial r}-\frac{1}{r}\frac{\partial \psi}{\partial r}\frac{\partial}{\partial z}-\frac{2}{r^2}\frac{\partial \psi}{\partial z}\right)\nabla^2\psi=\upsilon\nabla^4\psi \quad (6)$$

where $\nabla^2=\dfrac{\partial^2}{\partial z^2}+\dfrac{\partial^2}{\partial r^2}-\dfrac{1}{r}\dfrac{\partial}{\partial r}$ and $\nabla^4\psi=\nabla^2(\nabla^2\psi)$.

The boundary conditions are

$$\frac{1}{r}\frac{\partial \psi}{\partial r}=0=\frac{1}{r}\frac{\partial \psi}{\partial z} \text{ at } r=\pm R \text{ at (no-slip condition)} \quad (7)$$

and
$$\psi=\text{constant at } r=R \quad (8)$$

Normalization
Introducing the following non-dimensional quantities

$$z^*=\frac{z}{\lambda},\ r^*=\frac{r}{R_0},\ \psi^*=\frac{\psi}{\upsilon R_0},\ t^*=\frac{\upsilon t}{\lambda R_0},\ R^*=\frac{R}{R_0} \quad (9)$$

in Eqs. (6) to (8) and removing the asterisks, we get

$$\delta\left(\frac{\partial}{\partial t}+\frac{1}{r}\frac{\partial \psi}{\partial z}\frac{\partial}{\partial r}-\frac{1}{r}\frac{\partial \psi}{\partial r}\frac{\partial}{\partial z}-\frac{2}{r^2}\frac{\partial \psi}{\partial z}\right)E^2\psi=E^4\psi \quad (10)$$

where $E^2=\delta^2\dfrac{\partial^2}{\partial z^2}+\dfrac{\partial^2}{\partial r^2}-\dfrac{1}{r}\dfrac{\partial}{\partial r}$ and $\delta=\dfrac{R_0}{\lambda}$

The boundary conditions are

$$\frac{\partial \psi}{\partial r}=0=\frac{\partial \psi}{\partial z} \text{ at } \quad r=\pm R \quad (11)$$

and
$$\psi=\text{constant at } \quad r=R \quad (12)$$

Method of Solution
The stream function ψ can be written as an asymptotic power series in δ as

$$\psi=\psi_0+\delta\psi_1+\delta^2\psi_2+\ldots\ldots \quad (13)$$

Substituting Eq. (13) in Eqs. (10) to (12), and collecting the term of similar power of δ on both sides up to zeroth and first order, respectively.

Then, in the resulting equations further substitute

$$\psi_0 = \psi_{00} e^{i\omega t}, \quad \psi_1 = \psi_{10} + \psi_{11} e^{i\omega t} + \psi_{12} e^{2i\omega t} + \ldots \ldots \tag{14}$$

Equating like harmonic terms and solving, we get the stream function

$$\psi = R_1^2 \left(2 - R_1^2\right) e^{i\omega t} + \delta \left[\frac{4i\omega R^2}{192}\left(1 - 3R_1^2 + 2R_1^6\right) e^{i\omega t} + \frac{1}{36}\frac{1}{R}\frac{\partial R}{\partial z}\left(-5 + 16R_1^4 - 12R_1^6 + R_1^8\right) e^{2i\omega t}\right] \tag{15}$$

the non-dimensional axial velocity

$$u_z = -\frac{4}{R^2}\left(1 - R_1^2\right) e^{i\omega t} + \delta \left[\frac{i\omega}{4}\left(R_1^2 - R_1^4\right) e^{i\omega t} - \frac{1}{9R^3}\frac{\partial R}{\partial z}\left(36R_1^2 - 28R_1^4 - 8R_1^6\right) e^{2i\omega t}\right] \tag{16}$$

the non-dimensional wall shear stress

$$\tau_{rz} = \left(\frac{8}{R^3} - \frac{i\omega \delta}{2R}\right) e^{i\omega t} + \frac{40\delta}{9R^4}\frac{\partial R}{\partial z} e^{2i\omega t} \tag{17}$$

and the non-dimensional axial pressure gradient

$$\frac{\partial p}{\partial z} = e^{i\omega t}\left(\frac{16}{R^4} + \frac{i\omega \delta}{R^2} - \frac{112}{9}\frac{\delta}{R^5}\frac{\partial R}{\partial z} e^{i\omega t}\right) \tag{18}$$

where $R = 1 + \varepsilon \sin 2\pi z$, $\varepsilon = \dfrac{a}{R_0}$ (amplitude ratio) and $R_1 = \dfrac{r}{R}$

Result and Discussion

The numerical values of non-dimensional axial velocity profile, wall shear stress and pressure are obtained for different values of parameters such as heart beat and ωt. The numerical values of radius R_0 of the aorta and the amplitude ratio ε are taken 1.25 cm and 0.2, respectively, throughout the computation. Angular frequency ω is calculated by the relation $\omega = ((2\pi \cdot \text{heartbeat})/60)\ \text{s}^{-1}$. The non-dimensional pressure distribution variation with axial distance is shown in Fig. 3 for different values of heartbeat. It is clear from the figure that amplitude of pressure increases as the heartbeat increases. It is also clear that pressure is maximum at the constriction part (i.e. $(z/\lambda) = 0.75$). As stated earlier the sound level produced a constriction of blood vessels and an increase in heart rate, and from Fig. 3 heart rate affects the pressure in aorta.

Therefore it is clear that high noise affects the blood pressure. The amplitude and phase of the wall shear stress are shown in Table 3. The amplitude of wall shear stress is maximum at the constriction and minimum at the wider part of the aorta. The amplitude also increases as the heart rate increases. In Fig. 4 non-dimensional axial velocity profile variation with the non-dimensional radial distance is shown at different location of axial distance z. The maximum velocity occurs on the centerline and the velocity gradient is maximum at the constricted part.

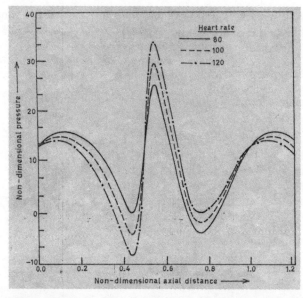

Fig. 3. Variation of pressure with axial distance at different values of heart rate.

Table 3

(i) $R_0 = 1.25$, Heart rate = 80 and $\delta = 0.83$

z	Amplitude			Phase		
	$\pi/4$	$\pi/2$	$3\pi/4$	$\pi/4$	$\pi/2$	$3\pi/4$
0.0	11.2928	8.0842	4.7932	0.7677	−1.4264	−0.8278
0.2	5.7849	5.2368	4.8833	0.3526	1.1338	−1.3053
0.4	6.2874	7.9696	8.8693	−0.0907	0.8026	−1.3613
0.6	11.0564	15.4554	18.0753	−0.0701	0.8530	−1.2654
0.8	17.5384	15.0961	12.8484	0.6742	1.5068	−0.9374
1.0	11.2928	8.0842	4.7132	0.7677	−1.4264	−0.8278

(ii) $R_0 = 1.25$, Heart Rate = 100 and $\delta = 1.04$

z	Amplitude			Phase		
	$\pi/4$	$\pi/2$	$3\pi/4$	$\pi/4$	$\pi/2$	$3\pi/4$
0.0	12.1877	8.0083	4.1109	0.6752	−1.5254	−1.1175
0.2	6.6724	6.0087	5.7007	0.1520	0.9102	−1.5507
0.4	7.8831	9.7585	10.5385	−0.3114	0.6278	−1.5136
0.6	13.1923	18.1641	20.7274	−0.3001	0.6957	−1.3834
0.8	18.4134	15.2802	12.6850	0.5787	1.4028	−1.0879
1.0	12.1877	8.0083	4.1109	0.8752	−1.5254	−1.1175

(iii) $R_0 = 1.25$, Heart Rate = 120 and $\delta = 1.25$

	Amplitude			Phase		
z	$\pi/4$	$\pi/2$	$3\pi/4$	$\pi/4$	$\pi/2$	$3\pi/4$
0.0	13.2614	8.0475	4.2304	0.5636	1.4621	−1.5464
0.2	8.0246	7.2818	7.0687	−0.0294	0.7097	1.3843
0.4	10.0990	12.0933	12.6764	−0.4660	0.4938	1.4984
0.6	16.2909	21.6173	23.9186	−0.4698	0.5687	−1.4897
0.8	19.6028	15.7760	13.0575	0.4662	1.2695	−1.2759
1.0	13.2614	8.0475	4.2305	0.5636	1.4621	−1.5464

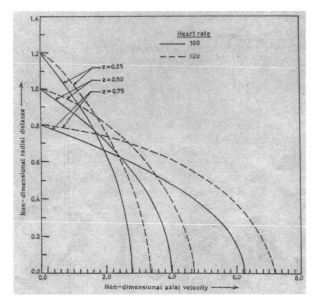

Fig. 4. Velocity profile at different positions of z.

The diastolic blood pressure of more than 60% subjects is found high enough (≥ 90.0 mm Hg) and the heart beat in the normal range (≤ 80 per min) is found in very few persons as can be seen from Table 1 and Jaipal et al. (1992). A raised diastolic pressure is thought to have more serious long term effects than a raised systolic pressure. This may seems strange when the systolic pressure is always the higher of the two. But the systolic pressure, however high is reached for only a fraction of second in each beating of the heart. On the other hand, the diastolic pressure is the lowest level reached between two heartbeats and therefore represents the minimum pressure the artery wall has to sustain continuously. The higher the diastolic pressure the greater the damage to the wall over the months or years that the patient has hypertension. We have tried to explain the nature of these effects through our model, however it is very difficult to establish a direct mathematical relationship between the noise pressure level and blood pressure level. It is concluded that repeated and prolonged exposure to a stressful stimulus may be a contributing factor to rise in blood pressure and heart beat through a mechanism involving structure adaptation of blood vessels in response to repeated acute episodes of raised blood pressure. It is also suggested that acoustic barriers with sound absorbing materials should be used regularly during fettling and grinding operations as well as ear muffs, ear plugs or both must be used by the workers working on grinders, splatter guns and in noisy sections such as foundry etc.

References

1. Ahrlin, U. and Öhrstorm. 1978. Medical Effects of Environmental Noise on Humans. **59**(1): 79-87.
2. Chen, C.J., Hiramatsu, K., Ooue, T., Takagi, K. and Yammamoto, T. 1991. Measurement of Noise Evoked Blood Pressure by Means of Averaging Method. *Journal of Sound and Vibration.* **151**(3): 383-394.
3. Dijk, F.J.H. Van. 1986. Non-auditory Effects of Noise in Industry. *Int. Arch. Occup. Environ. Health.* **58**: 325-332.
4. Dijk, F.J.H. Van, Souman, A.M. and Varies de, F.F. 1987. Non-auditory Effects of Noise in Industry. *Int. Arch. Occup. Environ. Health.* **59**: 133-145.
5. Harris, M.C. 1979. *Handbook of Noise Control.* McGraw Hill Book Co. ING. New York.
6. Jonsson, A. and Hansson, L. 1977. Prolonged Exposure to a Stressful Stimulus (Noise) as a Cause of Raised Blood Pressure in Men. *The Lancet.* **Jan. 8**: 86-87.
7. Jaipal, Katiyar, V.K. and Sharma, H.G. 1992. Effect of Noise on Blood Pressure. *Proceedings, Indian Society of Industrial and Applied Mathematics,* Feb. 4-7, 1993.
8. Jones, S.A. and Fronek, A. 1988. Effect of Vibration on Steady Flow Downstream of a Stenosis. *J. Biomechanics.* **21**(11): 903-914.
9. Moller, A. 1978. Review of Animal Experiments. *Journal of Sound and Vibration.* **59**: 73-77.
10. Miyakita, T. and Miura, H. 1991. Combined Effects of Noise and Hand-arm Vibrations on Auditory Organs and Peripheral Circulation. *Journal of Sound and Vibration.* **151**(3): 395-405.
11. Peterson, E.A., Augenstein, J.S., Tanis, D.C. and Augenstein, D.G. 1981. Noise Raised Blood Pressure Without Impairing Auditory Sensitivity. *Science.* **211**: 1450-1452.
12. Stephans, D.B. and Rader R.D. 1983. Effects of Vibrations, Noise and Restraint on Heart Rate, Blood Pressure and Renal Blood Flow in the Pig. *Journal of Royal Society of Medicine.* **76**: 841-847.
13. Talbott, E., James, H., Matthews, K. and Kuller, L. 1985. Occupational Noise Exposure. *American Journal of Epidemiology.* **121**(4): 501-514.
14. Yamamoto, T. and Osada, Y. (Eds.). 1980. Abstracts of the Collected Papers on the Effects of Noise on Man (Vol. II). Kyoto, Nakanishiya.

Biomechanics
R.K. Saxena and P. Mishra (Editors)
Copyright © 2005, Anamaya Publishers, New Delhi, India

37. Review of EMG Controlled Prosthetic Hand Control

A.S. Arora[1], S.K. Soni[1], P.K. Pathak[2], Vinay Gupta[3] and Santosh Kumar[3]
[1]EIE Department, [2]Instrumentation and Control Department, [3]Department of Instrumentation Engineering,
Sant Harchand Longowal Central Institute of Engineering and Technology, Longowal, Sangrur - 148106, India

Abstract: In this paper we have presented various works done by different researchers in the field of prosthetic hand and their comparative study. The purpose of this paper is to select a prosthetic hand that simulates fundamental dynamic properties of neuromuscular control system of human hand. The different designs can be classified on the basis of their price, complexity, degree of freedom and ease of use. EMG is the muscular signal used for the control of prosthetic hand.

Introduction

The word "prosthesis" comes via new Latin from the Greek "prostithenai" meaning "to add to, or to put in addition". The plural of prosthesis is prostheses. Electromyography is a seductive muse because it provides easy access to physiological processes that cause muscle to generate force, produce movement and accomplish countless functions, which allow us to interact with the world around us. Electrode potentials associated with muscle activity constitute the electromyogram abbreviated EMG. These potential may be measured at the surface of the body near a muscle of interest or directly from the muscle by penetrating the skin with needle electrodes. Since most EMG measurements are intended to obtain an indication of the amount of activity of a given muscle, or group of muscles, rather than of an individual action potential from the fibers constituting the muscle or muscles that are measured.

Characteristics of the EMG Signal

It is well established that the amplitude of the EMG signal is stochastic (random) in nature and can be reasonably represented by a Gaussian distribution function. The amplitude of the signal can range from 0 to 10 mV (peak-to-peak) or 0 to 1.5 mV (rms). The usable energy of the signal is limited to the 0 to 500 Hz frequency range, with the dominant energy being in the range 50-150 Hz. Usable signals are those with energy above the electrical noise level. An example of the frequency spectrum of the EMG signal is presented in Fig. 1.

Frequency spectrum of the EMG signal detected from the Tibialis Anterior muscle during a constant force isometric contraction at 50% of voluntary maximum.

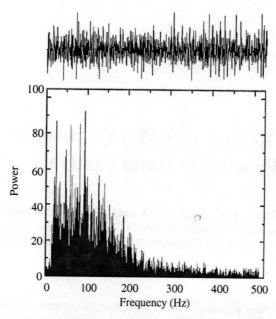

Fig. 1. Frequency spectrum of the EMG signal.

Prosthetic Hand

Prosthetic hands are the hands which give hope to recapture their ability to perform complicated physical movements of the lost human hands. EMG is a suitable approach for human-machine interface in the prosthetic hand control.

After World War 2, with the huge increasing young amputees, the need for better limb control became more apparent. This led to technology being concentrated on developing a new knee that would stabilize during weight bearing but swing freely during walking.

The design of body-powered upper-limb prostheses in particular has experienced few, if any, major breakthroughs since the early 1960s. Upper-limb prostheses are either hook or hand-shaped, and are actuated by body or external power. Persons with amputation frequently express dissatisfaction with the current state of upper-limb prosthesis technology, noting numerous deficiencies with their prostheses, for example functionality, reliability, ease of use, weight, and energy consumption.

The 1970's then saw the development of 'modular assembly prosthesis', which allowed the assembly of prosthesis from a series of stock components. Then, in the 1980's, with the development of materials in the aircraft industry, the world's first carbon fiber prosthetic system was made. This technology promoted high strength and lightweight system.

Then in the 1990's, development into the first commercially available microprocessor controlled prosthetic knee was carried out called the Intelligent Prosthesis (IP), the unit was programmed to each individual user during walking to achieve the smoothest, energy saving pattern. It reacted to speed changes but its intelligence did not extend to understanding environmental considerations such as stairs, ramps or uneven terrain.

Using EMG Data in Prosthetic Hand Control

Electromyography (EMG) is a suitable approach for human-machine interface in the prosthetic hands control. The first artificial hand controlled by EMG was developed in Russia by A.E. Kobrinski in 1961. The Otto Bock Orthopedic Industries developed an EMG controlled multi-finger hand in 1965. In 1980's and 1990's, high-tech solutions were emerging which used new material and miniaturized components. Users can comfortably wear the lightweight electrically powered prosthetic hands during an eight hour day. Integrated with modern computer system the future prosthetic hand probably can accept command direct from the human nerve system and perform multiple tasks.

Acquisition of EMG

As the brain's signal for contraction increases, it both recruits more motor units and increases the "firing frequency" of those units already recruited. All muscle cells within one motor unit become active at the same time. By varying the number of motor units that are active, the body can control the force of the muscle contraction. When individual motor contract, they repetitively emit a short burst of electrical activity known as the motor unit action potential (MUAP). It is detected by electrodes on the surface of the skin in proximity of the motor (Fig. 2).

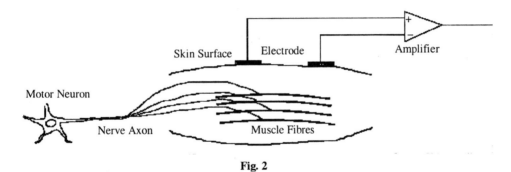

Fig. 2

Detection of the Motor Unit Action Potential (MUAP)

The motor unit action potential (MUAP) is the electrical response to the impulse from the axon. A MUAP looks like the following figure.

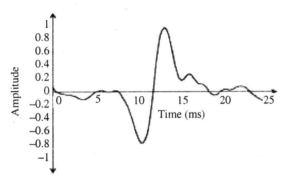

Fig. 3

Action Potential (AP) of One Motor Unit

The contraction of a muscle recruits a number of motors during a period of time. When several motor units are active (the timing of the electrical burst between distinct motor units is mostly uncorrelated), a random interference pattern of electrical activity results detection of EMG Signal.

The usable energy of the signal is limited to the frequency range of 0 to 500 Hz. Usable signals are those with energy above the electrical noise level. There are many factors that influence the EMG signals' detection. These include the electrode structure and its placement on the surface of the skin above the muscle.

Controlling Method of the EMG Operated Prosthetic Hand

Two control systems were implemented for the gripper:

(1) Electrical control to convert the amputee impulses into the gripper actions.
(2) Mechanical control to regulate the force exerted by the prosthetic fingers.

The control system requires an adaptation mechanism for each amputee's characteristics. EMG is a complex signal and is different for each person, making the design of the prosthesis more challenging. For practical use, the electrical control needed to be embedded in the prosthetic hand for the process in a timely manner.

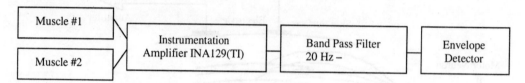

Various Works Done by Different Researchers

1. In this article the author used the adaptive control for the prosthetic limb using neural network. He used an adaptive inverse prosthetic architecture; it reduced the need for specialized external computing and measurement equipment during training. By this patient could directly control leaning of both efferent (control) and afferent (sensing function).
2. In this article the author proposed the prosthetic control method considering flexors and extensors in order to construct the multi-joint prosthetic system and to realize natural feeling of its control. In the experiments, the natural feeling of the prosthetic control similar to that of the human hand was realized using the neural networks and the biomimetic impedance control method.
3. In this article the author focused on development of a myoelectric discrimination system for a multi degree prosthetic hand. He proposed 8 type movement three jaw chuck, lateral, hook grasp, power grasp, cylinder grasp, centralized grip, flattened hand and wrist flexion. He also used the techniques of variance, bias zero crossings, auto regressive model and spectral estimation were employed for preprocessing of features also with 2nd order AR parameters for best performance for pattern recognitions.
4. The author proposed a new EMG control method of prosthetic hand, which could estimate the joint torque from the EMG signals using the neural network. The joint torque was estimated from EMG signals using artificial neural network. He showed the possibility to apply the proposed method to the prosthetic hand control under the conditions that the controlled object was the second order delay system.

5. The author has proposed the learning system utilizing the feedback error-learning schema. A direct torque control method for the prosthetic hand was estimated from an artificial neural network by the feedback error learning schema of EMG signals, 2-DOF motions, i.e. hand grasping/opening and ann. flexion/extension, were picked up.
6. The author presented development of a novel prosthetic hand based on biomechatronic design. Hhe presented the cylindrical grasp of a cylindrical object. The parallel force /position control was developed that ensured the stability of the grasp. This was useful to gain the information about the load and forces using the objects with different shapes and weight.
7. In this article the author describes an Evolvable Hardware (EHW) chip, and the application of this chip as a controller for myoelectric prosthetic hand. The chip consists of Genetic Algorithm (GA) hardware, reconfigurable hardware logic, a chromosome memory, a training data memory, and a 16-bit CPU core (NEC V30). For compact implementations of GA hardware, the following three criteria are necessary:

   a. An effective search method for small populations
   b. Steady state gas
   c. Crossover operations requiring less hardware for random number generation

8. The author discussed the necessity of learning mechanism for EMG prosthetic hand controller; and the real-time learning method was proposed and designed. He divided the controller into three units. Analysis unit extracts useful information for discriminating motions from EMG. Adaptation unit learns the relation between EMG and control command and adapts operator S characteristics. Trainer unit makes the adaptation unit learn in real-time, which has an ability of real-time learning, and performed experiments with the discriminating of 10 motions.
9. The author presented his EHW (Evolvable Hardware) design in contrast to conventional hardware where the structure was irreversibly fixed in the design process, which previously could not be fully realized due to the lack of autonomously reconfigurable digital hardware. In general, applications that tend to be time variant in nature and to had real-time constraints are suitable for EHW, because EHW enables adaptation at the hardware level, and fast execution, being accomplished by the hardware itself. Although reconfigurable hardware devices, such as FPGA and PLD, are spreading rapidly and the usefulness of reconfigurable hardware is being more widely recognized, reconfiguration in FPGA's is not autonomous and requires human intervention. Thus, EHW indicates a new direction in reconfigurable hardware beyond FPGA's. Rather than discarding chips that do not meet the specifications, minute adjustments can, however, be made with EHW architectures and evolutionary computation.
10. In this paper the author describes the design and fabrication of a novel prosthetic hand based on a "biomechatronic" and cybernetic approach. In this approach the integrated biomimetic mechanisms, sensors, actuators and control, and by interfacing the hand with the peripheral nervous system. He has used intelligent neural network for stimulation and recording from nerve. He used PNS with a telemetry connection with an external control system. It had two degree of freedom.
11. In this paper the author has presented a EMG operated biometric prosthetic hand. He used digital servo system of dc motor on a mechanical hand of one degree of freedom. The non-linear properties of the neuromuscular system were realized by position control sys.
12. In this paper the author has presented the electromyography (EMG) based signal analysis to discriminate 8 hand motions: power grasp, hook grasp, wrist flexion, lateral pinch, flattened hand, centralized grip, three-jaw chuck and cylindrical grasp. His analyzed the PC-based control system,

a three-channel EMG signal was used to distinguish 8 hand motions for the short below elbow amputee. Three surface electrodes were placed on palmaris longus, extensor digitorum and flexor capitularies. He also presented a controller based on digital signal processor (DSP) and implemented it in this discriminative system. The on-line DSP controller could provide 87.5% correct rate for the discrimination of 8 hand motions.

13. In this paper the author focused on the hand mechanisms design and presented preliminary results in developing the three fingered anthropomorphic hand prototype and its sensory system. He presented the design approach of the mechanical structure of the sensory system and of the socket of a cybernetic prosthesis. It was of two degree of freedom.

Conclusions

Through the comparative study of various researchers in the field of prosthetic hand control using EMG signals, we studied various types of techniques that are proposed for feature extraction from EMG signals and various optimization techniques used for optimizing the control system design depending on degree of freedom. Comparative study reveals that the most recent and dynamic approach is the use of EHW and neural network for the design part. In modern computer system the future prosthetic hand probably can accept command directly from the human nerves system and perform multiple tasks.

References

1. Elsley, R.K. 2001. Adaptive Control of Prosthetic Limb using Neural Network. *Proceedings of the 2001 IEEE/RSJ International Conference on Intelligent Robots and Systems.* pp. 1366-1375.
2. Tsuji, T., Fukuda, O., Shigeyoshi, H. and Kanekol, M. 2000. Bio-mimetic Impedance Control of an EMG-controlled Prosthetic Hand. *Proceedings of the 2000 IEEE/RSJ International Conference on Intelligent Robots and Systems.* pp. 377-373.
3. Huang, H.P. and Chen, C.Y. 1999. Development of a Myoelectric Discrimination System for a Multi degree Prosthetic Hand. *Proceeding of 1999 IEEE International Conference on Robotic and Automation,* May 1999. pp. 2392-2397.
4. Morita, S., Kondo, T. and Ito, K. 2001. Estimation of Forearm Movement from EMG Signal and Application to Prosthetic Hand Control. *Proceedings of the 2001 IEEE,* Seoul, Korea. May 21-26, 2001. pp. 3692-3697.
5. Morita, S., Shibatat, K., Zhengt, X-Z, Ito, K. 2000. Prosthetic Hand Control based on Torque Estimation from EMG Signals. *Proceedings of the 2000 IEEE/RSJ International Conference on Intelligent Robots and Systems.* pp. 389-394.
6. Scherilla, P. and Siciliano, B. 2003. Parallel Force/Position Control of a Novel Biomechatronic Hand Prosthetic. *Proceeding of 2003 IEEE/ASME International Conference on Advanced Mechatronic (AIM 2003).* pp. 920-925.
7. Kajitani, I., Murakawa, K.M., Nishikawa, D. Yokoi, H. and Kajihara, N. 1998. An Evolvable Hardware Chip for Prosthetic Hand Controller. *Proceedings of the 1998 IEEE.*
8. Nishikawa, D., Wenwei, Y., Yokoi, H. and Kakazu, Y. 1999. EMG Prosthetic Hand Controller Discriminating Ten Motions using Real-time Learning Method. *Proceedings of the 1999 IEEERSJ International Conference on Intelligent Robots and Systems.* pp. 1592-1597.
9. Higuchi, T., Iwata, M., Keymeulen, D., Sakanashi, H., Murakawa, M., Kajitani, I. and Takahashi, E. 1999. Real-World Applications of Analog and Digital Evolvable Hardware. *IEEE Transactions on Evolutionary Computation.* **Vol. 3, No. 3, September 1999**: 221-235.
10. Carrozza, M.C., Micera, S., Massa, B., Zecca, M., Lazzannt, R., Canelli, N., Dario, P. "The Development of a Novel Biomechatronic Hand – Ongoing Research and Preliminary Results". 2001 IEE/ASME International Conference on Advanced Intelligent Mechatronics Proceedings 8-12 July 2001 Como, Italy, pp. 249-254.

11. Ohno, R., Yoshida, M. and Akazawa, K. 1999. Development of Myoelectric Hand End Effector with Voluntarily Controlled Compliance. *Proceedings of The First Joint BMEEMBS Conference SMng Humanity, Advancing Technology 06 13.16, '99*, Atlanta, G.A., U.S.A. pp. 640.
12. Huang, H.P. and Chiang, C.Y. 2000. DSP-Based Controller for a Multi-Degree Prosthetic Hand. *Proceedings of the 2000 IEEE International Conference on Robotics 8 Automation*, San Francisco, CA April 2000. pp. 1378-1383.
13. Carrozza, M.C., Micera, S., Massa, B., Zecca, M., Lazzannt, R., Canelli, N. and Dario, P. 2001. The Development of a Novel Biomechatronic Hand-Ongoing Research and Preliminary Results. *2001 IEEU ASME International Conference on Advanced Intelligent Mechatronics Proceedings*, 8-12 July 2001. Como, Italy. pp. 249-254.

Biomechanics
R.K. Saxena and P. Mishra (Editors)
Copyright © 2005, Anamaya Publishers, New Delhi, India

38. Simulation of Normal Human ECG and Arrhythmias Using Advanced Electronics

Shahanaz Ayub
Department of Biomedical Engineering, Institute of Engineering and Technology,
Bundelkhand University, Jhansi - 284128, India

Abstract: The hospital coronary care unit is developed to provide the critical cardiac patient with specialized attention of highly skilled medical professionals, aided by cardiac surveillance equipment. Detection and treatment of arrhythmias has become one of the CCU's major functions. Arrhythmia/ECG simulator will allow the user to ascertain that arrhythmia/ECG monitor is recognizing normal ECG and typical arrhythmia, and properly responding when such signals occur. Today, the available simulators are for normal ECG or separately arrhythmia simulator detecting only lead II signal. This article describes the 17 different life threatening arrhythmias and their simulation to obtain lead I, lead II, lead III, avR, avL, avF, V1, V2, V3, V4, V5, V6 waveforms, simultaneously. Arrhythmias selected are Bradycardia, Tachycardia, Asystol, Fusion Beat, Missed Beat, Atrial Fibrillation, Ventricular Fibrillation, Bigeminy, Multifocal Ventricular Extrasystols, R on T wave, Ventricular Tachycardia, Heart Block, VPB 1 and VPB 2. The heart mechanics explains the change by showing abnormalities in normal ECG waveform as arrhythmia and the data for each arrhythmia calculated is explained. Microcontroller 8752 is used to control all the operations like detection of key press, performing the calculations and displaying the corresponding 9 waveforms for one arrhythmia. The other necessary circuits used like D to A converter, analog demultiplexer, low pass filter, sample and hold circuit are also explained. The results obtained and checked at Nair Hospital, Mumbai for satisfactory performance are given.

Introduction

Cardiac Monitoring [1] is now an essential and important medical field. It includes diagnosis and continuous monitoring of patients suffering from cardiac troubles. As the monitoring facility increases, it is also important to check whether the machines are functioning properly, so that they produce correct patient data. For this purpose simulators [6] were introduced, which gave reliable output representing normal or abnormal patient data. This data was given to the machine and checked whether the machine produced the expected waveforms. With further development, simulators which could be used as tool of comparison for intelligent monitor to diagnose the patient data and give directly the disease or arrhythmia the patient is suffering from could also be developed.

The hospital coronary care unit was developed to provide the critical cardiac patient with specialised attention of highly skilled medical professionals aided by cardiac surveillance equipment. Detection and treatment of arrhythmias has become one of the Cardiac Care Unit's (CCU) major functions.

A close correlation exists between innovative developments and enhanced prognosis and patient longevity. Today, the CCU patient may require constant and exact monitoring, which is provided by specially programmed, sophisticated arrhythmia/electrocardiograph (ECG) [2] monitors. These instruments can detect potentially life threatening arrhythmias with a high degree of accuracy [2, 3, 23], thus, alerting the hospital staff to the patient's need for immediate treatment.

Arrhythmia/ECG simulator is a device for quick and accurate testing of arrhythmia/ECG monitors. This unit allows the user to ascertain that the arrhythmia/ECG monitor is recognizing normal ECG and typical arrhythmia, and properly responding when such signals occur. Therefore, the instrument is an essential tool for CCU personnel.

Today, the available simulators are for normal ECG or separately arrhythmia simulator detecting only lead II signal and not the abnormalities of all points [6], i.e. RA, LA, LL, RL, V1, V2, V3, V4, V5 and V6.

Hence, the need is for ECG/arrhythmia simulator which simulates the abnormalities of all point waveforms, e.g. LA, RA, LL, V1, V2, V3, V4, V5 and V6. The present Holter Cardiography [21] does give the interpretations of the arrhythmias but the software is not accessible. Further, its use is restricted to the holter equipment only. Hence, one must know the surety of the interpretations given by any arrhythmia/ECG monitor before preparing the patient's report.

Till date no such simulator is designed to provide a means to confirm these interpretations.

Methodology

The research work consists of designing of the ECG/arrhythmia simulation which will generate the normal and abnormal sinus rhythms [2, 3] by appropriate conversion of the digital data [4, 5] stored in memory. Waveforms for all signals, i.e. avR, avL, avF, lead I, lead II, lead III and V1 to V6 [1] for a particular rhythm are generated simultaneously. The hardware circuitry should be considerably reduced by making efficient use of the microcontroller [4, 5].

To output the waveforms of the respective lead pin to the terminals, the individual lead is scanned to output the single data byte for each waveform at the respective pin. The waveform to be output is selected by pressing a particular key which is also indicated by appropriate LED glow on front panel. In order to simulate the arrhythmias, the physiological description of the most common and important life threatening arrhythmias are explained here. Such 17 arrhythmias [1-3] as VPB 1, VPB 2, multifocal run, missed beat, fusion beat, couplet, run, R on T wave, asystole, ventricular fibrillation, atrial fibrillation, heart block, paced rhythm, tachycardia, bradycardia, ventricular tachycardia, bigeminy are described. The description gives the details of how the heart mechanics shows the abnormalities in its functioning (Fig. 1). Some of the arrhythmias are:

1. Bradycardia: Extreme bradycardia is a critical reduction of heart rate. R-R interval is greater than 1.2 s (50 beats/min).
2. Tachycardia: It is a serious racing of heart with average heart rate exceeding 120 beats/min.
3. Asystol: There is a lack of conduction for an extended duration.
4. Ventricular Fibrillation: Unconditional, chaotic, uncoordinated fluttering of ventricles. No defined P-QRS-T signal is observed.

The actual designing of the various circuits involves the criteria for component selection from data sheets used in the design. The sections includes power supply design, digital section designing includes microcontroller and DAC (digital to analog converter) [20]. The analog [20] section design includes demultiplexer [19], sample and hold circuit [17, 18], low pass filter [17, 18], attenuator, keyboard [22] and display [22] section.

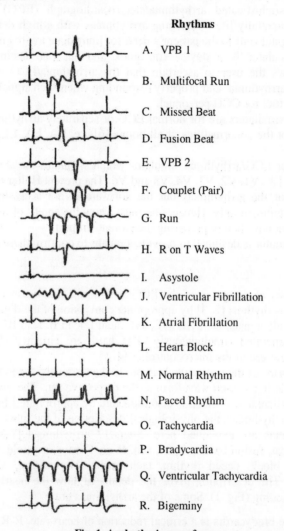

Fig. 1. Arrhythmia waveforms.

The software, contributing the major part of the project, includes programming of microcontroller to output the particular digital data from the memory on detection of key press, obtaining its keycode, initializing the memory pointer to point the required data and outputting the data at a fixed time period decided by the programmed timer/counter of microcontroller [4, 5]. Microcontroller 8752 is used because of its timer/counter and input/output port facilities. Also the memory capacity is selected taking into view the expanding of arrhythmias of interest of cardiac care unit in future, as and when required. Also, the hardware circuit is designed in such a way that in future it will provide the facility to introduce few more arrhythmias without disturbing the current 17 arrhythmias.

Data is collected and a normal ECG is simulated designing a circuit which will output this pattern. 10 patient's data is collected. Their ECG taken by an expert on an ECG machine is analyzed and then the normal ECG waveform is simulated. The electrode [1] applied on a patient's body gives individual signal of left arm, right arm, left leg and six chest points but the ECG machine gives the signal as avR, avL, avF, lead I, lead II, lead III, V1 to V6. These are the signals recognised as ECG waveforms and not the individual signal picked up by electrode. To extract the individual signals at electrode from the ECG waveforms obtained by ECG machine or any cardiac monitor, needs a careful calculation, which relates these signals by certain formulae [1-3]. Using this, the arrhythmias are simulated. The hardware circuit is designed to a great extent and a microcontroller selection has been done, taking into account its memory capacity and other facilities.

Experimental Work

Figure 2 shows block diagram of the entire simulation process. The individual signal at each electrode obtained is in the form of analog signal. In order to make it work with microcontroller for further treatment, it needs to be converted into proportionate digital data, which is done by digital to analog converter (DAC). In case of the above signals, 8 bit DAC is selected. All the analog data is converted into digital data. This data is stored in the memory of microcontroller in the form of 'Look Up Table' [22]. The data is grouped which forms a typical pattern that will be fetched frequently as and when required in order to minimise the utilisation of memory space.

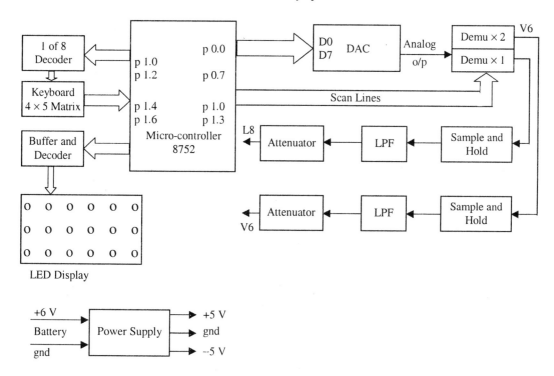

Fig. 2. Block diagram of ECG/arrhythmia simulator.

Microcontroller is used to control all the operations of the instrument like detection of the key press, get the key code, branch to the corresponding routine [22] and display the corresponding 9 arrhythmia waveforms at a time.

The digital data stored in the memory for any arrhythmia waveform detected is passed to the digital to analog converter which converts the signal in a bipolar range [9, 10].

An analog demultiplexer [9, 10] selected using four bits of microcontroller port will be used to generate nine waveforms. This is essential to output all 9 lead signals at a time with the restriction of being able to output only one digital data on the microcontroller port at a time.

The demultiplexer provides analog signal which is held by sample and hold circuit till the next data appears on it. This held signal is then passed through second order low pass filter with a gain of 40 dB/decade.

The waveforms are generated at amplitude levels of 1, 1.5 or 2 V and hence, an attenuator of 1:1000 is used to provide signals in the millivolt range.

The keyboard is scanned at the rate at which the 9 circuits are enabled. The keyboard matrix is formed using 1 of 8 decoder and 3 port lines of the microcontroller. In all, 20 keys will be used.

Display is provided to indicate the arrhythmia selected. On any key press, the corresponding Light Emitting Diode (LED) glows, indicating to the user the operation selected. The display consists of 18 LEDs corresponding to the 17 arrhythmias and one normal ECG using 1 of 8 decoder [7]. The LEDs are connected after the buffer to raise the current capability of the signal coming from the microcontroller.

Power supply [9, 10] is designed to generate +5/−5 for the analog and digital ICs used in the circuit. Depending on the current requirement of each IC, the total current being consumed by the circuit is calculated. The power supply will be DC to DC converter [9, 10].

To make the instrument portable and to avoid the 50 Hz noise signal, a 6 V lead acid battery (1.2 Ah) is used. Using a switching regulator and the appropriate transformer, the supply is generated and supplied all over the circuit.

The PCB layout and artwork is done using the package 'SMARTWORK'. Software is written in assembly language [4, 5, 22] of 8752 microcontroller using Norton Editor [12]. It is assembled using Cross 16 assembler [13] and using HEXOBJ software for obtaining the object code. This data is then loaded into the 8752 EPROM using the EPROM programmer.

8752 microcontroller has two software interrupts T0, T1. It is difficult to program using "polling" [22] method as keyboard scanning along with outputting display and waveforms has to be done. Therefore, the best method is to use the software timer interrupt provided by the microcontroller. Each timer interrupt can be used in mode 0, mode 1, mode 2 or mode 3 operation. The details of the modes of operation are provided in the appendix [4, 5]. Mode 1 is used for our purpose in which the timer register can be loaded as 16 bit data.

Complete data for each waveform is to be outputted in 2 ms. Since all the 9 waveforms need to be outputted at a time, the timing for each data byte is 2 ms/9, i.e. 222 µs. Hence, the timer is loaded with a count for 222 µs. Timer mode is selected using TMOD register [4, 5], timer interrupt is selected using TCON register [4, 5] and TH0, TL0 [4, 5] is loaded with the appropriate count. IE register [4, 5] is used to enable the interrupt, since it is started because of TCON register.

Software algorithm includes initialization of all ports of 8752 microcontroller, initialization of timer interrupt T0, starting of timer and enabling timer interrupt. It will wait till interrupt comes. In timer interrupt routine, the keyboard is scanned. If new key press is detected, keycode is found. If no key press is observed, it continues the previous operation. Depending on the key code, it goes to the respective routine. In arrhythmia routine, it checks for initialization flag [4, 5, 22] and if not set, it does the initialization of the memory pointer, counter for data bytes and scan lines. It fetches appropriate byte from the respective memory location and output to DAC port. It outputs scan line bits to enable the analog demultiplexer switches. It increments memory pointer and scan line [22] counter for next data. It checks for scan lines, if greater than 9 then reinitialize it to zero. Counter is decremented by 1 and if 0, the initialization flag is cleared. It returns from interrupt.

Results and Discussion

This simulation program is an effective tool for medical and biomedical engineering students to teach them about arrhythmias especially life threatening arrhythmias, so that the students will be aware of these and will learn it fast.

Holter Cardiography machine can make use of this simulation for regular check of its identification of arrhythmias.

The actual implementation of the circuit design is done on printed circuit board (PCB). The entire synchronised hardware circuit with software is tested and the assembled project is tested on different ECG machines and on Holter Cardiography at Nair Hospital, Mumbai.

Though at present, the scope of the project is limited to generation of only 17 arrhythmias and normal ECG, the number can be extended by using the remaining keys in the hardware circuit. The remaining memory space of the microcontroller can be utilized. Besides this, the project can be expanded further by interfacing it with a computer for diagnostic monitoring.

Acknowledgement

The author is grateful to Mr. R.A. Raichur, Director, Situ Electro Instruments Pvt. Ltd., Ghatkopar, Mumbai and Nair Hospital, Mumbai for providing necessary facilities.

References

1. Khandpur, R.S. 1987. *Handbook of Biomedical Instrumentation.* Tata McGraw-Hill, New Delhi.
2. Rowlands, D.J. 1980. *Understanding ECG.* Manchester.
3. Booklet Presented as a Service to Cardiology. Pharmaceuticals Division, Imperical Chemical Industries PLC, England.
4. Philips Microcontroller 8051 Data Book.
5. MCS- 51 Microcontroller Data Book.
6. Manual on ECG Simulator. BPL.
7. Motorola High Speed CMOS Data Book.
8. TTL Logic Data Book.
9. Intersil Linear Data Book.
10. National Semiconductor Linear Data Book.
11. Grabowsky, T. *Principles of Anatomy and Physiology, 17th edn.* Harper Collins.
12. Norton Editor Manual.
13. Cross 16 Assembler Manual.
14. EPROM Programmer Manual.
15. Ross and Wilson. *Anatomy and Physiology in Health and Illness.* ELBS.
16. Guyton. *Physiology of Human Body.*
17. Gayakwad, R. *Op-Amp and Linear Integrated Circuit Techniques.* Prentice Hall of India, New Delhi.
18. Botkar. *Integrated Circuit Designing.*
19. Jain, R.P. *Modern Digital Electronics.*
20. Cooper. *Modern Electronic Instrumentation.* Prentice Hall of India. New Delhi.

21. Holter Cardiography Manual.
22. Gaonkar, R. *8085 Microprocessor Programming*.
23. Pueyo, Smetana, Camina. 2004. Characterisation of QT Interval Adaption to RR Interval Changes and its Use as Risk-Stratifier of Arrhythmic Mortality in Amiodarone-Treated Survivors of Acute Myocardial Infarction. *IEEE Transactions on Biomedical Engineering.* **Vol. 51, Issue 9**: 1511-1520.

39. Effect of Physical Exercise on Three-Dimensional Centre of Gravity, Height, Weight and Ponderal Index of the Students of University of Delhi

Dhananjoy Shaw[1], Seema Kaushik[2] and Indu Kaushik[1]

[1]Biomechanics Laboratory, Department of Natural/Medical Sciences,
I.G.I.P.E.S.S., University of Delhi, New Delhi - 110018

[2]Department of Physical Education, Lakshmibai College, University of Delhi, Delhi - 110007

Abstract: The study was conducted with the purpose of studying the effect of physical exercise on three-dimensional centre of gravity and weight, height, ponderal index as an effect of exercise on male and female students of University of Delhi. 225 students (male = 135 and female = 90) of University of Delhi, age ranging between 17 and 25 years were selected randomly for the purpose of the study. The sample was classified into six groups viz. Conditioning male ($n_1 = 43$), non-conditioning male ($n_2 = 62$), sedentary male ($n_3 = 30$), conditioning female ($n_4 = 33$), non-conditioning female ($n_5 = 35$) and sedentary female ($n_6 = 22$). The conditioning program consisted of three meso cycles viz. M-l, M-2 and M-3 of six weeks duration each with the target training intensity (HR) of 130 ± 10, 150 ± 10 and 170 ± 10 beats/min, respectively, for a session of 1 h duration (45 min for general conditioning program and 15 min for warming up and cooling down) regularly for five sessions per week. The subjects were tested four times viz. T-l (at zero weeks of training), T-2 (after 6 weeks of training), T-3 (after 12 weeks of training) and T-4 (after 18 weeks of training). The variables were "Z" dimension of CG (% height of CG), "Y" dimension of CG (posterio-anterior aspect of CG), "X-RL" dimension of CG (CG from the right lateral aspect of the foot) and "X-LL" dimension of CG (CG from the left lateral aspect of the foot), body weight, height and ponderal index. Equipments included lever based weighing scale, a wooden board of $183 \times 100 \times 2$ cm (fitted within a steel frame and fixed fulcrum) and anthropometer etc. The data was analyzed using mean, standard deviation, and percentage etc. The study concluded that there was a significant relationship between three-dimensions of centre of gravity and selected variables.

Introduction

Human body is a system of bony articulations bound by muscle and tissue fibers. This system behaves like a structure when one stands erect or adopts a particular stationary posture and/or machine during locomotion or other physical activities. Structurally, the human body is in a state of unstable equilibrium as the body's centre of gravity lies near the navel, which is much above the supporting points on the ground, leading to instability in dynamic conditions.

The center of gravity of an individual, while standing in the anatomical position marks the intersection of the three primary planes (sagittal, lateral and transverse) and their axes. For solid masses of uniform density, the location of this point, often called the mass center, is at the geometric center and remains constant, no matter what position the object assumes. The human body's flexibility and its internal fluid structure create great problems in accurately locating the center of gravity, because, mass center can be determined for any given, momentarily fixed stance, any major movement is accompanied by a slight shift in the location of the center of gravity. Thus, in many sports skills, the mass center is constantly moving (Charles, 1985).

The resultant of centre of gravities (CG) of various bodily segments, depends upon its own structural dimension and mass which is likely to change due to certain changing factors, such as height of the individual, mass of various body segments, posture, width of bone, varying length dimension of each segment of the body, body composition (lean body mass, percentage of fat, percentage of water in the body) and type of physique etc., which are under the direct influence of nature of activities, nutrition, physical training/conditioning etc. (Charles, 1985; Burleigh et. al., 1994; Kawamura et. al., 1987; Kaushik and Shaw, 1995; Oyster, 1979), leading to various physical, physiological and psychological changes (Blomquist, 1983; Shephord, 1983; Monahan, 1987; Harris et. al., 1989; McHenery et. al., 1990).

Any such momentarily change is likely to deviate the location of CG at any of the three planes and axes. Due to the little deviation of centre of gravity, the unstable equilibrium of the body is markedly disturbed, leading to the postural disturbances and further may cause postural deformities or functional impairment. Hence, the present investigation was conducted to study the effect of physical exercise on the location of three-dimensional centre of gravity, height, weight and ponderal index of the male and female students of University of Delhi.

Material and Method

Sample
225 students (male = 135 and female = 90) of University of Delhi, age ranging between 17 to 25 years were selected randomly for the purpose of the study. The sample was classified into six groups namely conditioning male ($n_1 = 43$), non-conditioning male ($n_2 = 62$) and sedentary male ($n_3 = 30$), conditioning female ($n_4 = 33$), non-conditioning female ($n_5 = 35$) and sedentary female ($n_6 = 22$).

Selection of Variables
The following variables were selected for the purpose of the present study:

1. "Z" dimension of CG (% height of CG)
2. "Y" dimension of CG (posterio-anterior aspect of CG)
3. "X-RL" dimension of CG (CG from the right lateral aspect of the foot)
4. "X-LL" dimension of CG (CG from the left lateral aspect of the foot)
5. Body weight
6. Height
7. Ponderal Index

Experimental Protocol
The conditioning programme consisted of three meso cycles namely, M-1, M-2 and M-3 of six weeks duration each with the target training intensity (HR) of 130 ± 10, 150 ± 10 and 170 ± 10 beats/min respectively for a session of 1 h duration (45 min for general conditioning programme and 15 min for warming up and cooling down) regularly for 5 sessions per week where, conditioning group participated in the described conditioning programme of physical exercises along with the regular

physical education programme of Indira Gandhi Institute of Physical Education and Sports Sciences (IGIPESS) curriculum, however, non-conditioning group participated in the regular physical education programme of IGIPESS curriculum only, whereas, sedentary group did not participate in any kind of physical exercise for the period of experimentation. The detailed experimental protocol is presented in Table 1.

Table 1. Experimental protocol

S. No.	Meso Cycles	Target HR/Intensity (beats/min)	Target Components
1	Meso cycle one (M-1)	130 ± 10	Flexibility, cardio-respiratory endurance
2	Meso cycle two (M-2)	150 ± 10	Muscular endurance, strength
3	Meso cycle three (M-3)	170 ± 10	Speed, power (explosive strength)

Testing Protocol

Each subject was tested 4 times during 4½ month of physical exercise programme viz. T 1 (at zero weeks of training), T-2 (after 6 weeks of training), T-3 (after 12 weeks of training) and T-4 (after 18 weeks of training). The detailed testing protocol is exhibited in Table 2.

Table 2. Testing protocol

S. No.	Tests	Test Code	Time of Testing
1	Testing one (Pre-test)	T-1	Before the start of physical training/conditioning at zero weeks of training, i.e. 3rd and 4th week of July.
2	Testing two (First post test)	T-2	At the end of first meso-cycle (M-1) after 6 weeks of training, i.e. 1st and 2nd week of September.
3	Testing three (Second post test)	T-3	At the end of second meso-cycle (M-2) after 12 weeks of training, i.e. 4th week of October.
4	Testing four (Third post test)	T-4	At the end of third meso cycle (M-3) after 18 weeks of training, i.e. 2nd and 3rd week of December.

Collection of Data

For the purpose of collection of data, the method explained by Sen and Ray (1983), Das and Ganguli (1982), scientifically authenticated by Shaw, Kaushik and Kaushik (1998) and validated by Shaw et. al. (1998) was strictly administered. A lever based weighing machine (Avery) with a range from 0 to 100 kg with the balancing accuracy of ±10 grams, a wooden board of 183 cm in length, 100 cm in width and 2 cm in thickness were among the equipment used to collect the data for determination of centre of gravity. The partial body weight of the wooden board was determined (S_0). The two knife edged wooden blocks, one on the platform of the balance and other on the wooden box (placed opposite end to the balancing scale) below which the footrest was fabricated. Thereafter, the scale reading was recorded (S_0).

The subjects were instructed to be flat in supine position on the board in such a way that the imaginary vertical plane passed through the longitudinal axis of symmetry of the board and the sagittal axis of the body are coincident. The feet were kept flush at the end of the board (foot rest), opposite to the direction of the weighing scale was placed. The scale reading in this position of the body was recorded (S_1). Then, the subjects were asked to stand facing the scale to the opposite end of scale, placing the heel with the footrest and scale reading was recorded (PA). Thereafter, subjects were asked to stand while keeping the right lateral aspect of the right footrest placing both the feet together and scale reading was recorded (*RL*). With the same posture, left lateral aspect of the foot with the footrest, the reading was recorded (*LL*).

The following equations were used to calculate the three-dimensional centre of gravity:

1. $$CG(Z) = \frac{(S_1 - S_0)}{M} \times L$$

2. $$CG(Y) = \frac{PA - S_0}{M} \times L$$

3. $$CG(X) = \frac{(a) + (b)}{2}$$

   (a) $$CG(X - RL) = \frac{RL - S_0}{M} \times L$$

   (b) $$CG(X - LL) = \frac{LL - S_0}{M} \times L$$

where M = mass of the body, S_0 = scale reading of the system, S_1 = scale reading when subject is lying flat on the system, L = length of the reaction board, PA = scale reading while subject stands at opposite end facing the weighing scale, RL = scale reading while subject stands at opposite side keeping right side to the scale and LL = scale reading while subject stands at opposite side keeping left side to the scale.

The height of centre of gravity was converted to percentage height of centre of gravity by the following formula:

$$\text{Percentage Height of CG} = \frac{\text{Height of CG (m)}}{\text{Stature/Height (m)}} \times 100$$

Weight as a composite measure of total body size was measured in kg to the nearest 100 g. The subjects in light indoor clothing stood erect in the centre of the scaled platform. After taking each measurement, the scale was reset to zero.

Height (stature) as a major indicator of general body size and of bone length was measured in meteres to the nearest centimeter. Landmark was the maximum distance from the point vertex on the head to the ground. An anthropometeric rod was used to take measurement. The measurement was taken with the individual standing straight against the wall touching it with heel, buttocks, back and arms hanging naturally by the sides. The upward force under mastoid level was exerted to raise the subject. The subject inhaled deeply and was instructed to "look straight ahead" with the head positioning in the Frankfurt horizontal plane. Height was recorded to nearest of 1/10th of a centimeter.

The Ponderal Index (P.I.) was computed by using the following formula:

$$\text{Ponderal Index (PI)} = \frac{\sqrt[3]{\text{Weight}}}{\text{Height}} \times 100$$

Analysis of Data

The collected data was analyzed while computing mean, standard deviation and percentage. The analysis of data pertaining to the effect of physical exercise on three-dimensional centre of gravity, height, weight and ponderal index of the male and female students of University of Delhi have been presented in Tables 3 and 4 and illustrated in Figs. 1 and 2.

Table 3. Effect of physical exercise on selected variables of male students of University of Delhi

S. No.	Variable/Group	Mean ± S.D. (Pre-Test)	Mean ± S.D. (Post-Test)	% Change	Effect
1	CG (Z dimension) (m)				
	Conditioning male	1.032 ± 0.099	0.990 ± 0.097	4.07%	Decrease
	Non-conditioning male	1.012 ± 0.067	0.992 ± 0.074	1.88%	Decrease
	Sedentary male	1.004 ± 0.063	1.020 ± 0.058	1.49%	Increase
2	CG (Y dimension) (m)				
	Conditioning male	0.102 ± 0.017	0.092 ± 0.016	9.80%	Decrease
	Non-conditioning male	0.102 ± 0.017	0.099 ± 0.017	2.94%	Decrease
	Sedentary male	0.102 ± 0.014	0.108 ± 0.016	5.88%	Increase
3	CG (X-RL dimension) (m)				
	Conditioning male	0.093 ± 0.012	0.085 ± 0.017	8.60%	Decrease
	Non-conditioning male	0.091 ± 0.011	0.087 ± 0.087	4.40%	Decrease
	Sedentary male	0.091 ± 0.011	0.095 ± 0.011	4.40%	Increase
4	CG (X-LL dimension) (m)				
	Conditioning male	0.089 ± 0.012	0.091 ± 0.017	2.25%	Increase
	Non-conditioning male	0.090 ± 0.010	0.093 ± 0.087	3.33%	Increase
	Sedentary male	0.091 ± 0.011	0.087 ± 0.011	3.33%	Decrease
5	Height (m)				
	Conditioning male	170.16 ± 5.742	170.53 ± 5.753	0.22%	Increase
	Non-conditioning male	169.77 ± 4.730	170.00 ± 4.747	0.14%	Increase
	Sedentary male	168.01 ± 5.467	168.29 ± 5.485	0.17%	Increase
6	Body Weight (kg)				
	Conditioning male	61.16 ± 9.917	60.47 ± 8.966	1.13%	Decrease
	Non-conditioning male	58.92 ± 7.654	59.21 ± 7.743	0.65%	Increase
	Sedentary male	60.69 ± 10.028	62.41 ± 10.344	2.83%	Increase
7	Ponderal Index				
	Conditioning male	2.309 ± 0.085	2.297 ± 0.079	0.52%	Decrease
	Non-conditioning male	2.289 ± 0.095	2.286 ± 0.093	0.13%	Decrease
	Sedentary male	2.332 ± 0.088	2.350 ± 0.088	0.77%	Increase

The analysis of data in Table 3 pertaining to the effect of physical exercise on selected variables of male students of University of Delhi reveals that the CG at Z-dimension reduced in conditioning male (CM) and non-conditioning male (NCM) whereas, it was increased among sedentary male (SM); CG at Y-dimension reduced in CM and NCM whereas, it was increased among SM; CG at X(RL)-dimension reduced in CM and NCM whereas, it was increased among SM; CG at X(LL)-dimension increased in CM and NCM whereas, it decreased among SM; height improved in all the selected three groups; weight reduced in CM whereas, it increased among NCM and SM; ponderal index decreased for CM and NCM, whereas, it increased for SM.

Table 4. Effect of physical exercise on selected variables of female students of University of Delhi

S. No.	Variable/Group	Mean ± S.D. (Initial)	Mean ± S.D. (Later)	% Change	Effect
1	CG (Z dimension) (m)				
	Conditioning female	0.898 ± 0.062	0.864 ± 0.073	3.57%	Decrease
	Non-conditioning female	0.897 ± 0.073	0.879 ± 0.080	1.90%	Decrease
	Sedentary female	0.887 ± 0.061	0.912 ± 0.064	2.93%	Increase
2	CG (Y dimension) (m)				
	Conditioning female	0.091 ± 0.015	0.080 ± 0.014	12.09%	Decrease
	Non-conditioning female	0.088 ± 0.012	0.082 ± 0.013	6.82%	Decrease
	Sedentary female	0.090 ± 0.011	0.096 ± 0.016	6.67%	Increase
3	CG (X-RL dimension) (m)				
	Conditioning female	0.081 ± 0.019	0.074 ± 0.019	8.64%	Decrease
	Non-conditioning female	0.081 ± 0.015	0.077 ± 0.017	4.94%	Decrease
	Sedentary female	0.086 ± 0.016	0.091 ± 0.017	5.81%	Increase
4	CG (X-LL dimension) (m)				
	Conditioning female	0.081 ± 0.013	0.088 ± 0.015	8.64%	Increase
	Non-conditioning female	0.081 ± 0.016	0.086 ± 0.012	6.17%	Increase
	Sedentary female	0.087 ± 0.012	0.084 ± 0.014	4.55%	Decrease
5	Height (m)				
	Conditioning female	157.02 ± 4.340	157.31 ± 4.331	0.18%	Increase
	Non-conditioning female	156.85 ± 5.522	157.05 ± 5.522	0.13%	Increase
	Sedentary female	153.85 ± 3.457	154.03 ± 3.413	0.12%	Increase
6	Body Weight (kg)				
	Conditioning female	51.05 ± 9.063	50.45 ± 8.761	1.18%	Decrease
	Non-conditioning female	50.60 ± 7.927	50.35 ± 7.624	0.49%	Increase
	Sedentary female	50.64 ± 7.026	52.18 ± 7.453	3.04%	Increase
7	Ponderal Index				
	Conditioning female	2.355 ± 0.108	2.342 ± 0.105	0.55%	Decrease
	Non-conditioning female	2.354 ± 0.118	2.349 ± 0.114	0.21%	Decrease
	Sedentary female	2.401 ± 0.134	2.422 ± 0.134	0.87%	Increase

The analysis of data in Table 4 pertaining to the effect of physical exercise on selected variables of female students of University of Delhi reveals that the CG at Z-dimension reduced for conditioning female (CF) and non-conditioning female (NCF) whereas, it was increased among sedentary female (SF); CG at Y-dimension reduced in CF and NCF whereas, it was increased among SF; CG at X(RL)-dimension reduced in CF and NCF whereas, it increased among SF; CG at X(LL)-dimension increased in CF and NCF whereas, it decreased among SF; height improved in all the selected three groups; weight reduced in CF whereas, it increased for NCF and SF; and ponderal index decreased for CF and NCF, whereas, it increased for SF.

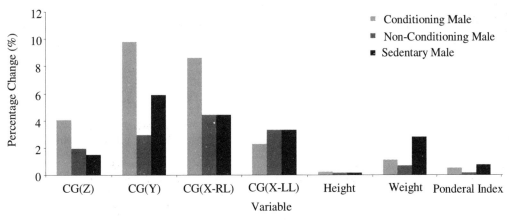

Fig. 1. Effect of physical exercise on selected variables of male students of University of Delhi.

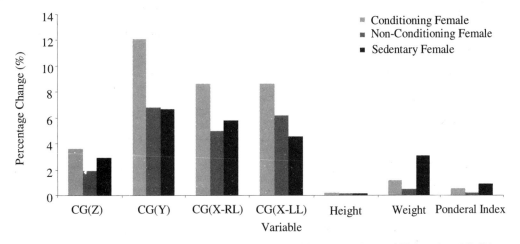

Fig. 2. Effect of physical exercise on selected variables of female students of University of Delhi.

Conclusions

The following conclusions have been drawn from the analysis of data:

1. The conditioning male reduced their CG(Z), CG(Y), CG(X-RL), body weight and ponderal index.
2. The conditioning male increased their CG(X-LL) and height.
3. The non-conditioning male reduced their CG(Z), CG(Y), CG(X-RL) and ponderal index.
4. The non-conditioning male increased their CG(X-LL), height and body weight.
5. The sedentary male reduced CG(X-RL).
6. The sedentary male increased their CG(Z), CG(Y), CG(X-RL), height, body weight and ponderal index.
7. The conditioning female reduced their CG(Z), CG(Y), CG(X-RL), body weight and ponderal index.
8. The conditioning female increased their CG(X-LL) and height.
9. The non-conditioning female reduced their CG(Z), CG(Y), CG(X-RL) and ponderal index.
10. The non-conditioning female increased their CG(X-LL), height and body weight.

11. The sedentary female reduced their CG(X-RL).
12. The sedentary female increased their CG(Z), CG(Y), CG(X-RL), height, body weight and ponderal index.

References

1. Shephord, J.R. 1983. Employee Health and Fitness – State of the Art. *Preventive Medicine.* **12**: 644-653.
2. Ray, G.G. and Sen, R.N. 1983. Determination of Whole Body Centre of Gravity in Indians. *Journal of Human Ergo.* **12**: 3-4.
3. Monahan, T. 1987. Is Activity As Good As Exercise. *The Physician and Sports Medicine.* **15(10)**: 181.
4. McHenery, P.L. et al. 1990. Statement of Exercise. *Special Report Circulation 51 (Jan., 1990).* p. 1.
5. Logan, G.A. 1976. *Adaptation of Physical Activities.* Prentice Hall, Englewood Cliffs, N.J.
6. Kawamura, T. et al. 1984. An Analysis of Somatotypes and Postures of Judoist. Bulletin of the Scientific Studies on Judo: Report VI, Kodokan, Tokyo. pp. 107-116.
7. Kaushik, R., Kaushik, S. and Shaw, D. 1995. Changes in the Location of Three-Dimensional Centre of Gravity as a Result of Six-weeks Conditioning Programme on Male Athletes. *Souvenir: 5th National Conference of NAPESS and GANSF*, October 28-29, Delhi. p. 33.
8. Johnson, P. and Strolberg, D. 1971. *Conditioning.* Prentice Hall, Englewood Cliffs, N.J.
9. Jensen, C.R., Schultz, Gordo W., and Bangerter, B.L. 1984. *Applied Kinesiology 3rd Edn.* McGraw-Hills, Singapore.
10. Hoeger and Hoeger. 1990. *Fitness and Wellness.* p. 3.
11. Harris, S.S., Caspersen, C.J., Defriese, G.H. and Estes, E.H. 1989. Physical Activity Counseling for Health Adults as a Primary Preventive Intervention in the Clinical Setting. *JAMA.* **261**: 3590-98.
12. Das, R.N. and Ganguli, S. 1982. Moment of Inertia of Living Body and Body Segments Geometrically Shaped Timber and Sleep Specimens. *Journal of the Institution of Engineers.* **62: IDGE-3**.
13. Das, R.N. and Ganguli, S. 1982. Mass and Centre of Gravity of Human Body and Body Segments. *Journal of the Institution of Engineers.* **62: IDGE-3**.
14. Charles, S. 1976. *Fundamentals of Sports Biomechanics.* Kendall/Hunt Publishing Company, Dubuque, Ia.
15. Burleigh L., Horak, F.B. and Malquin, F. 1994. Modification of Postural Responses and Step Initiation: Evidence for goal-directed postural interactions. *Journal of Neuro Physiology.* **72(6)**: 2892-2901.
16. Bunn, J.W. 1972. *Scientific Principles of Coaching 2nd Edn.* Prentice Hall, Inc., Englewood Cliffs, N.J.
17. Bunn, J.W. 1957. *Basket Methods.* The Mc Millan Company, New York.
18. Blomquist, C.G. 1983. CV Adaptation to Physical Training. *Annual Review of Physiology.* **45**: 169.

Biomechanics
R.K. Saxena and P. Mishra (Editors)
Copyright © 2005, Anamaya Publishers, New Delhi, India

40. Handmade Peristaltic Pump

Abhishek Bakshi and A. Bhanu Prakash
Joginpally B.R. Engineering College, Yenkapally,
Moinabad (Mandal), Hyderabad - 500075, India

Abstract: One of the fastest-growing types of positive displacement pumps in the market is the peristaltic pump, but many people are unfamiliar with the technology behind this type of pump and its advantages. In the past, peristaltic pumps, also called tubing pumps, were used almost exclusively in the laboratory. However, as pumps have been redesigned to allow higher flow rates and pressures, many peristaltic pumps have demonstrated successful operation in a number of chemical process control and production applications.

The popularity of the peristaltic pump can be attributed to its design, which uses a handmade pump to turn a set of rollers. The rollers compress and release a flexible tube as they pass across the tube. This squeezing action creates a vacuum, which then draws fluid through the tubing to achieve the pumping action. Because the flexible tubing is the only wet part, maintenance and cleanup are simple and convenient.

This design results in a number of benefits, making peristaltic pumps suitable for a variety of challenging applications. The most important of these benefits are:

- Non-contamination
- Simple operation
- Low maintenance
- Self-priming capabilities
- A variety of available tubing materials
- Gentle pumping action
- Liquid, gas, mixed-phase and viscous-phase-fluid pumping capabilities
- Cost-effective operation

Peristalsis
The wavelike muscular contractions of the tubular structure that propel the contents onward by alternate contraction and relaxation are called as peristalsis, also called vermicular movement. It is a common phenomena by which the contents are forced onward toward the opening.

Peristaltic Pump

Principle
A peristaltic pump is a type of positive displacement pump used for pumping a variety of specialized fluids. The fluid is contained in a flexible hose. A rotor with a number of cams (also called 'shoes' or 'wipers') attached to the external circumference compresses the flexible tube. As the rotor turns, the part of tube is compressed thus forcing the fluid to be pumped to move through the tube. This process is called peristalsis and is used in many biological systems such as the gastrointestinal tract.

Materials

Metal base plate - 1
Rods - 5
Toothed wheels - 4
Wooden piece - 1
Latex tube - 1
Seed chamber - 1
Fermenter - 1
Cams - 3
Wheel with cams - 1
Chains - 2
Handle - 1

Design and Working

Design

On the base plate of 24 inches × 12 inches, two 26 inches long rods are erected. These rods are placed at a distance of 17 inches. The main wheel of radius 9 cm which is fixed on a rod is connected to the second wheel of radius 4.5 cm on another rod via a chain. The third wheel of radius 2.5 cm is placed behind the second wheel. The third wheel is connected to the fourth wheel of radius 3.5 cm on the same rod at a distance of 16 inches with a chain. In front of the fourth wheel there is a wheel of radius 2 cm on which 3 (cams) rollers of radius 1.75 cm are welded. Below the rollers there is a wooden piece to hold the tube in place (Fig. 1). The seed chamber should be kept above certain height so that the flow rate of the liquid will be at faster rate.

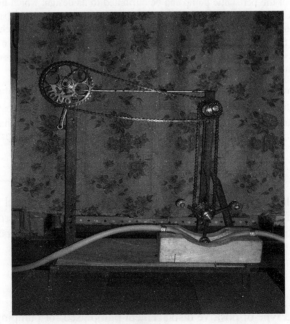

Fig. 1

Working

The working of the peristaltic pump is quite simple. When the main wheel is rotated with the help of a handle it leads to the synchronized motion of the other wheels (wheels 2, 3 and 4). Rollers, which are fixed in front of the fourth wheel, compress the latex tube at regular intervals, which creates vacuum in the tube. Due to this vacuum both the Newtonian and non-Newtonian fluids move into the tube from the seed chamber.

These fluids are directed towards the opening of the tube in the fermentor in which positive pressure is being maintained. A bit of experience in rotating handle will lead a great amount of the fluid getting transferred from seed chamber to fermentor.

Advantages

In the past, peristaltic pumps, also called tubing pumps, were used almost exclusively in the laboratory. However, as pumps have been redesigned to allow higher flow rates and pressures, many peristaltic pumps have demonstrated successful operation in a number of chemical process control and production applications.

The popularity of peristaltic pumps can be attributed to its design that uses a handmade pump to turn a set of rollers. The rollers compress and release a flexible tube as they pass across the tube. The squeezing action creates a vacuum, which draws fluid through the tubing to achieve pumping action. The flexible tubing is the only wet part, hence maintenance and cleanup are simple and convenient.

This design results in a number of benefits, making peristaltic pumps suitable for a variety of challenging applications. The most important of these benefits are:

- Non-contamination
- Simple operation
- Low maintenance
- Self-priming capabilities
- Cost-effective operation

Non-contamination

The peristaltic pump design keeps the fluid being pumped inside the tubing at all times. This is important for two reasons: The fluid cannot contaminate the pump, and the pump cannot contaminate the fluid. This is a distinct advantage in transferring chemically aggressive or high-purity fluids. Because the fluid never touches any gears, seals, diaphragms or other moving parts, finding a pump compatible with a particular solution is as simple as choosing a chemically compatible tubing.

Simple Operation

Peristaltic pumps are simple to operate, and many are ready to use right out of the box. Some models have interchangeable pump heads that can be mounted in minutes using finger-tightened screws. Engineering advances have yielded designs that allow the pump tubing to be loaded or replaced in just seconds. In addition, most pumps feature simple controls, with a dial or keypad for speed control and straightforward menus for programming more complex tasks.

Low Maintenance

Most peristaltic pumps require very little maintenance to keep them in peak operating condition. The tubing must be replaced regularly to prevent leakage or poor flow performance. This procedure typically takes just seconds, which can be an advantage when using the same pump to transfer or dispense different chemicals. When solutions are changed, only the tubing needs to be changed, and the pump can be up and running within minutes. This saves time and labour, and allows the same pump to be used in a variety of processes, potentially contributing to significant cost savings.

Self-Priming Capabilities
When starting, peristaltic pumps generate sufficient suction to draw fluid from the source into the tubing and through the pump, eliminating the need for pump and suction line priming before use. This ends the need for flooded suction–locating the pump below the level of the source tank–providing added flexibility during pump installation and allowing the pump to be moved as needed. In addition, peristaltic pumps do not lose their prime when air is introduced, preventing related downtime.

Cost-Effective Operation
Peristaltic pumps can be very cost-effective to operate. This benefit is especially apparent when dealing with aggressive chemicals that would otherwise require a pump with a number of chemically resistant wetted parts. Instead of selecting a pump with expensive, specially made components, it may be wise to purchase a peristaltic pump with compatible tubing to transfer or dispense these chemicals.

Because the same peristaltic pump can handle a variety of solutions if the tubing is changed, the same pump can be used in multiple applications, eliminating the need for a large number of dedicated pumps. This benefit, combined with lower labour and maintenance costs than most other pump types, reduces the overall cost of ownership in many applications tubing or reinforced hose to achieve significantly higher flow rates. Some industrial hose pumps offer maximum flow rates as high as 108 gallons/min, satisfying many high-volume fluid handling needs. This makes peristaltic pumps a viable alternative for companies needing a non-contaminating low-maintenance pump.

Some people have found the pulsation inherent in peristaltic pumping as undesirable in their applications. This pulsation can be reduced by increasing the number of rollers, which helps smooth out the pulses. A pulse dampener also helps.

Results
The following results were obtained from pump designed by us:
Flow rate - 4000 ml/min
Drive speed - Dependent upon person
Chemical inertness - Tubing
Corrosion resistance - Tubing
Dead volume - None
Contact with the mechanical parts of the pump - No
Dry running - No problem
Maintenance (time and cost) - Very low (only tubing exchange)
Self priming - Yes

Conclusion
Peristaltic pumps are gaining acceptance in more applications and industries than ever. They have made the transition from the lab to become a critical and reliable part of a number of processes, from chemical transfer to pharmaceutical processing to wastewater treatment. Their versatility and ease of use has helped them meet a variety of fluid-handling challenges. As worldwide demand for pumps continues to increase, the future usefulness of peristaltic pumps will be limited by only the number of applications invented.

Acknowledgement
The authors are very thankful to Dr. A.B. More for his support which has made this project possible.

41. Effect of Translations Mobilization on the Rotational ROM of the Normal Shoulder Done at the End Range of Abduction

K. Ramanathan, Jince Thomas, Patitapaban Mohanty, Monalisa Pattnaik, Sharada Lakshmana Nayak and Raja Bhattacharya
Swami Vivekanand National Institute of Rehabilitation Training and Research, Cuttack, India

Abstract: The glenoid cavity is too shallow to hold the humeral head. Therefore, during normal physiological movements, translations occur to keep the head on the shallow glenoid cavity. Mobilization techniques are frequently used by physical therapists as an intervention for treating limited ROM in periarthritic shoulders. They advocate gliding movements in direction of limited joint glide in accordance with concave-convex rule as proposed by Kaltenborn. But experimental evidence for this rule is lacking. Studies have predicted that concave-convex rule may not be applicable in the end range of the shoulder and few cadaver studies have concluded that translations for rotations can occur in violation to this rule in non-pathological shoulder like ventral translations during internal rotation and dorsal translation during external rotation. So, the need for clinical study was felt. The purpose of this study was to evaluate the efficacy of translatory mobilization in improving rotational ROM in non-pathological shoulders. 20 shoulders between age group of 20 and 30 yrs were chosen, and ventral and dorsal translations were done for each shoulder in random order. The changes in rotational range of motion before and after translations were measured using a 180° goniometer. There was increase in ROM in external rotation when ventral translations were done and internal rotation increased when dorsal translatory mobilization was done. This was in accordance with Kaltenborn concave-convex rule.

Introduction

Glenohumeral joint articulation is between the incongruent glenoid cavity and the head of the humerus. The glenoid cavity is too shallow to hold the humeral head and during normal physiological movements translations occur to keep the head within the shallow glenoid cavity [1]. These translations are believed to occur in accordance with concave-convex rule as proposed by Kaltenborn [2, 3]. This rule states that the sliding occurs in the opposite direction of the angular movement of the bone if the convex joint surface moves over a concave surface and sliding occurs in the same direction as the angular movement of the bone, if concave surface moves over a convex surface [2, 3]. When there is a limited range of motion of shoulder joint, this biomechanical principle is used to increase joint range, i.e. if internal rotation is limited, posterior translation is done and if external rotation is limited anterior translation is done. But experimental support for this has been lacking [4].

Some studies have predicted that movement in shoulder does not follow concave-convex rule at end ranges. Harryman et al. [5] have stated that anterior superior translations occur during flexion, posterior translations during extension, and external rotation and anterior translations during internal rotation, and cross body movement during passive glenohumeral motion. This was attributed to capsuloligamentous tissue, i.e. in a joint that has tightened capsule, translations away from the side of the joint (capsular constraint mechanism).

Harryman et al. [5] and Itio et al. [6] have suggested through cadaver studies that the translations partly obey the concave-convex rule, i.e. not throughout the range of motion. But the study report of Howell et al. [7] was in violation of the concave-convex rule. The study by Hsu et al. [4] suggested that internal rotation increases after ventral translatory motion procedure in resting position and internal rotation increases after dorsal translatory motion at the end range which supported concave-convex rule. But internal rotation after ventral translatory motion at end range and medial rotation after dorsal translatory motion in resting position did not improve. This study was done on 14 fresh frozen cadaver elderly shoulder specimen. However, all these studies were done on cadaver and reproduction of these results on clinical situation is questionable.

Use of mobilization at end range to increase range of motion suggested by Maitland [8] was substantiated in cadaver study of Harryman et al. [5] and Hsu et al. [4]. So the present study was conducted to find out the effect of translatory mobilization in the end range of abduction on internal rotation and external rotation of shoulder. Since we believed that the above studies could not be applicable in clinical situation, we chose young normal shoulder for the study. The aim was to find the effect of anterior and posterior translatory mobilizations on the internal and external rotational range of motions of the shoulder.

Maitland [8] has proposed Grade III and Grade IV in phased manner for a period of 2 min to increase the range of motion. The authors have also proposed that normal tissue could be stretched because of viscoelastic properties. So we chose normal shoulder, which had the capability for stretching to test the hypothesis.

Method

20 young men, with mean age of 21 yrs (ranging between 18 and 24 yrs) participated in this study. The inclusion criteria were that they should not have had any pathology or surgery of the shoulder in the past. If there was any neurological or orthopaedic problem, the subjects were excluded. This was conducted on students of the institution after obtaining oral consent from each individual.

Measurements

Complete examination of the shoulder was done to rule out any pathology of the shoulder and external rotation and internal rotation range of motion was measured using universal goniometer 180 degrees.

The subjects were positioned in prone with arm supported over the plinth. The proximal arm of the goniometer was aligned perpendicular to the ground and the distal arm was aligned parallel to the ulna. The forearm was in mid-prone. Internal rotation and external rotation was done and measurements taken [9]. The passive range of motion was limited when the gross scapula movements or discomfort or any substitution mechanism were seen, to prevent any measurement errors. The reliability of goniometer for measurement of internal rotation and external rotation was established [9].

Procedure

After the measurements, the subjects were allotted randomly to either Group I or II. Group I was given ventral translations to dominant shoulder and dorsal translations to non-dominant shoulder. Group II was given dorsal translations to dominant shoulder and ventral translations to non-dominant shoulder.

These mobilizations were done at the end range of abduction (90° of abduction) so that the inferior capsule is not stretched.

Randomization for the translatory mobilization (anterior or posterior) of the dominant and non-dominant shoulder was done to prevent any variations in range of motion with regard to dominance. Initially, grade II was done to relax the muscles and then, grade III was done as per the group allocation for the translatory glides. This mobilization was done for 2 min for 3 sessions with a rest period of 1 min in between the mobilization for each shoulder. This was done for 5 consecutive days. This protocol was chosen as per the recommendation of Maitland [8]. However, for the treatment sessions, there is a lack of experimental evidence.

Measurements were done using goniometer before the start of mobilization (pre-test measurement) and the post-test measurements were taken at the end of 3rd day and 5th day of treatment. Measurements were also taken one month after the completion of the mobilization. These measurements were done by an examiner who was blinded to the group allocation for mobilization.

Data Analysis
For investigating the effect of translations over the internal and external rotations of the shoulder, ANOVE 2×3 was used.

Results
There was a main effect for the international rotation range of motion in dorsal translations, $F = 4.189$ and $P = 0.048$. There was also main effect for time $F = 68.869$ and $P = 0.000$. The main effect was also found in the interaction for dorsal translations $\times$ time, for internal range of motion, $F = 28.639$ and $P = 0.000$. For dorsal translations, the mean increase in internal range of motion was 14.75° and mean increase in external rotation was 3.9°. This is illustrated in Table 1.

Table 1.

Mobilization	Range of Motion	Pre-Test (Mean degrees ± SD)	Post-Test Measurement I (3rd Day) (Mean degrees ± SD)	Post-Test Measurement II (5th Day) (Mean degrees ± SD)
Dorsal translations	Internal rotation	81.65 ± 11.76	87.95 ± 11.17	96.4 ± 9.61
	External rotation	78.95 ± 14.62	81.35 ± 13.56	82.25 ± 13.03
Ventral translations	Internal rotation	95.25 ± 11.29	97.25 ± 9.93	97.75 ± 10.06
	External rotation	91.25 ± 10.11	99.55 ± 7.46	107.85 ± 6.87

For ventral translation, main effect was found for external rotation, $F = 0.981$ and $P = 0.038$ and there was also main effect for time $F = 69.152$ and $P = 0.000$. The main effect was also found in the interaction for ventral translation $\times$ time, for external rotation range of motion $F = 37.750$ and $P = 0.000$. For ventral translations, the mean increase in external range of motion was 16.59° and mean increase in internal rotation was 2.5°. This is illustrated in Table 1.

Discussion
Thus, these results imply that there is a significant increase in internal rotation range of motion of shoulder when dorsal translations are done and external rotation range of motion increases significantly when ventral translations are done.

The middle posterior capsule restricted motion for dorsal translations of the glenohumeral head in the neutral position, whereas, both the middle and inferior capsule were involved in limiting the dorsal translations of the head in the abducted position. Studies done on the cadavers have suggested that these ligaments restrict humeral head translations [10]. So, by stretching these ligaments, range of motion could be improved. Translations may accompany passive movement of glenohumeral joint and this could be influenced by asymmetrical tightening of the capsule. The results of this study show that ventral translatory mobilization increases external rotation of the shoulder when done at the end range of abduction, which is in accordance with the arthrokinematics of the joint, i.e. concave-convex rule by Kaltenborn and increases the range as proposed by Maitland [11].

In graded Maitland [8] technique, stretching of the capsule is questionable because oscillations were done. Low intensity long duration stretching has been proved to be effective in enhancing the length of the soft tissue than the cyclic stretching. So, this increase could be attributed to the arthrokinematic influence than the soft tissue. Further, anterior mobilization done at the end range of abduction increases external rotation. Hsu et al. [4] found improvement when ventral mobilizations are done in resting position and not at the end range of abduction. The results are contrary to the findings of Howell et al. [7]. The results of this study may not be compared with the other studies, since studies done by other authors have been done on cadavers [4-7].

In our study, one subject felt extreme numbness and decrease in vascular supply (coldness of the digits) by the end of the session. No other complaints were noticed during or after the mobilization. Follow up measurement was done to prevent stretch instability (weakness) after one month. The results showed that the range of motion returned back to the pre-treatment ranges.

Clinical Implications

Our study provides evidence that ventral mobilization technique performed at the end range of abduction increases external rotation and dorsal mobilization technique increases internal rotation. For example, in Periarthritis with external rotation limited, ventral mobilization technique could be done to increase arthrokinematic of the joint, which increases the external rotation of the shoulder. In the same way, dorsal mobilization technique could be applicable for increasing internal rotation.

Limitation of the Study

The data obtained through the study in normal subjects may not be applicable in shoulder with limited range of motion, since the pathological shoulder may not follow the arthrokinematics of the normal joint, but the results of this study could pave way for further evaluation of the arthrokinematics of the pathological joint. The translation force used in mobilization was not quantified.

Conclusion

Ventral translations done at the end of abduction range improves the external rotation of the shoulder and dorsal translations done at the end of abduction range improves the internal rotation of the shoulder in healthy individuals, as suggested by Kaltenborn and Maitland.

References

1. Halder, A.M., Itio, E. and An, K.N. 2000. Anatomy and Biomechanics of the Shoulder. *Orthopedic Clinics of NA.* April, **31**: 2, 159-176.
2. Edmond, S.L. 1993. *Manipulation and Mobilization: Extremities and spinal techniques.* Mosby, St Louis, Mo.

3. Hertling, D. and Kessler, R.M. 1996. *Management of Common Musculoskeletal Disorders: Physical therapy principles and methods, 3rd edn.* Lippincott, Philadelphia, Pa.
4. Hsu, A.R., Hedman, T., Chang, J.H., Vo, C., Ho, L., Ho, S. and Chang, G.L. 2002. Changes in Abduction and Rotation Range of Motion in Response to Simulated Dorsal and Ventral Translational Mobilization of the Glenohumeral Joint. *Phys. Ther.* **82**: 544-556.
5. Harryman, D.T., Sidles, J.A., Clark, J.M., McQuade, K.J., Gibb, T.D. and Matsen, F.A. 1990. Translation of the Humeral Head on the Glenoid with Passive Glenohumeral Motion. *J. Bone and Joint Surg.* **72A**: 9, 1334-1343.
6. Itoi, E., Motzkin, N.E., Morey, B.F. and An, K.N. 1994. Contribution of Axial Arm Rotation to the Humeral Head Translation. *Am. J. Sports Med.* **22**: 499-503.
7. Howell, S.M., Galinat, B.J., Renzi, A.J. and Marone, P.J. 1988. Normal and Abnormal Mechanics of the Glenohumeral Joint in the Horizontal Plane. *J Bone and Joint Surg.* **70A**: 227-232.
8. Maitland, G.D. 1992. *Peripheral Manipulation, 3rd edn.* Butterworths, Boston.
9. Norkin, C.C. and White, D.J. 1998. *Measurement of Joint Motion: A guide to goniometry, 2nd edn.* Jaypee Publications.
10. Turkel, S.J., Panio, M.W., Marshall, J.L. et al. 1981. Stabilizing Mechanisms Preventing Anterior Dislocation of the Glenohumeral Joint. *J. Bone and Joint Surg.* **63A**: 1208-1217.

Biomechanics
R.K. Saxena and P. Mishra (Editors)
Copyright © 2005, Anamaya Publishers, New Delhi, India

42. Proprioceptive Neuromuscular Facilitation in Restoring Altered Pulmonary Mechanics Post Mid-sternotomy

Neha Gupta[1], V.P. Gupta[1], Sajad Ali[2], Jeyasundar[2], Sundar Kumar[1] and Thiruvarangan[1]

[1]Department of Cardiothoracic and Vascular Surgery, AIIMS, New Delhi - 110029, India

[2]Department of Rehabilitation Sciences, Jamia Hamdard, New Delhi - 110062, India

Abstract: The objective of this paper is to investigate the efficacy of PNF in restoring altered pulmonary mechanics post mid-sternotomy. Mid-sternotomy is a procedure with elaborate surgical exploration causing an insult to diaphragmatic excursion and rib cage movements. Altered rib cage mechanics (objectively noted as decreased chest expansion) and impaired respiratory mechanics (noted as decreased PEFR) give a gross presentation of a restrictive breathing pattern and reduced lung volumes. 20 subjects were selected after fulfilling inclusion and exclusion criteria. Randomized control group trial was used. In the experimental group, patients were getting conventional postoperative management that includes incentive spirometry coughing, huffing maneuvers and a set of therapist guided PNF exercises. In the control group the patients were getting only conventional post operative management. Readings were to be taken pre and post intervention. The results will be extrapolated after appropriate statistical analysis of data collected.

Introduction

Over decades mid-sternotomy has been an elective procedure for wide exploration of thorax [1]. Restrictive ventilatory pattern follows mid-sternotomy [2-7]. There is discoordination of chest wall movement and presence of Frank paradox [2]. There is decreased chest wall stability, pain and muscle spasm leading to decreased respiratory excursion and gross reduction in lung volumes, especially in CABG patients with IMA graft. Muscle weakness is more likely to contribute to altered pulmonary mechanics [8] by reducing the ability to reach high lung volume, and thus, decreasing cough effectiveness [9, 10].

Respiratory muscle stretch (special technique of proprioceptive neuromuscular facilitation) has been found to decrease chest wall stiffness and increase chest expansion in COPD patients [11]. Intercostal muscle stretch has been found to facilitate inspiratory neurons and improve lung volumes [12]. There is no scepsis to role of conventional chest physiotherapy (CPT including diaphragmatic breathing, forced expiratory technique, percussion and vibration). In respiratory lung functions but benefits on chest wall mechanics remain to be clarified. So, this study was undertaken to compare effects of respiratory muscle stretch (RMS) as a special PNF technique and conventional CPT (being routinely followed) in restoring altered pulmonary mechanics post mid-sternotomy.

Methods

Patients
18 patients with mid-sternotomy were involved in the study. The subjects were well informed and their consent was taken. Subjects were age matched/surgical procedure matched. The entry criteria were as follows:

1. Haemodynamically stable.
2. No significant lung disease preoperatively.
3. No chest drains.

   Pre-determined drop out criteria were as follows:

1. Acute exacerbation of pulmonary condition.
2. Prolonged mechanical ventilation > 24 h.
3. Patients needing intropic support.

Design
There was a single-blind pre- and post-experimental control group design. The patients were randomly assigned to either RMS group or conventional CPT group. One therapist gave the treatment session and the other therapist, blind of procedure took pre- and post-treatment readings of the following variables:

1. Chest wall expansion: Chest circumference measured with standard plicometer at 3 levels– axillary, nipple and xiphisternum [13].
2. PEFR – measured with wright peak flow meter: Subjects were encouraged to produce maximal effort [14, 15].

Respiratory Muscle Sketch
Position and stretch technique were chosen to adequately stimulate respiratory muscle groups like the intercostals and accessory muscles [13].

Conventional Chest Physiotherapy
This included a combination of diaphragmatic breathing, force-expiratory maneuvers, percussion and vibration [16, 17, 18].

Analysis
Statistical analysis was performed using SPSS 10 package. Comparison was done of paired T test for pre- and post-treatment readings and two sample T tests with equal variance for inter-group comparisons. Significance level for this study was 95% ($p < 0.05$).

Results and Discussion

18 subjects were enrolled in the study with mean age of 42.2 ± 16 in the control group (Group 1) and 41.8 ± 15 in the experimental group (Group 2), so age was comparable with p value 0.96 (not significant).

Chest Expansion

As seen from the data chart, there is significant improvement in chest expansion of both axillary and nipple level in both groups ($p < 0.05$) but more significant in Group 2.

Data Chart

		Pre-Test	Post-Test	p Value
Chest expansion at axillary level	Group 1	1.52 ± 0.40	1.65 ± 0.39	0.03
	Group 2	1.87 ± 0.45	2.16 ± 0.43	0.001
Chest expansion at nipple level	Group 1	1.37 ± 0.53	1.48 ± 0.56	0.01
	Group 2	1.46 ± 0.43	1.75 ± 0.42	0.008
Chest expansion at xiphisternum level	Group 1	1.28 ± 0.56	1.32 ± 0.56	0.19
	Group 2	1.11 ± 0.24	1.22 ± 0.26	0.007
PEFR	Group 1	223 ± 41	231 ± 40	0.008
	Group 2	246 ± 50	273 ± 41	0.07

At xiphisternum level, there was significant improvement in Group 2 but not in Group 1. It reflects that RMS is more effective in increased lower chest diameter as compared to CPT.

PEFR

There was no significant change in values of PEFR. It is also possible that PEFR is not adequate estimator of respiratory functions for patients undergoing CABG [19]. Patient's effort and motivation also hold important for PEFR, so this could possibly explain why we encountered large variance in PEFR values in our patients sample.

Ventilatory responses in this study are most likely due to activation of muscle spindles included by muscle stretch.

Conclusion

This is the first study that highlights the comparison between rehab protocols targeting towards pulmonary mechanics. The results suggest that RMS may have unique effects on restoring respiratory mechanics post-sternotomy by increasing chest expansion as compared to CPT. Future studies are also necessary to further examine the clinical significance with larger sample size.

References

1. Connors, A.F. 1980. The Immediate Effect on Oxygenation in Acutely Illed. *Chest.* **78 (4)**: 559-584.
2. Locke, T.J. 1990. Ribcage Mechanics after Median Sternotomy. **45**: 465-468.
3. Braun, S.R., Birhbaum, M.L. and Chopra, P.S. 1978. Pre and Postoperative Pulmonary Function Abnormalities in Coronary Artery Revascularisation Surgery. *Chest.* **73**: 316-320.

4. Burgess, G.E., Cooper, J.R., Marino, R.J. et al. 1978. Pulmonary Effect of Pleurotomy during and after Coronary Artery Surgery with Internal Mammary Artery vs. Saphenous Vein Grafts. *J. Thorac. Cardiovasc. Surg.* **76**: 230-234.
5. Stock, M.C., Downs, J.B., Weaver, D. et al. 1986. Effect of Pleurotomy on Pulmonary Function after Median Sternotomy. *Am. Thorac. Surg.* **42**: 441-444.
6. Estenne, M., Yernault, J.C., DeSmet, J.M. and De Troyer, A. 1985. Phrenic and Diaphragm Function after Bypass Grafting. *Thorax.* **40**: 293-299.
7. Berrizbeita, L.D., Tessler, S., Lenora, R.A.K. et al. 1988. Effect of Sternotomy and Coronary Bypass Surgery on Postoperative Pulmonary Mechanics. (Abstract) *Am. Rev. Respir. Dis.* **137**: 248.
8. Clergue, F. 1995. Interference about Respiratory Muscle Use after Cardiac Surgery from Compartmental Volume and Pressure Measurements. *Anesthesiology.* **82**: 1318-1327.
9. Alejan dro suarez, A. 2002. Peak Flow and Peak Cough Flow in the Evaluation of Expiratory Muscle Weakness and Bulbar Impairment in Patients with Neuromuscular Disease. *Am. J. Phys. Med. Rehabil.* **81**: 506-511.
10. VanBelle, A.F. 1992. Postoperative Pulmonary Function Abnormalities after Coronary Artery Bypass Surgery. *Respiratory Medicine.* **86**: 195-199.
11. Minoguchi, H. 2002. Cross-over Comparison between Respiratory Muscle Stretch Gymnastics and Inspiratory Muscle Training. *Internal Medicine.* **41**: 805-812.
12. Puckree, T. 2002. Does Intercostals Stretch after Breathing Pattern and Respiratory Muscle Activity in Conscious Adult? *Physiotherapy.* **88**: 89-97.
13. Anderson, J.M. 1987. Assessment of Chest Function by the Physiotherapist. In *Cash's Textbook of Chest, Heart and Vascular Disorders for Physiotherapists, 4th edn.* P.A. Downie, D.M. Innocenti and S.E. Jackson (eds.). J.B. Lippincott Company, Philadelphia. pp. 318-324.
14. Suarez, A.A. et al. 2002. Peak Flow and Peak Cough Flow in the Evaluation of Expiratory Muscle Weakness and Bulbar Impairment in Patients with Neuromuscular Disease. *Am. J. Phys. Med. Rehabil.* **81**: 506-511.
15. Fink, J.B. and Hunt, G.E. 1999. *Clinical Practice in Respiratory Care.* Lippincott Williams and Wilkins.
16. Jenkins, S.C. 1989. Physiotherapy after Coronary Artery Surgery. Are breathing exercises necessary? *Thorax.* **44**: 634-639.
17. Oikkonen, M. 1991. Comparison of Intensive Spirometry and Intermittent Positive Pressure Breathing after Coronary Artery Bypass Graft. *Chest.* **9**: 60-65.
18. Howell, S. 1978. Chest Physical Therapy Procedures in Open Heart Surgery. *Physical Therapy.* **58(1)**: 1205-1214.
19. Stein, M. 1963. Pulmonary Evaluation of Surgical Patients. *JAMA.* **181(9)**: 765-770.